BASIC BIOINFORMATICS

BASIC BIOINFORMATICS

Manju Bansal

ATLANTIC
PUBLISHERS & DISTRIBUTORS (P) LTD

Published by

ATLANTIC

PUBLISHERS & DISTRIBUTORS (P) LTD

7/22, Ansari Road, Darya Ganj,
New Delhi-110002
Phones : 2327 3880, 23275880, 23280451
Fax : +91-11-23285873
Web : www.atlanticbooks.com
E-mail : info@atlanticbooks.com

Branch Office
5, Nallathambi Street, Wallajah Road,
Chennai-600 002
Phones : +91-44-64611085, 32413319
E-mail : chennai@atlanticbooks.com

ISBN: 978-81-269-1043-4

Printed in India at Nice Printing Press, A-33/3A, Site-IV, Industrial Area, Sahibabad, Ghaziabad, U.P.

Dedicated to

My husband Mukesh Gupta
and
My dear son Akshat
on his thirteenth birthday
on 5th July 2008 with love

ACKNOWLEDGEMENTS

It is perhaps unfair to say that this textbook is the result of the sole efforts of mine since I truly feel that the writing of this book has been an enormous collaborative effort. It took me four long years to finish this work. Each chapter went through several rigorous revisions and was reviewed by many scientists and editors. I wish to acknowledge them all; in particular I acknowledge the contributions of Dr. Devapriya Choudhury, Center for Biotechnology, JNU for his critical comments and rigorous revisions. Several of the pages in this book include copy of the slides with which he teaches in the class. Every time I incorporated the changes something new was waiting to be added. I am truly grateful to him for the assistance rendered during the writing of this book. I am also thankful to Dr. Naveen Kumar, Computer Science Department, Delhi University and Dr. Saurabh Raghuvanshi, University of Delhi, South Campus for their valuable comments.

I am thankful to Dr(s) Julie Thompson, Dennis M. Ritchie, Russel Doolittle, and Joe Felenstein for providing valuable pictures which have been used to supplement the text. I am also grateful to Dr(s) Steven Henikoff, William R. Pearson, Amos Bairoch and Des Higgins for pictures provided through web. I am also thankful to Wikipedia, the free encyclopedia, for providing the pictures of Frederick Sanger, Max Perutz, E.O. Wilson, John Kendrew, James Watson, Margaret Dayhoff, Craig Venter and Hamilton O. Smith. Contributions of these scientists have been so immense that the book would have been incomplete without their pictures.

Experiments included in the book are those repeated by the students in the classroom or submitted by them as projects. Many students made very useful suggestions and helped me in finding errors. In addition, I am grateful to Sh. Harjeet Singh who worked meticulously for the creation of this work and showed all his patience for four years. I would also like to thank Ms. Ritu for skillful crafting of the book.

I wish to express my gratitude to my husband Sh. Mukesh Gupta. His support has been remarkable. He really helped me to come out of the mess which my life had become at one stage and he almost revived this book. Two great souls which left me halfway before this book was finished are: my first husband late Sh. Atul Gupta and my father-in-law late Sh. I.M. Gupta. I cherish their memories and treasure their blessings. My parents, specifically the critical input of my father was immensely valuable that shaped the book's final contents and organization. My son Akshat and my mother-in-law Mrs. Rajeshwari Gupta also helped by motivating and encouraging me to complete this book. I extend my sincere gratitude to them all. Daughterly affections from my sister Sheeba and sister-in-law Aashu are deeply appreciated.

I also wish to thank all the difficult people whom I came across during the completion of this work. They served as a source of strength and inspiration to me. God Almighty has been always looking after me.

Manju Bansal

PREFACE

The idea to write this textbook on bioinformatics actually stemmed from my interest in computers and computer programming at the University of Western Australia many years ago. During a refresher course at JNU in 2001, while interacting with several speakers on biotechnology and bioinformatics, I got the opportunity to discuss this topic and work in this direction. Soon I realized that bioinformatics is very different from the perspective of a mathematician, a computer scientist or a biologist. As a teacher of genetics and genomics and biotechnology, I felt that the continuous addition of bioinformatics to the curricula was making the teaching community and the students really uncomfortable in the absence of basic books on bioinformatics.

In 2004, during major restructuring of undergraduate courses in life sciences, further addition of several bioinformatics topics in the syllabi was made. Students were required to do practicals and submit projects on bioinformatics. However, from their previous course work, they were not trained to handle computers. They had a very poor background of mathematics and physics. On the positive side, it was impressive to see how students improved their ability to handle computers and practically visualized application of results. On the negative side, however, it left me with a disappointing sense that most college science courses do not provide the students with enough background to work in integrated disciplines like bioinformatics. I felt that lack of availability of textbooks in bioinformatics was to be further blamed. At that point I decided to write a textbook that could incorporate the scientific methods to each chapter to introduce students with basic computer concepts and simplify mathematical equations along with genetics and genomics applications.

When you use this book you will see that each chapter has one or two experiments which have been analyzed according to scientific methods.

Similarly, each chapter has one or two historical profile(s) of distinguished scientists relating to that particular branch of bioinformatics. I hope this book will help students to see the fundamental connection between scientists, scientific analysis and principles. There is a set of questions at the end of each chapter to test the students' knowledge and help them explore the relationship between concepts and experiments. All the latest information relating to the topics has been given in the boxes in every chapter.

Having an academic background of biology helped me to write a book on bioinformatics for biologists, although all scientific community must benefit from this. I hope this book will go a long way in solving the problems of students and research scholars in the field of bioinformatics. Constructive comments and valuable suggestions are invited from the students and learned teachers to further improve this book in subsequent editions.

Manju Bansal
Sri Aurobindo College
University of Delhi
Email: mbansalsac@rediffmail.com

CONTENTS

ABBREVIATIONS

ANSI	American National Standards Institute
ASP	Active server pages
BLAST	Basic Local Alignment Search Tool
BLOSUM	Blocks Amino Acid Substitution Matrix
CDS	Coding Sequence
COGS	Clusters of Orthologous Groups
DBCAT	The Public Catalog of Databases
DDBJ	DNA Databank of Japan
EBI	European Bioinformatics Institute
EMBL	European Molecular Biology Laboratory
EST	Expressed Sequence Tags
ExPASy	Expert protein analysis system
GBIF	Global biodiversity Information Facility
GOLD	Genomes Online Database
HMM	Hidden Markov Model
HSP	High Scoring Segment pair
IMMs	Interpolated Markov Models
KEGG	Kyoto Encyclopedia of Genes and Genomes
IPNI	International Plant Names Index
ISCB	International Society for Computational Biology
LANL	Les Alomos National Laboratory
MALDI	Matrix Assisted Laser of Desorption/Ionisation
MARS	Matrix Attachment Region
MDM	Mutation Data Matrix
MHC	Major Histocompatibility Complex
NBRF	National Biomedical Research Foundation
NCBI	National Centre for Biotechnology Information
NHGRI	National Human Genome Research Institute
NLM	National Library of Medicine
ORF	Open Reading Frame
OS	Operating System
PAM	Percent Accepted Mutation
PCFs	Potential Coding Fragments
PDB	Protein Data Bank
PHI-BLAST	Pattern HIT Initiated BLAST
PIR	Protein Information Resource
PSI-BLAST	Position Specific Iterative BLAST

PSSM	Position Specific Scoring Matrix
RFLP	Restriction Fragment Length Polymorphism
SARS	Scaffold Associated Regions
SIB	Swiss Institute of Bioinformatics
SNP	Single Nucleotide Polymorphisms
SQL	Structured Query Language
SSLP	Simple Sequence Length Polymorphism
UPGMA	Unweighted Pair Group Method using Arithmatic Averages
UTR	Untranslated Regions
WGS	Whole Genome Shotgun

1

OVERVIEW OF BIOINFORMATICS

In bioinformatics we aim at uncovering the wealth of biological information hidden in the mass of data and obtain a clearer insight into the fundamental biology of organisms. This new knowledge has profound impact on fields as varied as human health, agriculture, environment, energy and biotechnology.

The genomic era has seen a massive explosion in the amount of biological information available due to huge advances in the fields of molecular biology and genomics. Bioinformatics involves the study of biological information as it passes from its storage site in the genome to the various gene products in the cell. Generally speaking, bioinformatics involves the creation and development of advanced information and computational technologies for problems in molecular biology. As such, it deals with methods for storing, retrieving and analyzing biological data, such as nucleic acid (DNA/RNA) and protein sequences, structures, functions, pathways and interactions. Bioinformatics uses computer software tools for database creation, data management, data warehousing, data mining and global communication network. Functional genomics, biomolecular structure, proteome analysis, cell metabolism, biodiversity, downstream processing in chemical engineering, drug design, vaccine design are some of the areas in which bioinformatics is an integral component.

Advancements in information technology has revolutionized the science of bioinformatics. The knowledge of multimedia databases, tools to carry out data analysis and modeling of molecules and biological systems on computer systems has become essential for any student of bioinformatics. Advances in bioinformatics have made it possible to maintain biodiversity databases. Indeed, bioinformatics has evolved into a discipline of its own. Therefore large bioinformatics research and development divisions are being established in many pharmaceutical companies, biotechnology companies and even in other conventional industries. Bioinformatics has thus potential to provide unlimited career opportunities.

1.1 MILESTONES IN BIOINFORMATICS

Bioinformatics has emerged as a multidisciplinary subject that encompasses developments in information technology as applied to biotechnology and biological sciences. The study of bioinformatics started with the collection of protein sequences as databases. In 1960s, Margaret Dayhoff, pioneer in bioinformatics (Historical Profile 1) along with her collaborators was first to assemble the protein sequences atlas and their collection eventually became known as PIR (Protein Information Resource). She and her coworkers also organised protein into families and superfamilies based on degree of similarity. She and her colleagues also gave PAM tables for evolutionary selection. Listed in

Table 1.1 are some of the major events in bioinformatics over the last several decades. Most of the events in the list occurred long before the term, "bioinformatics" was coined. Science of bioinformatics gained importance after 1995 when first genome of a bacterium *Hemophilus influenza* was sequenced.

Table 1.1: Milestones in Bioinformatics

1951	Development of protein sequencing methods by Sanger and Tuppy.
1962	Margaret Dayhoff's developed Atlas of Protein Sequences.
1970	Needleman-Wunsch algorithm for optimal alignment of sequences.
1977	Bacteriophage ϕ X-174 sequenced, the first complete genome.
1977	DNA sequencing and software to analyze it developed by Staden.
1979	Walter Goad conceived idea of Gen Bank.
1981	Smith-Waterman algorithm developed.
1981	Human mitochondrial-DNA sequenced: 16569 base pair.
1982	Gen Bank Release 3 made public.
1983	Sequence database searching algorithm given by Wilbur and Lipman which subsequently was called ENTREZ.
1984	Epstein-Barr virus genome sequenced: 1, 72, 281 base pair.
1988	Fast sequence similarity searching method called FASTA developed by Pearson and Lipman.
1988	National Center for Biotechnology Information (NCBI) created.
1988	EMBnet network for database distribution.
1990	BLAST: fast sequence similarity searching developed by S. Altschul.
1991	J.C. Venter and colleagues identified active genes via express sequence tags (EST).
1993	1st genome database called ACEDB (a *Caenorhabditis elegans* database) was developed by Mike Cherry and his colleagues.
1995	First bacterial genomes of *Hemophilus influenza* completely sequenced.
1996	First fungal genome of yeast completely sequenced.
1997	PSI-BLAST developed.
1998	Worm (multicellular) genome of *Caenorhabditis elegans* was completely sequenced.
1999	Fly *Drosophila melanogaster* genome completely sequenced.
2000	Human genome chromosome number 20 sequenced.
2000	Genome of a plant *Arabidopsis thalliana* sequenced.
2000	Mouse genome sequenced.
2003	Human genome completely, sequenced and gene count estimate changed to 20,000 from 25,000.
2005	Chimpanzee genome sequenced.
2007	Personal human genome of Dr. C. Venter and Dr. James D. Watson sequenced.
2007	Neanderthal a fossil human genome to be completely sequenced within 2 years.

1.2 THE SCIENCE OF BIOINFORMATICS

The discipline of bioinformatics involves molecular, cellular, organismal and population biology and computers. Scientists interested in bioinformatics need to have an understanding of the supporting discipline such as biochemistry, biophysics, cell biology, mathematics, computer science, microbiology, population biology, ecology, agriculture and medicine.

The study of bioinformatics in the present book in organised in four parts: software tools in bioinformatics, techniques in bioinformatics, organism specific bioinformatics and application.

1.2.1 Software Tools in Bioinformatics

The development of sequence analysis methods has largely depended on the computational databases and softwares. In the beginning, the databases were developed mainly for biological sequences of DNA and proteins. Margaret Dayhoff (1972, 1978) and her collaborators at National Biomedical

Research Foundation (NBRF) Washington DC assembled databases of protein sequences into a protein sequence atlas in 1960s. Their collection was eventually known as Protein Information Resource (PIR). Similarly in 1978 Walter Goad and colleagues at Les Alomos National Laboratory, (LANL) New Mexico conceived the idea of developing nucleic acid databases. These computer based biological databases proved to be of indispensable for the study of molecular biology. In chapter 2, several concerned computer databases are cited, which give the readers a clear understanding about functioning of common computer databases. In early days a sequence entry in the database included a computer file name and DNA or protein sequence name. These were eventually expanded to include much more information about sequences such as functions, mutations, encoded proteins, regulatory sites and references. The information is now placed along with the sequence into a database format that could be readily searched for many types of information. Protein and nucleic acid databases are discussed in chapters 3 and 4, respectively. The merger of computational databases with molecular biology revolutionalised the field. Indeed, now we have very specific databases like EST databases or antibody databases which have immense applications.

Although biological databases first began to be used in molecular biology, over a period of time, they find use in various branches of biology for example, a database of complex reactions in the body is called Metabolic pathway databases (chapter 5). Pathway databases are complex networks of reactions, study of which has become easy with the use of computers. In Biodiversity databases started in 1990s (chapter 6) not only the correct scientific name of an organism could be searched but all details about species distribution can also be obtained. Bibliographic databases are collections of references which are widely used by scientists (chapter 7). In the last chapter, in this section, important sequence formats used in the different biological databases are given. Knowledge of the file formats in databases is important for submitting and retrieving information.

1.2.2 Methods in Bioinformatics

As the name implies the goal of this section is to study how to apply the bioinformatics tools to the field of biology. Besides databases, bioinformatics deals with several softwares which function as tools for databases search, sequence function identification, structural analysis of biological molecule and analysis of sequences. Once all of the biological data is stored consistently and is easily available to the scientific community, the requirement is to provide methods for extracting the meaningful information from the mass of data. The software programs are designed to carry out this analysis step. These software must be user friendly keeping in view the end users i.e. biologists.

The application of biological data analysis techniques fall into the following four major categories:

1. **Sequence analysis:** This set of tools allows us to carry out furthermore detailed analysis on the query sequence including evolutionary analysis, identification of mutations, identification of introns, exons, etc. The identification of these and other biological properties are clues that aid the search to elucidate the specific function of the sequence at hand. This is discussed in chapter 9 along with an introduction to computer programing.
2. **Sequence alignment:** Sequence alignment is a procedure for comparing two or more sequences by searching for a series of individual character patterns that are in the same order in a given sequence. It is useful for discovering functional, structural and evolutionary information contained in biological sequences. It is discussed in chapters 10 and 11 on sequence alignments concepts and algorithms and pairwise and multiple sequence alignment respectively.
3. **Similarity searching:** This technique can be used to identify similarities between novel query sequences of unknown structure and function and database sequences whose structure and function have been elucidated (chapter 12). Homologous sequences are sequences that are

related by divergence from a common ancestor. Thus the degree of similarity between two sequences can be measured while their homology is a case of being either true of false. The software allows you to compare your protein sequence to the secondary (or derived) protein databases that contain information on motifs, signatures and protein domains. Highly significant hits against these different pattern databases allow you to approximate the biochemical function of your query protein. Whether a sequence is coding a gene or not, is studied in details in chapter 13 on gene prediction. In chapter 14 it is discussed in the context of homology, evolution and phylogenetic prediction.

4. **Structural analysis:** The set of softwares for structural analysis allow you to compare the given molecular structure with the other known structures contained in structure databases. The function of a protein (chapter 15) or nucleic acid (chapter 16) is more directly a consequence of its structure rather than its sequence. The determination of 2D/3D structure of a proteins and nucleic acid is crucial in the study of their function.

1.2.3 Organism Specific Bioinformatics

In earlier sections, the basic unit of study was a sequence either a stretch of nucleotide or a protein sequence. When we study all the DNA sequences of an organism, the branch of study is called genomics and the nucleic acid content is called the genome. Similarly proteome refers to all protein sequences contained in an organism and proteomics is the study of these proteins as a unit. Application of bioinformatics tool and techniques at the level of genome and proteome is studied in Section 3. An introduction to genomes and proteomes is given in chapter 17 and application of different bioinformatic tools is discussed in this context. Studies on different forms of life i.e. bacteria, fungi and animals are given in chapters 18, 19, 20 and 21 respectively.

1.2.4 Applications of Genomics and Bioinformatics

Some current and potential applications of bioinformatic resources and genome research for human welfare are discussed in the last chapter (chapter 21). Applications of this technology in DNA sequence analysis can detect the absence of a particular gene or a mutation. The identification of specific gene sequence associated with diseases will lead to fast and reliable diagnosis of disease and also help in development of therapeutic procedure. Diagnosis of disease and disease risks i.e. identification of specific gene sequence is associated with diagnosis of disease, can be done even when symptoms are not present. Similarly in case of in utero diagnosis of potential abnormalities and for genetic counselling of couples contemplating children identification of a particular gene becomes very useful. Genetic responses to therapy are also predetermined with bioinformatic tools. Sequence analysis helps in selecting drugs and dosage optimal for individual patients. Identification of drug targets and gene therapy are other potential areas.

1.3 IMPORTANT BIOINFORMATICS WEB SITES

There are various bioinformatics web sites and search launchers which provide easy access and information to bioinformatics resources. It is a good idea to be familiar with many of the bioinformatics centers, because they tend to emphasize the methodology that they have had a hand in developing, and therefore, each institution offers different tools. Most of the databases share information, particularly the Japanese, European and North American databases, so the databases essentially mirror each other. However, it is worth exploring different bioinformatics web sites to take advantages of the different tools for data analysis that they have to offer. Selected bioinformatics sites are mentioned in Table 1.2.

Table 1.2: Important Bioinformatics Web Sites

1. National Centre for Biotechnology Information (NCBI) hosts the nucleic acid database called Genbank. Also known for its sequence search program ENTREZ and BLAST for fast similarity searches.
2. EMBL (The European Molecular Biology Laboratory) is one of the Top Research Institutions in Molecular biology. The European Bioinformatics Institute (EBI) is a non-profit organisation which forms a part of EMBL.
3. National Human Genome Research Institute (NHGRI) is known for bringing genomic revolution to public. It is working on identifying genetic basis of common diseases.
4. ExPASy proteomics server refers to expert protein analysis system. It is a server of Swiss institute of Bioinformatics and is dedicated to the analysis of protein sequences and structure.
5. The DBCat database is a public catalogue of biological databases, maintains statistical record of all biological databases.
6. The International Society for Computational Biology (ISCB) is advancing applications of computational biology.
7. The Research Collaboratory for Structural Bioinformatics. (RSCB) is known for Protein Data Bank and Nucleic Acid Structural Data Bank.
8. The PROSITE database is a site maintained and developed by Swiss Institute of Bioinformatics. It is a secondary protein database with a special value of protein families.
9. The PIR is international protein resource web site is the most comprehensive and latest catalogue of information on proteins.

Established in 1988 as a National Resource for Molecular Biology Information, NCBI creates public databases (*www.ncbi.nlm.nih.gov*), conducts research in computational biology, develops software tools for analyzing genome data, and disseminates biomedical information—all for the better understanding of molecular processes affecting human health and disease. NCBI provides integrated access to DNA sequence database, the human genome database and more than 40 related molecular biology database services, and the scientific literature. Sequence analysis tool available at NCBI is given in Table 1.3 Web view of NCBI is shown in Fig. 1.1.

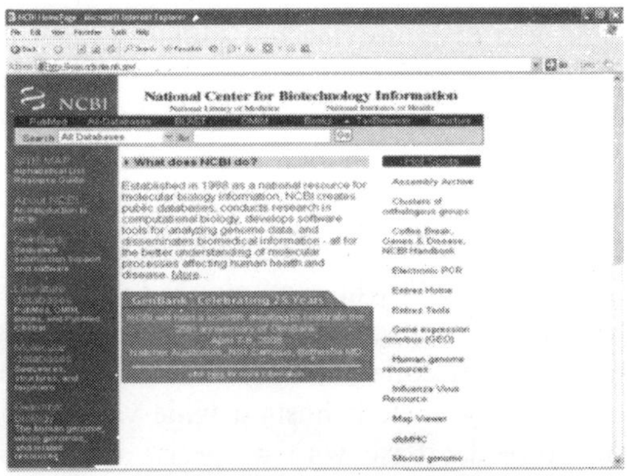

Fig 1.1: Web view of NCBI

EMBL (The European Molecular Biology Laboratory) is one of the top research institutions in the world. EMBL was founded with a fourfold mission: to conduct basic research in molecular biology, to provide essential services to scientists to provide high-level training to its staff, students, and visitors, and to develop new instrumentation for biological research. These core functions are combined with significant outreach activities in the areas of technology transfer, science and society and training for science teachers. Over its 30-year history, the Laboratory has had a deep impact on European science in all of these areas.

The European Bioinformatics Institute (EBI) is a part of EMBL. The EBI is the bioinformatics arm of the EMBL, and it functions in Europe to maintain and create databases and bioinformatics tools in the same way NCBI does in the United States. EBI maintains the EBI toolbox, a selection of bioinformatics software that can be accessed on the internet. The Toolbox web page has a menu in a grey border on the left of the page that lists five different categories of tools: 1. Homology and similarity, 2. Protein functional analysis, 3. Structural analysis, 4. Sequence analysis and 5. Miscellaneous tools.

The Research Collaboratory for Structural Bioinformatics (RCSB) is a non-profit consortium dedicated to improving our understanding of the function of biological systems through the study

Table 1.3: Sequence analysis tools available at NCBI

1. BLAST	Sequence similarity search tool are available for comparing a nucleotide or protein sequence against a collection of databases. BLAST is discussed in details in chapter 12.
2. ORF Finder	ORF Finder is a graphical analysis tool which finds all open reading frames in a sequence.
3. Clusters of Orthologous Groups	(COGs) were delineated by comparing protein sequences encoded in seven complete genomes representing five major phylogenetic lineages. A tool for comparing your sequence to the COG database is available.
4. Bankit	Bankit is a web sequence submission tool that uses a simple forms-based approach to formatting and submitting a sequence to GenBank. Accession numbers are assigned within 24 hours. New submissions become GenBank records after extensive quality assurance processing by NCBI staff and consolation, if necessary, with the author.
5. Sequin	Sequin a stand-alone software tool for submitting DNA sequences to GenBank, can be downloaded from the NCBI Web site. Sequin is designed to simplify multiple sequence submissions, provide graphical viewing and editing options, and accommodate long sequences and complex submissions.

The Research Collaboratory for Structural Bioinformatics (RCSB) is a non-profit consortium dedicated to improving our understanding of the function of biological systems through the study of the 3-D structure of biological macromolecules. RSCB is famous for two major structural databases: the protein data bank (PDB) and nucleic acid database (NDB). RCSB members work cooperatively and equally through joint grants and subsequently provide free public resources and publications to assist others and further the fields of bioinformatics and biology.

The SIB (Swiss Institute of Bioinformatics) is an academic non-profit foundation established on March 30, 1998 whose mission is to promote research, the development of databanks and computer technologies, teaching and service activities in the field of bioinformatics, in Switzerland with international collaborations. It provides two famous servers: ExPASy proteomics server and Swiss node of EMBnet. It hosts a wide variety of databases including PROSITE is a secondary protein structure database with a special value of maintaining an extensive documentation on protein families.

A very large selection of tools is available on ExPASy (Expert Protein Analysis System), the proteomics server of the Swiss Institute of Bioinformatics (SIB). The ExPASy tools and databases are numerous, and relatively easy to locate and use. ExPASy is also unusually good about documentation, including in their databases "SeqAnalRef", one of the few good listings of bioinformatics is bibliographic references. Take a moment to inspect this web site and the array of tools it provides.

1.4 BIOINFORMATICS IN INDIA

India is a hub of scientific and academic research and is one of the first countries in the world to establish a nation wide bioinformatics network. Bioinformatics is a very essential component in the biotech revolution in India. The Department of Biotechnology (DBT), Government of India, initiated the program on bioinformatics way back in 1986-87. The Biotechnology Information System Network (BTIS), a division of DBT, has covered the entire country by connecting to the 57 key research centers. University of Delhi South Campus is among them. BTIS was established with the following objectives in mind:

HISTORICAL PROFILE 1

Dr. Margaret Oakley Dayhoff: Pioneer in Bioinformatics

Dr. Margaret Oakley Dayhoff

Dr. Margaret Oakley Dayhoff (1925-1983) was a pioneer in the use of computers in chemistry and biology, beginning with her PhD thesis project in 1948. Her work was multi-disciplinary, and she used her knowledge of chemistry, mathematics, biology and computer science to develop an entirely new field. She is credited today as a founder of the field of Bioinformatics. Dr. Dayhoff was the first woman in the field of Bioinformatics. She was also the first woman to hold office in the Biophysical Society, serving first as Secretary and later as president.

Among Dr. Dayhoff's many contributions to science, the one she is most known for is the creation of computerised protein and DNA sequence databases, which were developed because of her research interests in bio-molecular evolution.

In 1960 she along with her collaborators at National Biomedical Research Foundation (NBRF) was first to assemble protein sequences into a protein sequence atlas and their collection eventually became known as Protein Information Resource (PIR). She organised the protein into families and super families based on the degree of sequence similarity. She developed a method for predicting phylogenetic relationships and PAM table for performing protein sequence comparisons. These are routinely used in performing sequence alignment and database similarity searches. The work allowed advances in our understanding of biological relationships, disease processes, and evolution. Her work produced statistical and computational methods for studying protein and nucleic acid structures that were critical to a large number of practical applications in the Biological sciences, including the production of useful drugs.

1. To prepares databases on biotechnology and to develop relevant handling tools and techniques.
2. To provide a national bioinformation network designed to bridge the inter-disciplinary gaps in biotechnology information.
3. To establish link among scientists involved in R&D and manufacturing activities in biotechnology.
4. To organize creation of necessary infrastructure and to provide information and computer support services to the national community of users.
5. To establish linkages with international resources in biotechnology information (like databanks on genetic materials, published literature, patents and other information of scientific and commercial value).
6. To evolve and implement programs on education of users and training of information scientists responsible for handling of biotechnology information.

BTIS is providing an easy access to huge database to the scientists. Six national facilities on interactive graphics are dedicated to molecular modeling and other related areas. More than 100 databases on biotechnology have been developed. Two major databases namely coconut biotechnology databases and complete genome of white spota syndrome of shrimp has been released for public use.

Several major international databases for application to genomics and proteomics have been established in the form of mirror sites under the National Jai Vigyan Mission.

Bioinformatics Society of India (Inbios) has been working since August 2001 with an objective to help and rise bioinformatics sector to the world map. The Inbios already has over 270 members in a short span of one and half years. It has become a common informal platform for the younger generation to learn and contribute to this sun rising field in India.

Bioinformatics as private sector, is the quickest growing field in the country. The promising start-ups are already there in Bangalore, Hyderabad, Pune, Chennai, and Delhi. There are over 200 companies functioning in these places. IT majors such as Intel, IBM, Wipro are getting into this segment spurred by the promises in technological developments. The real money is in discovering new drugs for ourselves and not in supplying genetic information and data to the foreign companies, who would then use this information to discover new molecules. There are several problems in this direction. There is a lack of the records of clinical information about the patients, sequence date without clinical information will have little meaning. And hence partnership with clinicians is essential. Because of large number of bioinformatic institutes, there is a problem of finding talented and trained individuals in this industry. Also many of them have a superficial knowledge and a certificate, India lacks true professionals in this area. Most people, who opt for bioinformatics are from the life sciences areas that do not have exposure to the IT side of bioinformatics, which is very important. Another issue is that some companies face shortage of funds and infrastructure. The turn around time for an average biotech industry to breakeven would be around three to five years.

SUMMARY

Bioinformatics is the applications of computer technology to the management and analysis of data. Madam Margaret Dayhoff is a known pioneer scientist in bioinformatics (Historical Profile 1). Science of bioinformatics gained importance after 1995 when first genome of a bacterium *Hemophilus influenza* was sequenced. Now we have reached the era of personal genome sequencing. Tools in bioinformatics are the several types of databases and techniques include sequence analysis, similarity search, protein structure and function analysis, nucleic acid structure studies and sequence alignment. Genomic databases whether bacterial, fungal, plant or animal are impossible without bioinformatic tools. Important bioinformatics websites discussed in the chapter are: NCBI, EMBL, EBI, RCSB and SIB.

EXERCISE

1. These are only a few of the many bioinformatics sites available via the internet. Perform a Google search with "bioinformatic tools" as your query. How many results were returned from this search? List at least two sites other than an NCBI, PDB, EBI, ExPASy or BCM associated site that contain a collection of bioinformatic tools, and list the sponsoring organisations.

SECTION-1

TOOLS IN BIOINFORMATICS

2

BIOLOGICAL DATABASES: AN INTRODUCTION

Databases and database programs for analysis of data are the routine tools of modern biologists, better called as bioinformaticians today. All modern biological data whether molecular genetic data in the form of DNA, RNA or amino acids sequences or taxonomic information about a organism, are stored in computer databases. These databases are classified into several types based on what information they store or the source of this information. Computers not only effectively store this data but could also analyze this data at a rate of millions or billions of operations in a second. In this present introductory chapter on biological databases, we explore what are these databases, their types and how they are useful to molecular biologists. In the last segment we would discuss about challenges in front biologists using these databases and their management.

2.1 DATA AND DATABASES IN COMPUTERS

Data...it's everywhere. Everywhere you go or work, your senses are bombarded with various forms of data that your brain can only take a fraction of the data flowing into your sense at a given time. In a normal routine we write it down and keep it in filing cabinet. To keep track of, sort and find information from this data in paper files is very difficult, time consuming and unmanageable. A very efficient way to manage this information is to build databases in computers to do this job in minutes. Database is often used interchangeably with the term table. Technically, they're different: A table is a single store of related information; a database can consist of one or more tables of information that are related in some way. Databases excel at managing and manipulating structured information. What does the term 'structured information' mean? Consider that most ubiquitous of databases—the phone book. The phone book contains several items of information—name, address and phone number—about each phone subscriber in a particular area. Each subscriber's information takes the same form. In database parlance, the phone book is a table, which contains a record for each subscriber. Each subscriber record contains three fields: name, address, and phone number. The records are sorted alphabetically by the name field, which is called the key field. In Fig. 2.1 an example of a computer database, its relation with table, records, fields, key field, etc. is depicted. This database has three tables named customers, orders and overdue. Record in a table refers to a row of the table and field refers to a column. Table customers contain information name and addresses of customers. Table orders contain information about the orders and table overdue contain overdues data.

CUSTOMERS

Cid	Last_name	First_name	Shipping_address
1	Doe	John	505Som estreet
2	Doe	Jane	505Somestreet
3	Palmer	John	200on street
4	Johnson	Richard	1000Another street

ORDERS

Cid	Order-num	Stock_num	Priority	Shipped	Date
1	1002	100,252,342	3	Y	8/3/01
1	1221	200,353	1	N	10/2/01
3	1223	200,121	2	Y	10/2/01
2	1225	221,152	1	N	10/3/01

OVERDUE

Cid	Order-num	Days_overdue	Action
1	1002	32	Sentletter

Fig. 2.1: An example of using three related tables to track customers, orders and outstanding invoices

The computer databases are of two basic forms: a single file and multi-file database. A flat file is simple table with a space, tab or some other character determining different parts of information much similar to a table created in a word file for Microsoft windows. Multi-file databases have many such files. All the files/tables in multi-file databases are related to each other, because of this they are called as relational databases. Example shown in Fig. 2.1 is an example or a relational database. Relational databases are made up of two or more tables of information that are connected in some way.

To create and maintain a computer database, you need a database program, often called a database management system, or DBMS. A computer program is defined series of operations that analyze data in a desired way. Just as databases range from simple, single-table lists to complex multi-table systems, database programs, too, range in complexity. Some, such as the database component of Microsoft office, are designed purely to manage single-file databases. With such a product you cannot build a multi-table database. You can certainly create numerous tables for storing different types of information, but there's no way to link information from one table to another. Such programs are sometimes called flat-file databases, or list managers. Other database programs, called relational database programs or RDBMSs, are designed to handle multi-file databases. Oracle is a relational database that's easy to use and fairly inexpensive.

The most popular relational databases are offered from the big three software companies. Lotus, Corel and Microsoft each produces a full-featured relational database application available both as a stand alone program and as part of its integrated suite. Stand-alone program is works on a single PC while others work on computers connected as a network. Lotus has Approach, Corel has Paradox and Microsoft has access. In Table 2.1 commonly used database softwares are listed along with the company manufacturing them and the operation system on which they work. A brief knowledge about them is a must for biologists who want to create and maintain a database.

Table 2.1: Common examples of databases softwares

Name	*Working platform*	*Company*
Oracle	Windows	Oracle
Access	Windows	Microsoft
DB2	Windows	Microsoft
MYSQL	Linux	Red Hat, Inc.
POSTSQL	Linux	Red Hat, Inc.
SQLserver	Windows	Microsoft

2.2 BIOLOGICAL DATABASES ARE A SIMPLE TYPE OF COMPUTER DATABASES

Biological databases store the data files relating to biological macromolecules. These biological databases do not have anything different from routine computer databases except that they store biological information. The information contained in these databases range from single nucleic acid sequences to whole genome projects. The primary data of biological macromolecules usually contain information regarding nucleic acid sequences amino acid sequences and whole genome proteomes, protein and nucleic acid structures, protein functions, expression of genes, metabolic pathways, taxonomic information of biological organisms and publications of research articles. In beginning when databases were being tried to be assembled, the major form of biological data was protein and DNA sequences.

BOX: 2.1

Breakthrough Research: Automation of Sequencing Techniques

In the past few years, automation for genomics has enabled a 43-fold increase in the total finished human genomic sequence in the world. Three major milestones: the invention of sequencing reactions the Polymerase Chain Reaction (PCR) and automated fluorescent DNA sequencers have made it possible to streamline and automate most of the processes required for DNA analysis.

Today, there are instruments that can automate nearly every step of the large-scale sequencing process: isolating DNA, cloning or amplifying DNA, preparing enzymatic sequencing reactions, purifying DNA, and separating and detecting DNA fragments with fluorescent labels to obtain DNA sequence. Most large-scale sequencing facilities employ the consensus strategy, which combines a random shotgun phase with a directed finishing phase. Other centers use a whole-genome shotgun approach: random fragments of total human genomic DNA are subcloned directly, and high-throughput sequencing is used to generate random sequences that provide at least ten-fold coverage of the genome. Yet another approach is to sequence each end of every BAC in a genomic library. Matching a finished sequence to one or more BAC ends permits selection of the net clone.

Similarly protein-sequencing technique as given by Edman is now fully automated. Protein sequences are now available which does this job very fast and efficiently. Automated Edman (N-terminal) sequencing is performed with a Hewlett Packard G1005A automated protein sequencer.

In 1971, Frederick Sanger discovered conventional DNA sequencing technique. It was in the same year Alan Maxam and Walter Gilbert gave a different method for DNA sequencing. In 1977, Edman gave the conventional and modern protein sequencing technique and also he described the method for automation of protein sequencing. Craig Venter (Historical Profile 19) gave modern DNA sequencing methods in 1991. Efforts of these and several other scientists lead to development of fully

HISTORICAL PROFILE 2

Dr. Frederick Sanger for Discovering Conventional DNA Sequencing Methods

Dr. Frederick Sanger

Dr. Frederick Sanger was born on August 13, 1918 and since 1940 he carried out research in the Department of Biochemistry at Cambridge. While working on structure of insulin, he gave the dideoxy method of DNA sequencing in 1974. In the same year Alan Maxam and Walter Gilbert developed another method of DNA sequenced called as cleavage method that involved base specific cleavage of DNA. Sanger's method is more popular method of determination of DNA sequences and is used for routine purposes.

Sanger's approach involved copying DNA strands with dNTP (deoxyribonucleic acid triphosphates and terminating them with tagged ddNTPs (dideoxyribonucleic acid triphosphates), which would show the location of the nucleotides in the strands though the use of X-ray machines. This technique is very slow and tedious, usually taking many years to sequence only a few million letters in a string of DNA that often contain hundreds of millions or even billions of letters. However, because of this process's accepted accuracy and the lack of new techniques, many researchers still use a similar method today.

automated procedures for DNA and protein sequencing (Breakthrough Research box 2.1 and Historical Profiles 2 and 3). As a result of these rapid and automated DNA and protein sequencing techniques during last decade molecular biology has witnessed an information revolution. As a result of this and corresponding progress in computer based technologies enormous number databases are emerging which contain biological information.

Biological databases now have reached to the stage of databanks where enormous amount of information is stored. According to DBCAT, the Public Catalog of Databases, as on today there are about 511 databases that contain biological information. Of these protein databases are largest in number. For the sake of convenience the databases in the present book are divided into protein databases, nucleic acid databases, metabolic pathway databases and miscellaneous type databases, which include taxonomic information of biological organisms and publications. Databases for organisms of various category i.e. prokaryotes, eukaryotic lower organisms, higher plants and animals are discussed separately in Section III.

2.3 STRUCTURE AND FUNCTIONING OF BIOLOGICAL DATABASES

Biological data have indeed frequently been held in simple files often with tags to indicate the aning of a particular piece of data, for example Genbank file. Files may also have fixed record format as in case of a PDB file in the protein structure database. The biological data that traditionally been held in a file format are in reality now managed by a DBMS and familiar file formats are retained as one way of making such data available to the scientists. Each of the entry in the database

HISTORICAL PROFILE 3

Dr. Pehr Victor Edman 1916-1977—To have given Protein Sequencing and Automation Technique

Dr. Pehr Victor Edman was born in Stockholm, Sweden, in April 1916 and died in Munich, FRG, in March 1977. He was born into a lawyer's family and received his schooling in Stockholm. In 1935 he began medical studies at the Karolinska Institute and graduated with his primary medical qualifications in 1938. He became interested in research and, following graduation, continued to work at the Karolinska Institute, largely in the laboratory of Professor Eric Jorpes. He was awarded the degree of Doctor of Medicine in 1946. The subject of his thesis was the purification and analysis of angiotensin from bovine blood. His earlier published studies concerned heparin and secretin, which were interests of his mentor Jorpes.

At Princeton he began experiments to find a way to chemically decode the amino acid sequence of proteins. Edman synthesized the PTH (Phenylthioisocyanate) derivatives of all the amino acids found in proteins and developed chromatographic systems to identify and quantify them; After two years work Edman was able to publish the chemical details of a method capable, in theory, of solving the problem of primary structure of proteins and of providing the essential information about innumerable proteins essential for further advances in protein biochemistry. The method became widely known and was given the eponym 'Edman degradation'.

In 1957 Edman accepted the offer of the position of first Director of Research at St Vincent's School of Medical Research in Melbourne. His work here fell into three phases: improvements in the conditions for the degradation largely focused on the elimination of side reactions; 'automation' of the reaction sequence; and application of the degradation to various sequence problems. By 1960-61 the three-stage degradation reaction had been essentially perfected. Its universal application and repetitive nature suggested to G.S. Begg, Edman's Australian technical assistant, that it would be suitable for automation. Edman realized that the number of existing proteins (about ten million) made manual sequencing an impossible task and was quickly converted to the idea of automation. Edman and Begg worked as a team on the sequence automation project, with no sustained input from other workers. By 1967 in the first issue of the European Journal of Biochemistry with Begg as coauthor he published his definitive paper demonstrating an unbroken automated determination of the amino-terminal sixty amino-acids of humpback whale myoglobin at the rate of one residue per hour.

---

refers to one record and every database has different format for entering records. Different formats of records of popular databases and methods for their inter-conversion from one format to other are discussed in details in chapter 8.

The databases, that we search, could be divided into two ends: one is front end that interacts with us and the other is back end that does all the work in the back and is not visible to us. Whenever we desire to make a submission or retrieval from these databases, the form where we enter our information is connected to the RDBMS functioning at the back end with the help of a computer program. The program inserts the information in the database as a new record if it is a case of submission. In case of retrieval however it reads the information stored in the databases and brings it back to the front end and makes it visible to us.

A common example of form being used by the user on internet to submit/retrieve information from internet is shown in Fig. 2.2. The form usually is written in front end languages like HTML, DHTML, etc. As soon as we press submit button after filling in all the information it is read at 2 levels: at front end and at the back end. Front end languages like Javascript or VB script examine the information before sending it to the back end program. For example in the text box for e-mail address it would check for presence of symbol '@'. Similarly if the program wants to check if all the boxes in the form are filled or not, it would first be checked by Java script or VB script at the front

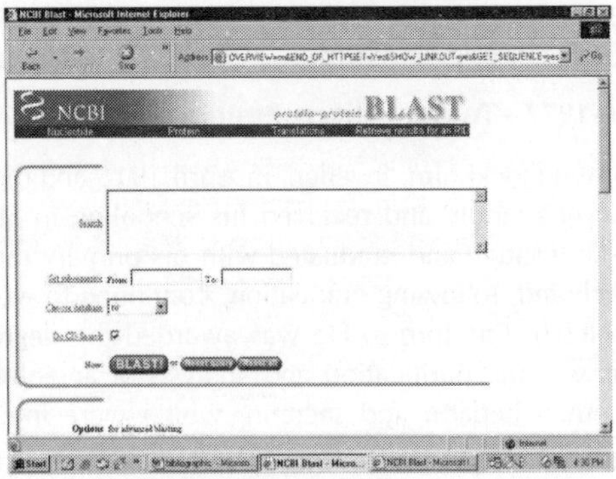

An example form used by the user
↓
As soon as Blast button is pressed it is read at two levels
↓
Front end level check up if all boxes are filled or not
↓
Back end level check up, if all required information present
↓
Programming languages search databases and retrieve the result
↓
Convert it back to user friendly form
↓
Viewer gets the Information

Fig. 2.2: Various steps of computer protocol in submitting and retrieving information in a database

end. All the information given in form is then collected by back end programming languages like active server pages (ASP) and CGI perl and converts this to database language like SQL. SQL statements, search the database and give the result. The same back end language like CGI perl or ASP convert this information back to HTML and display the information in user friendly way.

We would understand this process with the help of Experiment 2A with the help of 10 hypothetical sequences.

EXPERIMENT 2A

To Create a Small Biological Database and Perform the Similarity Searches using this Biological Database

The purpose of this experiment is to increase the practical understanding about relational databases and their functioning. Proper understanding of the experiment would not be possible without explanation to the three components i.e. LINUX, MySQL and SQL.

LINUX: Linux is an Operating System (OS) for computers like Windows from Microsoft. This was first developed and released to the World by LINUX Benedict Torvalds in 1991. Quite simply, the operating system provides users with the means to manipulate files and execute programs by providing access to the computer's hardware. The OS manages the computer's various programs and the memory they use so they don't interfere with each other and governs each program's access to the CPU so that one application doesn't hog the machine. The OS also manages individual users running programs on the same machine and keeps them from interfering with each other while governing each user's access to various system files and other user's directories through a series of permissions set by the system administrator. Also, the OS provides the means for programmers to access the hardware of the computer and the networks it is connected to. By providing an abstraction of the hardware, programs can take advantage of hardware features through a standard Applications Programming Interface, API. With Linux's aim of POSIX compliance (an IEEE standard for operating systems) and the use of the GNU C library, it provides programmers with an excellent platform to write programs that are portable to other flavors of UNIX. Using ANSI C and one of several portability GUI toolkits, programs can also be written for both UNIX and Windows platforms.

Gradually Linux is emerging as a very popular operating system. While most Linux distributions include the popular command line system administration utilities from the GNU Project, many developers around the world are working to enhance these tools to make them easy to use by desktop users. The goal is a desktop solution easily administered from a workstation's GUI for Intel x86, Motorola PPC, Sun Sparc, Compaq Alpha, Intel Strong Arm, and a host of other microprocessor based computers. Linux is a Free Software.

MySQL: is a RDBMS type of database that works very well on Linux platform. It is a SQL database available free of cost with Linux. It can handle very complex data hence it is very commonly being used in commercial organizations government agencies (e.g. NASA uses MySQL), research organizations and educational institutions.

SQL: Structured Query Language (pronounced *sequel* in the US; *ess-queue-ell* elsewhere). SQL is an ANSI (American National Standards Institute) standard computer language for accessing and manipulating database systems. SQL statements are used to retrieve and update data in a database. SQL works with database programs like MS Access, DB2, Informix, MS SQL Server, Oracle, Sybase, etc. Unfortunately, there are many different versions of the SQL language, but to be in compliance with the ANSI standard; they must support the same major keywords in a similar manner (such as SELECT, UPDATE, DELETE, INSERT, WHERE, and others).

Hypothesis: It would be possible to create and search a small RDBMS database of given sequences using MYSQL in Linux operating system.

TESTING THE HYPOTHESIS

Computer set up: A normal PC with Linux as operating system. Please check that the system administrator has already activated Mysql database on this system.

Starting material: Ten DNA sequences belonging to five genes in 10 different organisms are saved as experiment 2A.doc is given in data box 2.1. Taxonomic information about these organisms is given in data box 2.2.

1. Log into using password and username as provided by the system administrator. You would get a screen like shown in Fig. 2.3.
2. Create a database named sequence by passing the statement 'Create database sequence, in the consol for Mysql as shown in data box 2.3.
3. Create tables named Table 1 and taxonomy using SQL statements given in data box 2.3.
4. Insert the sequences in Table 1 along with the name of organisms by using insert SQL statements given in data box 2.3. Similarly insert the data of table taxonomy.
5. Query the database using SQL statements given in data box 2.4.

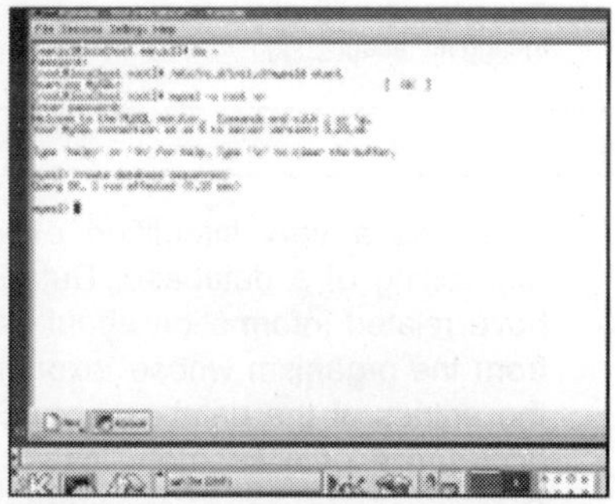

Fig. 2.3: Web view of MySQL

DATA

Data Box 2.1: Nucleotide sequence from ten different organisms for chloroplast rbcL gene for ribulose-1, 5-bisphosphate, carboxylase/oxygenase large subunit.

	Sequence	Organism
1.	tgcaggattt aaagcggggg taaaagacta tcgtctaact tattacacac	*Bryopsis plumosa*
2.	aggtgctgga tttaaagctg gtgtaaaaga ttatcgttta acatattata	*Caulerpa brachypus*
3.	tggtgctgga tttaaagccg gtgtaaaaga ttatcgttta acatattata	*Caulerpa okamurae*
4.	aggtgctgga tttaaagctg gtgtaaaaga ttatcgttta acatattata	*Caulerpa racemosa*
5.	tggtgctgga tttaaagctg gtgtaaaaga ttatcgttta acatattata	*Caulerpa racemosa*
6.	tggagctgga tttaaagctg gtgttaaaga ttatcgttta acatactata	*Halimeda discoidea*
7.	cggagctgga tttaaagctg gtgttaaaga ttatcgttta acgtattata	*Halimeda opuntia*
8.	aggagcagga tttaaagcag gtgtaaaaga ttatcgttta acctattaca	*Dichotomosiphon tuberosus*
9.	tgcaggattt aaagcaggag ttaaagatta tcgactaact tattatacac	*Codium lucasii*
10.	tgccggcttt aaagcagggg taaaagatta tcgtttaact tattatactc	*Pedobesia ryukyuensis*

Data Box 2.2: Taxonomic classification of these organisms

1. *Bryopsis plumosa,* Byopsidaceae, Caulerpales, Ulvophyceae, Chlorophyta, Viridiplantae, Eukaryota
2. *Caulerpa brachypus,* Calerpaceae, Caulerpales, Ulvophyceae, Chlorophyta, Viridiplantae, Eukaryota
3. *Caulerpa okamurae,* Calerpaceae, Caulerpales , Ulvophyceae, Chlorophyta, Viridiplantae, Eukaryota
4. *Caulerpa racemosa,* Calerpaceae, Caulerpales, Ulvophyceae, Chlorophyta, Viridiplantae, Eukaryota
5. *Caulerpa racemosa,* Calerpaceae, Caulerpales, Ulvophyceae, Chlorophyta, Viridiplantae, Eukaryota
6. *Halimeda discoidea,* Udoteaceae, Caulerpales, Ulvophyceae, Chlorophyta, Viridiplantae, Eukaryota
7. *Halimeda opuntia* Udoteaceae, Caulerpales, Ulvophyceae, Chlorophyta, Viridiplantae, Eukaryota
8. *Dichotomosiphon tuberosus,* Dichotomosiphon, Caulerpales , Ulvophyceae, Chlorophyta, Viridiplantae, Eukaryota
9. *Codium lucasii,* Byopsidaceae, Caulerpales, Ulvophyceae, Chlorophyta, Viridiplantae, Eukaryota
10. *Pedobesia ryukyuensis,* Byopsidaceae, Caulerpales, Ulvophyceae, Chlorophyta, Viridiplantae, Eukaryota

Data Box 2.3: SQL Statements passed for creating the database

1. Create database sequence.
2. Create table_1 [id int not null auto_increment primary key, sequence varchar (50), organism varchar (50)].
3. Insert into table_1 values ('tgcaggattt aaagcggggg taaaagacta tcgtctaact tattacacac', *'Bryopsis plumosa'*).
4. Insert into table_1 values ('aggtgctgga tttaaagctg gtgtaaaaga ttatcgttta acatattata', *'Caulerpa brachypus'*);
   and so on for all the ten entries.
5. commit..
6. create table taxonomy [id int not null auto increment primary key, organism_name varchar (50), classification varchar (100)].
7. Insert into taxonomy values (*'Bryopsis plumosa',* 'Byopsidaceae, Caulerpales, Ulvophyceae, Chlorophyta, Viridiplantae, Eukaryota').
8. Insert into taxonomy values (*'Caulerpa brachypus',* Calerpaceae, Caulerpales, Ulvophyceae, Chlorophyta, Viridiplantae, Eukaryota').
   And so on for all the ten entries.
9. Commit.

Data Box 2.4: SQL statements for querying the results

Select sequence from sequence where organism = *'Bryopsis plumosa'*.
This gives the result .
tgcaggattt aaagcggggg taaaagacta tcgtctaact tattacacac.

INTERPRETATION OF THE DATA

This was a very simplified experiment and helps in understanding about the structure and functioning of a database. Database here is made of 2 tables. Table 1 and taxonomy and both have related information about same organisms. Table 1 has information about sequence of DNA from the organism whose taxonomy is given in table taxonomy. Using different SQL statement all the entries of the database could be queried in a desired way. For example if we want to search occurrence of a segment 'aaa' in the given sequence this could be done with SQL statement

Select organism, sequence from Table 1 where sequence like '%aaa%'.

This statement would extract all the entries with aaa in their sequence structure and first display the name of the organism and then the complete sequence. Once this database is transferred to World Wide Web it could be accessed from anywhere in the world.

One important concept, which needs to be clarified at this stage, is how we get information in a user-friendly language rather than SQL, when we search any biological database available on Internet. The various steps of computer protocol in this process i.e. from submitting a form to retrieving information are depicted in Fig. 2.2. As depicted in the figure we make entry into a form, which represent the front end of the database. As soon as we submit this form the submit button, it invokes a program that reads all the information entered into the form, converts it to SQL and sends to the database at the back end. These programs are written in several languages like VBscript, cgi-perl, ASP etc. Search statements created by the program retrieves the desired information and deliver it to the program. Program again converts this SQL result into a user-friendly language and brings it to front end. Topic of programming is taken up once again in chapter 9 on sequence analysis with programming.

2.4 USES OF MAINTAINING BIOLOGICAL DATABASES

Biological databases maintain facilities for a variety of information retrieval and analysis functions. These databases are really useful for the following purposes:

1. Retrieval of sequence for a known gene or protein.
2. Sequence comparison in all cases like for a new sequence, whether it exists earlier can easily be determined with these databases and sequences similar to this can be studied.
3. Translation of DNA sequences to protein sequences.
4. Macromolecule structure analysis and prediction.
5. Determination of patterns in the sequences.
6. Gene annotation.
7. Gene functioning.
8. Phylogenetic and evolutionary relationships among genes and organisms.
9. Study of interactions in biological molecules.
10. Study of taxonomic classification and ecological distribution of species.
11. Bibliographic searches.

2.5 DIFFICULTIES WITH PRESENT DAY BIOLOGICAL DATABASES

Managing biological data and extracting useful information from these databases poses particular problems for the following reasons:

1. The data about any given biological molecule is held at many different sites each with a different approach.
2. Even if appropriate biological information is retrieved there are many different and incompatible formats and nomenclatures for data about the object and it's functioning.
3. Biological data is in different forms: structured as in Genbank or PDB file; unstructured as in biomedical literature abstracts; sequence in genomic or protein sequences or more complex in biochemical pathways.
4. There is no single object to which the biological data refers: individual have their own perspective depending on science they are carrying out ranging from atomic level to that of whole organisms.

SUMMARY

All modern biological data whether molecular genetics data in the form of DNA, RNA or animo acid sequence or taxonomic information about an organism all are stored in typical computer databases. These databases are of two basic forms; a single file and multiple files. All files in a multiple file databases are related to each other, because of this they are called as relational databases. The program used to create and maintain a computer database is called Database Management System (DBMS) and Relational Database Management System (RDBMS) when it is used for creating and maintaining relational databases. There are different types of database softwares and programs available which work on different platforms.

Biological databases and data accumulation has three milestones: the invention of sequencing reaction, the polymerase chain reaction and automated DNA sequencers. This all is a result of break through research of Dr. Frederick Sanger, Dr. Alan Maxam, Dr. Walter Gilbert, Dr. Craig Venter (Historical Profile 19) and Dr. Pehr Victor Edman (Historical Profile 3). Structure and functioning of biological databases is understood by means of 10 hypothetical sequences in Experiment 2A. This gives clear insight into the process of front end activities and back end processes on various web sites which have a form for filling information and retrieve information from databases.

EXERCISE

1. What is a database? What types of information are stored within a database? Where does the information come from?
2. What are the contributions of followings to biological sequence databases?
   (*a*) Sanger
   (*b*) Gilbert
   (*c*) Edman
   (*d*) Pauling
   (*e*) Mullis
   (*f*) Smith
   (*g*) Maxam
3. What are different types of databases other than MySQL which could have been used in Experiment 2A.
4. Enter following entry to the database sequence created earlier in Experiment 2A?
   Caulerpa racemosa, Caulerpales, Ulvophycease, Chlorophyta, Viridiplantae, Eukaryota
   Search the above database for how many entries for sequences belonging to *Caulerpa*?
5. How many types of biological databases exist on world wide web?
6. Write a short one on the followings:
   (*a*) submitting information to a database
   (*b*) retrieving information
   (*c*) back end
   (*d*) front end
   (*e*) biological database management.
7. Understand the functioning of Linux operating system. Create a database of books using MySQL in Linux operation system. It should contain information about title, authors, year of publication, and contents. Care is to be taken that there are no duplicate entries.
8. Log in to web site of SRS and find out how many biological databases are existing today.

3

PROTEIN SEQUENCE DATABASES

The protein databases are the most comprehensive source of information on proteins. Determination of protein/peptide sequence is basic requirement for biomedical research including cancer research. It is absolutely essential for characterizing and identifying proteins or peptides. Imagine if you are a biologist who has discovered a new peptide sequence, either theoretically translated from a nucleotide sequence or isolated from a gel. The first step in this process is to look for homology and similarity with already discovered peptide/protein. Homologus sequences are being related by evolutionary process and homologus is not synonymous of similarity. This is accomplished by comparing the novel sequence with those contained in protein databases. Besides deriving a conclusion that it is an unknown protein, lots of features of this protein/peptide including physical properties, secondary structure, protein family, protein fold etc. may be judged using these protein databases.

Protein sequences were the first sequences to be collected and stored in the form of databases. The development of protein sequencing methods by Sanger and Tuppy in 1951 led to the sequencing of human hormone insulin in 1955 and several other proteins in later years. In 1965, Margrett Dayhoff (Historical Profile 1) and her collaborators were the first to assemble these sequences in to a protein sequence atlas known as Protein Information Resource (PIR). With time with simplification and automation of sequencing techniques, we have uncountable protein sequences and protein sequence databases ranging from universal databases covering proteins from all species and specialized data collections storing information about a specific families or groups of proteins or about the proteins of a specific organism (proteomes).

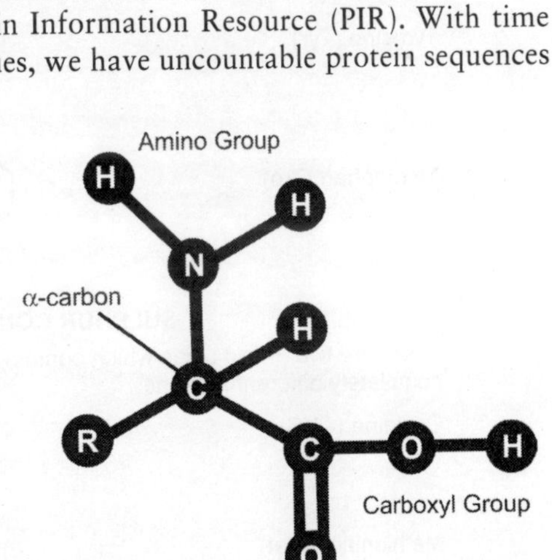

Fig. 3.1: Structure of an amino acid

Protein sequences are written as linear alphabets denoting each amino acid. A polypeptide in a protein is the immediate product of translation and its amino acid sequence is determined by the nucleotide sequences in the mRNA. The 20 amino acids utilized in translation differ in their chemical properties. Amino acids have a common structure consisting of a central carbon with a carboxyl group, an amino group and a hydrogen atom bound to it (Fig. 3.1). This leaves one binding site to

ALIPHATIC AMINO ACIDS

Aliphatic amino acids are hydrophobic. They do not like to be in contact with water molecules in an aqueous solution. For this reason, they are often located in the core of the protein, surrounded by the rest of the protein, and " shielded" by them from the aqueous surroundings.

Glycine (Gly)

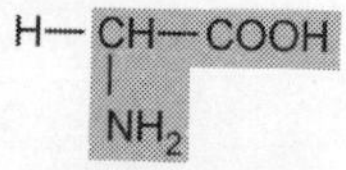

Alanine (Ala)

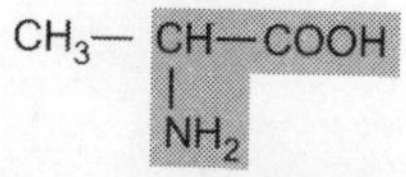

Valine (Val) (CH_3)

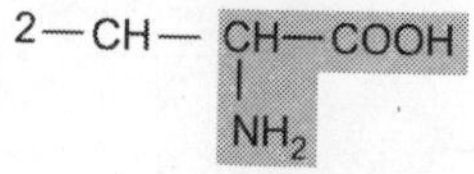

Leucine (Leu) (CH_3)

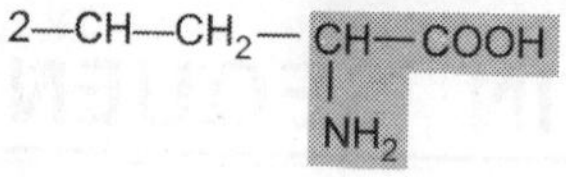

Isoleucine (Ile;)

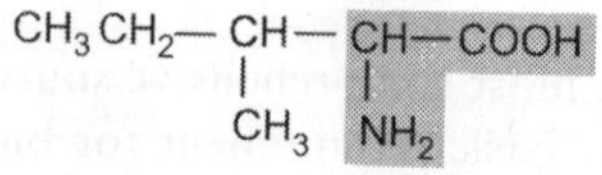

Proline (Pro)

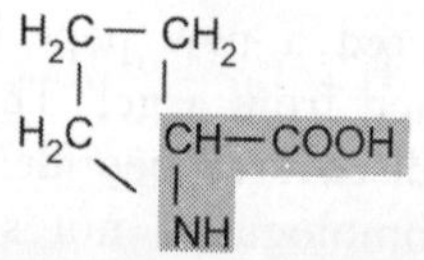

AROMATIC AMINO ACIDS

These are the amino acids which contain an aromatic ring as part of their side chains. Because of the hydrophobic nature of aromatic rings these amino acids are highly hydrophobic

Phenylalanine (Phe)

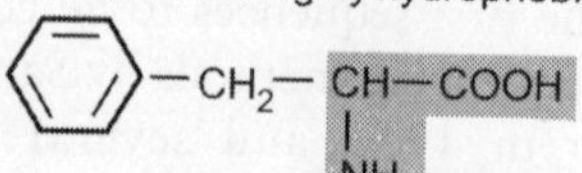

Tyrosine (Tyr)

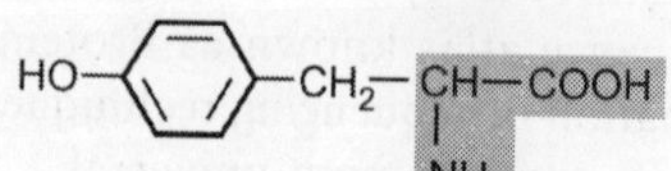

Tryptophan (Trp)

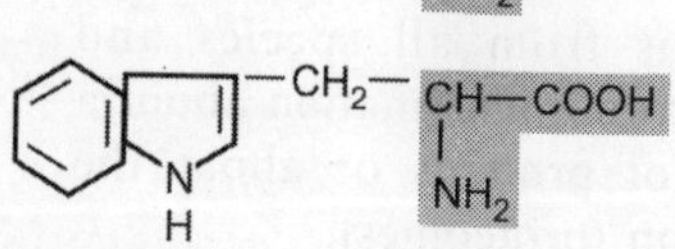

SULPHUR CONTAINING AMINO ACIDS

There are two amino acids which contain a sulphur atom. They are both very special but for completely different reasons.

Cysteine (CYS)

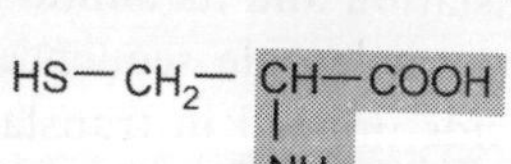

Methionine (Met)

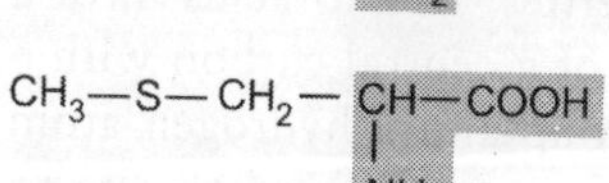

HYDROPHILLIC AMINO ACIDS

These amino acids can be split into a several groups. There are neutral; acidic basic.

A. Acidic Amino Acids

These amino acids are highly polar, and are nearly always negatively charged at physiological pH.

Aspartatic Acid (Asp)

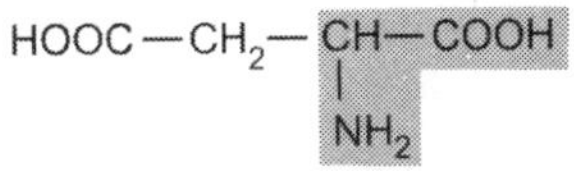

Glutamic acid (Glu)

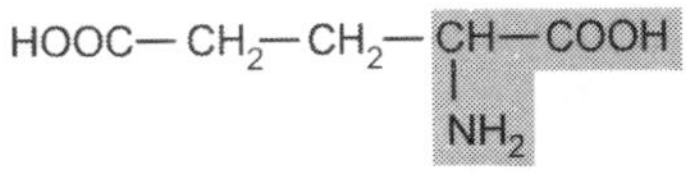

B. Basic Amino Acids

These amino acids contain side chains which are positively charged at physiological. pH. The pKa of Histidine, which is between 6 and 7 in proteins means that it is able to accept or donate proteins at physiological pH. For this reason histidine is often found at the active site of enzymes.

Lysine (Lys)

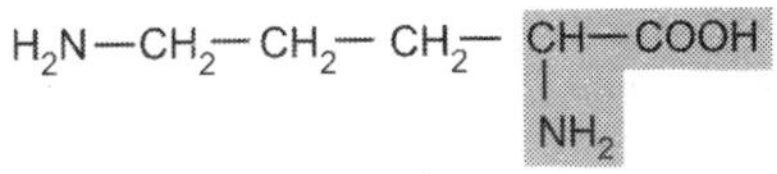

Arginine (Arg)

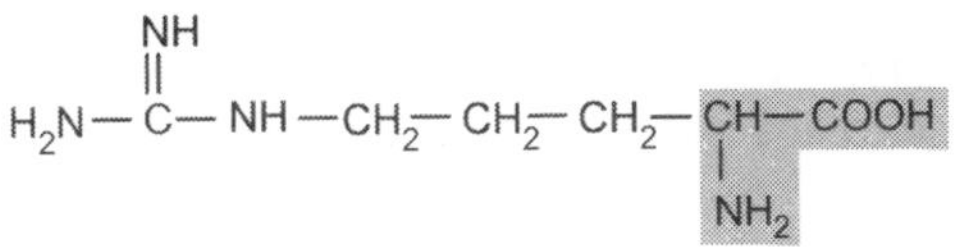

Histidine (His)

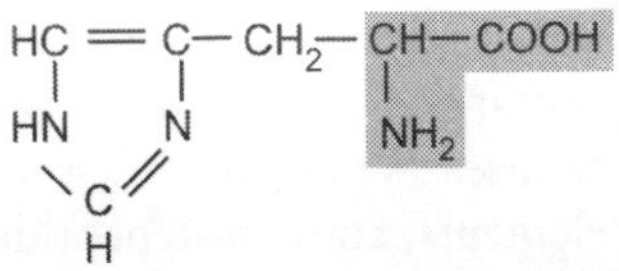

C. Neutral Polar Amino Acids

These amino acids are not charged at physiological pH. However they all have groups on their side-chains which are polar and can from

Serine (Ser)

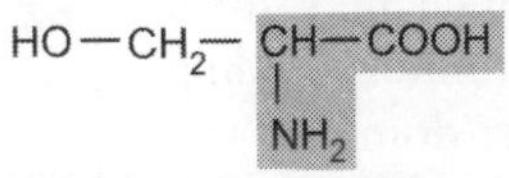

Threonine (Thr)

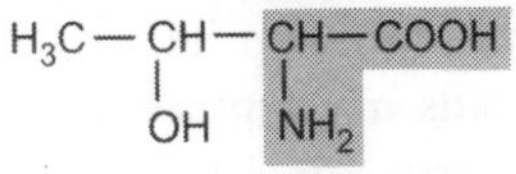

Asparagine (Asn)

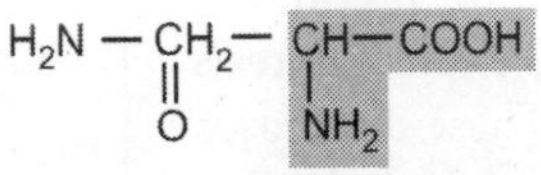

Glutamine (Glu)

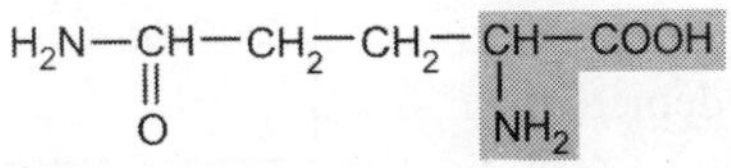

Fig. 3.2: Classification of amino acids based on different R group

which side groups may bind. The side group of an amino acid is denoted by R and is the only part of the molecule that differs from one amino acid to other. Based on R group, amino acids are categorized in to several groups as shown in Fig. 3.1. The proportion and location of an amino acid and the chemical properties of R group determine the structure and function of this protein.

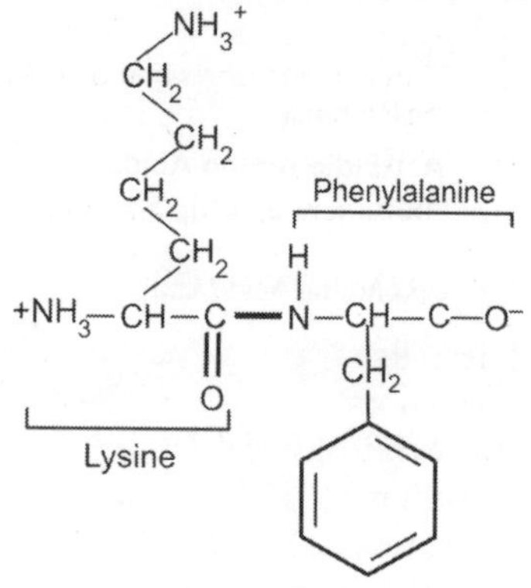

Fig. 3.3: A peptide bond

Amino acids can be joined together to form a peptide or polypeptide. They are called peptides because when the carboxyl group of one amino acid joins to the amino group of another, a peptide bond is formed. Chemically, this is an amide bond but when it occurs in proteins it is given the name peptide bond (Fig. 3.3). In a polypeptide, there are many peptide bonds. These bonds are rigid and planar due to electron sharing between the carboxyl carbon and the amide nitrogen which contribute to the bond and give it a partial double-bond character.

Classification of the types of protein databases is similar to the structural classification of proteins. As the protein structure is classified into primary, secondary, tertiary and quaternary structures, the protein sequence databases are classified in the present book into primary, secondary, structural and specialized protein databases. Like the primary structure of protein mainly deals with the amino acid sequence, the primary databases stores the amino acids sequences as linear alphabets that denotes the constituent amino acid residue. The secondary structure of a protein corresponds to spatial arrangement of amino acid residues that are near to each other in a linear sequence. α-helix, β-strands, coils etc. all are elements of secondary structures. One or more secondary structural elements very often follows a regular pattern like α-β-α- or β-α-β etc. These structures are called as motif and the level of structural organization is called as super-secondary structure. These are stored in secondary protein databases as patterns. As a result of interactions between amino acids contained in secondary structure elements, some polypeptide chains fold into two or more compact regions that may be joined by a flexible segment of polypeptide chain rather like pearls in a string. These compact units are called as domains. Tertiary structure arises from the packing of it's secondary structure elements, which may form discrete domains with a fold, autonomous folding units or modules; complete folds, domains and modules are stored in structural databases as set of atomic coordinates. Protein structure and its classification are discussed in details in chapter 15. Lastly, the properties of proteins like antibodies, enzyme reactions and metabolic pathway, mass spectrometry or gel electrophoresis are used to construct databases. These databases are studied under specialized databases. In Fig. 3.4, various levels of protein structure and their organization into databases are depicted.

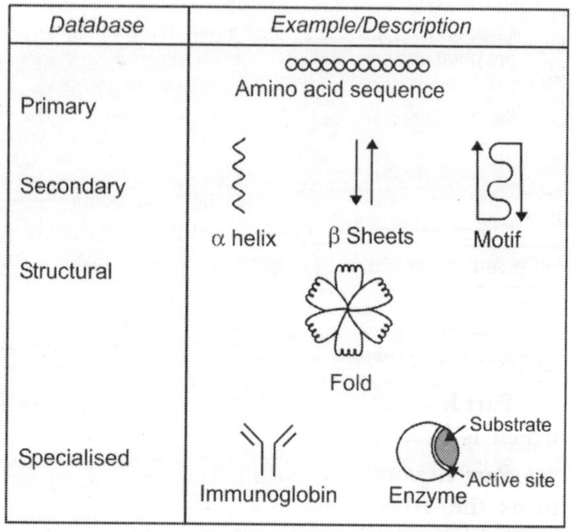

Fig. 3.4: Different types of protein databases based on structure of proteins

The various types of protein sequence databases are classified into four types:

1. Primary protein sequence database
2. Secondary sequence database
3. Protein structure database
4. Specialized protein database

3.1 PRIMARY PROTEIN SEQUENCE DATABASE

These usually have information about amino acids sequence in a peptide, which usually have been derived as a result of original research. Each amino acid is represented as an alphabet. Single word abbreviation for each amino acid is given in Table 3.1. UniProt (Universal Protein resource) is the world's most comprehensive and latest catalogues of information of proteins. It is central to repository of protein sequence and function created recently by joining the information contained in Swiss-Prot, TrEMBL and PIR, which existed independently earlier. PIR (Protein Identification Resource) was developed in 1960 by Late Dr. Margaret Dayhoff as a collection protein sequences for investigating evolutionary relationship among proteins. From 1988 onwards it was maintained collectively by NBRF, MIPS (Munich Center for Protein Sequences and JIPID (Japan International Protein Information Database) as PIR–International Protein Sequence Database. Department of Medical Biochemistry at University of Genava produced Swiss-Prot protein sequence database in 1986. Amos Bairock (Historical Profile 4) deduced the patterns of amino acids in proteins. His collection of protein sequences later became known as Swiss-Prot. The database endeavoured to produce a high level annotation, including description of the function of the proteins and the structure of the domains, it's post translational modifications and so on. TrEMBL was created in 1996 as a computer supplement to SwissProt and contained translations of all coding sequences in EMBL, a DNA sequence database.

Table 3.1: Abbreviation used for each amino acid in a protein sequence

One letter code	*Full name of Amino acid*	*One letter code*	*Full name of amino acid*
A	ala	R	arg
N	asn	D	asp
B	asx	C	cys
Q	gln	E	glu
G	gly	Z	glx
H	his	I	ileu
L	leu	K	lys
M	met	F	phe
P	pro	S	ser
T	thr	W	trp
Y	tyr	V	val
–	deletion	*	nonnsense
X	(stop)	?	X

* If given at the end means the end of the sequence and must be removed before further processing the sequence.

The current UniProt Release 12 (24 Jul 2007), consists of 4,949,164 entries. It consists of two parts:

Part I: containing fully manually annotated records, referred to as the Swiss-Prot Knowledgebase, which is having 2,76,256 entries.

Part II: containing computationally analyzed records awaiting full manual annotation, referred to as the TrEMBL Protein Database that is having 4,72,908 entries. UniProt databases can be accessed from the web at http://www.uniprot.org and downloaded from http://www.uniprot.org/database/download.html. UniProt database is studied in details in Experiment 3A where, it is used to study the enzyme ribulose bisphosphate carboxylase.

3.2 SECONDARY PROTEIN DATABASES

Secondary databases are called secondary not that they talk about the secondary structure but also because they contain the results of analysis of the sequences in primary sources. They are also called as pattern databases. Very often the sequence of an unknown protein is too distantly related to any of the known sequence but it can be identified by the occurrence of a particular cluster of secondary structure elements (eg. α-β-α or β-α-β) that is commonly called as a motif. Concept of motif was given in 1981 by Russel D. Doolittle (Historical Profile 5) and has turned out to be invaluable in studying secondary structure of a protein.

An amino acid sequence in a protein is divided into conserved and variable region. Conserved region shows little or no variation in homology sequences and are typically around 10 to 20 amino acids in length. This conserved region refers to motifs or regular expressions, sequence pattern or sequence motif. Motifs usually correspond to core structure and functional elements of a protein and are crucial for normal structure and function of a protein. Their conserved nature, allows them to be used for identification of the protein family to which a sequence belongs, through a range of sequence analysis techniques given in details in section II. Secondary protein databases are classified into three types based on single motif or multiple motif or full domain alignment is used for construction of the database. Examples of each type are given in Table 3.2.

A motif match finds matches of short sequence in one or more regions internal to a long one. In motif matching, one mismatching character is allowed. Alternatively, one could demand perfect matches, or could allow more mismatches or even gaps. An example for regular expression or motif region is:

C-x(2,4)-C-x(3)-[LIVMFYWC]-x(8)-H-x(3,5)-H

A sequence matches this pattern, if it contains a C followed by 2-4 arbitrary characters followed by C and 3 arbitrary letters and any one of L, I, V, M, F, Y, W or C and so on. Most classical zinc-finger DNA binding domain sequences match this pattern.

Table 3.2: Types of secondary protein databases

Type	*Example*	*Basis*
Single motif	Prosite	Patterns
	Identify	Patterns
Multiple motifs	Prints	Finger prints
	Blocks	Blocks
Full Domain	Profiles	Profiles
	pfams	Hidden Markov Models

In single motif databases, amino acid sequence in one motif is compared. The amino acid sequence, usually three to four amino acid in length, within a motif is reduced to a single, short consensus expression, which is expected to repeat at regular interval and follow a regular pattern. Such motifs where it is possible to identify regular expressions, are also called as patterns. Prosite and identify are examples single motif databases and are based on patterns.

Sequence A	AKSHAT	The	LAST	FA-T	BAY
Sequence B	AKSHAT	The	FAST	BA-T	- - -
Sequence C	AKSHAT	The	VERY FAST		B A T
Sequence D	- - - - - -	The	- - - -	FA-T	- - -

Fig. 3.5: Study of patterns in four sequences A, B, C, D

In prosite presence of a pattern is determined by construction of multiple alignment and by manual

HISTORICAL PROFILE 4

Dr. Amos Bairoch for Protein Databases and Softwares Tool

Dr. Amos Bairoch

His main work is in the field of protein sequence analysis and more particularly in the development of databases and software tools for this purpose. While working on PC/Gene he started to develop an annotated protein sequence database, which became SWISS-PROT. It was first released in July 1986. From 1988 onward it has been a collaborative project with the Data Library group of the EMBL. SWISS-PROT has grown both in size, in its scope and in the amount of work and people necessary to produce it. In 1988, he started to develop Prosite, a database of protein families and domains. A little while later he created Enzyme, a nomenclature database on enzymes as well as SeqAnalRef, a sequence analysis bibliographic reference database. In collaboration with Ron Appel he initiated in 1993, the first molecular biology WWW server, ExPASy. What was intended as a prototype grew rapidely into a major site that provides access to the many databases produced partially or completely in Geneva as well as many tools for the analysis of proteins.

He was involved with his colleagues in Geneva and Lausanne in setting up the Swiss Institute of Bioinformatics (SIB), whose mission is to establish in Switzerland a center of excellence in the field of bioinformatics with an emphasis on research, education, services and the developments of databases and tools. He is currently one of the six group leaders of the SIB. He has also founded GeneBio (Geneva Bioinformatics SA), a company involved in biological knowledge. IGeneProt (Geneva Proteomics), a high throughput proteomics company with his other colleagues.

inspection. In the four sequences A, B, C, and D given in Fig. 3.5 occurrence of alphabets 'THE' forms a pattern or motif. Similarly presence of 'Akshat' forms another pattern or motif.

In Identify on the other hand, a consensus sequence is generated by the program called eMOTIF. Here rather than encoding exact information observed at each position in a motif, eMOTIF adopts a 'fuzzy' approach in which alternative residues are tolerated according to a set of prescribed groupings such as small, small hydroxyl, basic, aromatic, small hydrophobic, etc.

Prosite introduced in 1968 provides a carefully crafted database of sequence patterns which are suitable for searching. Prosite, is a secondary database with a special value i.e. an extensive documentation on protein families, as defined by sequence domains and motifs is available. It was the first secondary database that is developed and maintained by the Swiss Institute of Bioinformatics. It contains biologically significant sites and patterns formulated in such a way that with appropriate computational tools it can rapidly and reliably work out to which family of proteins this new sequence belongs.

Most of the protein families are characterized not by one but several conserved motifs. When instead of one, several motifs are used to build a database; it is called as multiple motif secondary databases. The group of motifs, used as diagnostic signatures of a protein family, is called a fingerprint. Fingerprints have a better diagnostic reliability over single motif analysis methods because of obvious reasons. Prints is an example of fingerprint database. Blocks is another example of multiple sequence motif database. Here motifs or blocks are created in a given sequence automatically by detecting the most highly conserved region of each family. Steven Henikoff and Jorga Henikoff examined the alignments of protein sequences and gave the concept of blocks (Historical Profile 6). They are also known for producing the blossum matrices.

The full domain type of secondary protein database is based on the philosophy that the variable region of a protein also contains valuable information. Here, complete alignment is taken into consideration rather than the conserved regions. Profile is the unit of discrimination here. Profile is defined as a position specific scoring table that encapsulates sequence information within complete alignment. The profiled is weighed to indicate where insertion and deletion are allowed and what type of residues are allowed and at what position and where are the most conserved regions. In addition to data implicit in alignment the scoring system may include evolutionary weights and results from structural studies. Variable penalties are specified to weigh against insertions and deletions occurring in secondary structure elements.

Pfam is another example of full domain type of secondary database. It is based on Hidden Markov Model (HMM) a probabilistic model consisting of a number of interconnecting states. Like profiles HMMs encodes full domain alignments. They are linear chains of match, delete or insert states. A match state denotes a conserved column in an alignment and insert state allows insertion relative to match states, and delete state allows positions to be skipped. Pfam is used for identification of protein family of ribulose bisphosphate carboxylase in experiment 3A. At the time of writing of this book Pfam has 1344 entries.

3.3 STRUCTURAL PROTEIN DATABASES

Proteins may be classified according to both structural and sequence similarity. For structural classification, the sizes and spatial arrangement of secondary structures are compared in known three-dimensional structural databases. Study of structure of protein based on similarity searches from structural protein databases is very important because of following reasons:

1. Two entirely different protein sequences from different evolutionary origins may fold into a similar structure. Conversely, the sequence of an ancient gene for a given structure may have diverged in different species while at the same time maintaining same basic structural features.
2. Two proteins that share a significant degree of sequence similarity with each other and also share an evolutionary origin, should also share some structural features also.
3. There exist proteins with similar 3D structure however different amino acid sequence.

While studying the tertiary structure of a protein, we talk about protein family, folds, domains and modules. It would really be difficult to study protein structure without knowing what exactly is meant by these terms. Protein family is defined originally by Dayhoff and coworkers in 1978, is a group of proteins with similar biochemical function, which are more than 50%, aligned. A protein family comprises proteins with same function in different organisms (orthologus sequences) but may also include protein in the same organism (paralogus sequences) derived from gene duplication and rearrangement. Fold, in general, refers to somewhat larger combination of secondary structural units in the same configuration. Thus, the proteins sharing the same fold, have same combination of secondary structural elements. An example is a Rosman fold, comprising of several alternating α helices and β strands. Module is a region of conserved amino acids patterns, comprising of one or motifs and considered to be fundamental unit of structure and function. Presence of a module has

HISTORICAL PROFILE 5

Dr. Russel F. Doolittle for Giving the Concepts of a Sequence Motif in 1981

Dr. Russel F. Doolittle

Dr. Russel F. Doolittle is Professor at University of California San Diego in 1981 he reported that the systemic comparison of every newly determined amino acid sequence with all other known sequences may allow a complete reconstruction of the evolutionary events leading to contemporary proteins. He gave the concept of motif in sequence analysis.

Among the questions his group trying to resolve at present are: When did prokaryotic organisms diverge from eukaryotes? How many rudimentary families of proteins are there? How are new proteins invented? All of these questions can be answered, given enough sequences to compare. Other field of his present work is how clotting works and from where some of the components in this process come. Fibrinogen is a multidomained mosaic protein, a key part of which is found in numerous other animal proteins. His group also crystallized a native fibrinogen molecule (chicken fibrinogen = 320 KDa) and solved its X-ray structure.

HISTORICAL PROFILE 6

Dr. Steven Henikoff for Giving the Concepts of Blocks

Dr. Steven Henikoff

Dr. Steven Henikoff along with Jorga Henikoff examined the alignment of proteins sequences that make up each motif and discovered additional patterns in aligned sequences called BLOCKS. These patterns expanded ability to determine whether or not unknown protein possessed a particular biochemical activity. The changes that were in each column of these aligned patterns were counted and a new set off amino acid substitution matrices called BLOSSUM matrices were produced. Of these BLOSSUM62 is most often used for aligning protein sequences and searching databases for similar sequences.

been used to classify proteins into a family. Some polypeptide chains fold into two or more compact regions that may be joined by a flexible segment of polypeptide chain rather like pearls in a string. These compact units are called as domains. Please see chapter 15 for further information on these topics.

Three-dimensional structure of a protein is usually determined by making crystals of the proteins and using X-ray diffraction to determine the position of molecules that are fixed within the crystals. The technique of nuclear magnetic resonance (NMR) is also used to obtain the protein structure. The three-dimensional coordinates of each atom in the protein molecule are stored as a pdb file, which can be easily viewed with a molecular viewer, for example rasmol. The number of known protein structures is increasing very rapidly and these are available through the Protein Data Bank (PDB). A detailed description of protein structure study and prediction is given in chapter 15. PDB gives the atomic coordinates of amino acids in proteins, protein fragments or protein bound to substrate or inhibitors. The Cambridge Crystallographic Data Center (CCDC) provides a database of structures of small molecules, of interest to biologists concerned with protein-ligand interactions. SRS3D is an integrated environment that allows the end-user to quickly and easily retrieve/visualize sequence structure and also feature data from primary, secondary and tertiary protein databases.

3.4 SPECIALIZED PROTEIN DATABASES

There are many specialized protein sequence databases, some of them are quite small and only contain a handful of entries, and others are wider in scope and larger in size. As this category of databases is quite changeable, any list provided here would soon be outdated.

A brief description of some of specialized protein sequence databases follows:

1. **Enzyme databases:** These include databases which highlight the Enzyme properties of proteins. Database enzyme database is an annotated extension of Enzyme commissions publication linked to uniprot/Swiss-prot. There are also databases for enzyme properties such as BRENDA, ligand databases for enzyme reactions, such as LIGAND, and the database of enzymes and metabolic pathways, EMP.
2. **Two-dimensional get electrophoresis data database:** this is available from Expasy and Danish Center for human genomic research.
3. **Mass spectrometry protein data:** This is a useful resource which includes protein cleavage products, is maintained at Rockefeller University.
4. **Proteome database:** The Proteome Analysis database has been set up to provide comprehensive statistical and comparative analyses of the predicted proteomes of fully sequenced organisms. Such databases would revolutionize the science of biology (Software Update 1). The yeast protein database, also known as YPD, is a database for the proteins of *S. cerevisiae.* Based on the detailed curation of the scientific literature for the yeast *Saccharomyces cerevisiae*, YPD contains more than 50,000 annotation lines derived from the review of 8,500 research publications. The information concerning each of the more than 6,000 yeast proteins is structured around a one-page format, the Yeast Protein Report, with additional information provided as pop-up windows. Protein classification schemas are defining each protein's cellular role, function and pathway. YPD provides the user with a summary of the protein's function and its place in the biology of the cell. The first transcript profiling data has been integrated into the YPD Protein Reports, providing the framework for the presentation of genome-wide functional data. Altogether YPD is a very useful data collection for all yeast researchers and especially for those working on the yeast proteome.

SOFTWARE UPDATE 1

Human Protein Database could Change Biology

A human protein database called as Human Protein Reference Database has been developed that could change the world of biology by providing extensive information on proteins and allowing scientists to quickly create testable hypotheses. The Human Protein Reference Database currently contains about 3,000 human proteins and is expected to hold comprehensive information on 10,000. It includes known interactions between proteins, including their roles in health and disease.

This is the real beginning of systems biology in the human and research would go faster and available information could be easier to find and easier to organize. The database contains only experimentally proven or widely accepted facts about proteins involved in diseases and proteins involved in key pathways. Biologists looking for information about a protein can search the database for any of the protein's names and get a single entry that details everything from its alternative names to its whereabouts in cells to links for papers that describe it.

With this single database, biologists now will be able to quickly review what is known about the proteins and how they interact, speeding the creation of new hypotheses to test in the lab. Online, a user can access a visual web of protein-to-protein interactions. The database is expected to become indispensable as proteome database for several human pathogens viz are coming up. This really is going on facilitate the study of host- pathogen interaction studies and drug development. The entries have been critically reviewed, making the information in the database as accurate and complete as possible, Scientists can even link directly to the scientific paper behind an item, to judge for them its validity. The database is reported in the October issue of the journal *Genome Research,* 2003.

5. **MEROPS:** Families of proteins are thought to have common evolutionary origins and are known or expected to have similar tertiary folds are grouped into clans. The MEROPS database provides sets of files called FamCards and ClanCards, describing the individual families and clans. Each FamCard document provides links to other databases for sequence motifs and secondary and tertiary structures, and shows the distribution of the family across the major taxonomic kingdoms.

   This database provides a catalogue and structure-based classification of peptidases (i.e. all proteolytic enzymes). An index of the peptidases by name or synonym gives access to a set of files termed PepCards, each of which provides information on a single peptidase. Each card file contains information on classification and nomenclature, and hypertext links to the relevant entries in other databases. The peptidases are classified into families on the basis of statistically significant similarities between the protein sequences in the part termed the 'peptidase unit' that is most directly responsible for activity.

6. **GPCRDb:** This is a database of sequences and other data relevant to the biology of G-protein coupled receptors (GPCRs), a very large protein family of critical components of many different signaling systems in animals. The information available in a GCRDb entry is not much more extensive than that found in the EMBL nucleotide sequence entry from which it is derived. What make this database useful are not the entries themselves, but the analysis (e.g. multiple alignments, classification into subfamilies) which has been made on the data and which are available from the GPCRDb database. It is a good example for a specialised database adding value by offering an analytical view on data that a universal sequence database is unable to provide.

EXPERIMENT 3A

Study of a Primary Protein Database the UniProt

Purpose of this experiment is to introduce internet, world-wide web and global network of biological information to the students.

Hypothesis: It would be possible to extract all possible information regarding sequence and structure of enzyme, Ribulose bisphosphate carboxylase large chain 2 of *Thiobacillus ferrooxidans* using UNIPROT (the Universal Protein Resource). Information regarding protein family to which it belongs and domain and active site would also be gathered.

TESTING THE HYPOTHESIS

Computer set up: A normal PC with internet connection in windows environment

1. Log on to website of UniProt at (www.uniport.org). This takes to the welcome page of the web site (Fig. 3.6).
2. Select text search and enter Ribulose bisphosphate carboxylase in the text box for search.
3. Observe the results obtained and select only the entries for large chain 2 of *Thiobacillus ferrooxidans*.
4. Study the result and retrieve all possible information about the protein. Save the amino acid sequence as experiment-3A.doc. Entry for *Thiobacillus ferrooxidans* is given in databox 3.1. Go for identification of the motif and protein family using links to secondary protein databases provided in the results.

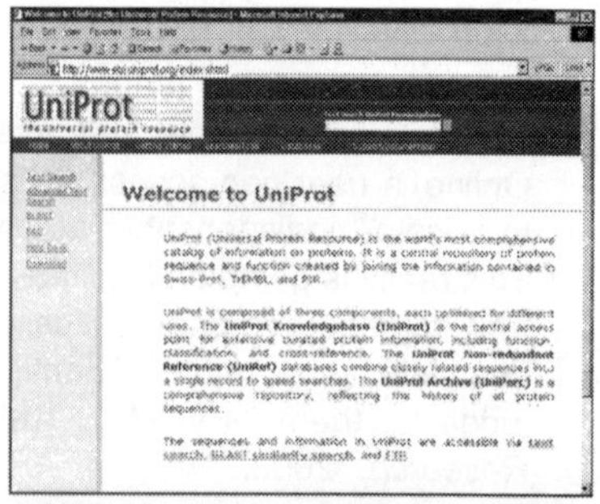

Fig. 3.6: Web view of UniProt

DATA

Data Box 3.1: A portion of UniProt entry for *Thiobacillus ferrooxidans* Hobacprot: RBL2_THIFE

```
ID   RBL2_THIFE    STANDARD;    PRT;    473 AA.
AC   Q07087;
DT   01-NOV-1995 (Rel. 32, Created)
DT   01-NOV-1995 (Rel. 32, Last sequence update)
DT   15-DEC-1998 (Rel. 37, Last annotation update)
DE   RIBULOSE BISPHOSPHATE CARBOXYLASE LARGE CHAIN 2 (EC 4.1.1.39)
DE   (RUBISCO LARGE SUBUNIT).
GN   CBBL2 OR RBCL2.
OS   Thiobacillus ferrooxidans.
OC   Bacteria; Proteobacteria; gamma subdivision; Acidithiobacillus.
OX   NCBI_TaxID=920;
RN   [1]
RP   SEQUENCE FROM N.A.
RX   MEDLINE=92070528; PubMed=1959634;
RA   Pulgar V., Gaete L., Allende J., Orellana O., Jordana X., Jedlicki E.;
RT   "Isolation and nucleotide sequence of the Thiobacillus ferrooxidans
RT   genes for the small and large subunits of ribulose 1,5-bisphosphate
RT   carboxylase/oxygenase.";
RL   FEBS Lett. 292:85-89(1991).
CC   -!-    FUNCTION: RUBISCO CATALYZES TWO REACTIONS: THE CARBOXYLATION OF
CC          D-RIBULOSE 1, 5-BISPHOSPHATE, THE PRIMARY EVENT IN PHOTOSYNTHETIC
CC          CARBON DIOXIDE FIXATION, AS WELL AS THE OXIDATIVE FRAGMENTATION
```

```
CC              OF THE PENTOSE SUBSTRATE IN THE PHOTORESPIRATION PROCESS. BOTH
CC              REACTIONS OCCUR SIMULTANEOUSLY AND IN COMPETITION AT THE SAME
CC              ACTIVE SITE.
CC    -!-       CATALYTIC ACTIVITY: D-RIBULOSE 1,5-BISPHOSPHATE + CO(2) = 2 3-
CC              PHOSPHO-D-GLYCERATE.
CC    -!-       CATALYTIC ACTIVITY: D-RIBULOSE 1,5-BISPHOSPHATE + O(2) =
CC              3-PHOSPHO-D-GLYCERATE + 2-PHOSPHOGLYCOLATE.
CC    -!-       SUBUNIT: 8 LARGE CHAINS + 8 SMALL CHAINS.
CC    -!-       SIMILARITY: BELONGS TO THE RUBISCO LARGE CHAIN FAMILY.
CC    -!-       CAUTION: IN T.FERROXIDANS TWO SIMILAR SET OF GENES CODE FOR
CC              RUBISCO LARGE AND SMALL CHAINS: THE RBCL1-RBCS1 AND THE RBCL2-
CC              RBCS2 SETS.
CC    -!- GENE_FAMILY: HBG014837 [ FAMILY / ALN / TREE ]
DR    EMBL; X70355; CAA49814.1; -.
DR    HSSP; P00880; 1RBL.
DR    InterPro; IPR000685; RuBisCO_large.
DR    Pfam; PF00016; RuBisCO_large; 1.
DR    Pfam; PF02788; RuBisCO_large_N; 1.
DR    PROSITE; PS00157; RUBISCO_LARGE; 1.
DR    PRODOM; Q07087.
DR    SWISS-2DPAGE; Q07087.
KW    Carbon dioxide fixation; Photorespiration; Lyase; Oxidoreductase;
KW    Monooxygenase; Calvin cycle; Multigene family.
FT    DOMAIN   33     85     PRODOM:2001.3:PD417314    493
FT    DOMAIN   90     130    PRODOM:2001.3:PD422231    98
FT    DOMAIN   131    174    PRODOM:2001.3:PD416183    515
FT    DOMAIN   175    266    PRODOM:2001.3:PD000044    532
FT    DOMAIN   267    301    PRODOM:2001.3:PD430602    421
FT    DOMAIN   302    363    PRODOM:2001.3:PD416302    514
FT    DOMAIN   382    421    PRODOM:2001.3:PD429431    426
FT    DOMAIN   422    468    PRODOM:2001.3:PD416521    510
FT    ACT_SITE 194    194    BINDING OF CO(2) ACTIVATES THE ENZYME
FT                           (BY SIMILARITY).
SQ    SEQUENCE   473 AA;  52810 MW;  E78CCC2744DA37C6 CRC64;
      MAVKTYNAGV KDYRNTYWEP DYSVKDTDIL AVFKITPQAG VDREEAPAAV AAESSTGTWT
      TVWTDLLTDL DYYKGRAYRI EDVPGDDTCF YAFIAYPIDL FEEGSVVNVF TSLVGNVFGF
      KAVRALRLED VRFPIAYVKT CGGPPHGIQV ERDIMNKYGR PLLGCTIKPK LGLSAKNYGR
      ACYEGLRGGL DFTKDDENVN SQPFMRWRQR FDFVMEAIQK AEAETGERKG HYLNVTAPTP
      EEMYKRAEYA KEIGAPIIMH DYITGGFCAN TGLANWCRDN GMLLHIHRAM HRVLDRNPHH
      GIHFRVLTKI LRLSGGDHLH SGTVVGKLEG DREATLGWID IMRDRFIKED RSRGIFFDQD
      WGSMPGVMPV ASGGIHVWHM PALVTIFGDD SVLQFGGGTL GHPWGNAKGA AANRVALEAC
      VEARNRGVAI EKEGKAVLTE AAKHSPELKI AMETWKEIKS EFDTVDKLDV AHK
//
```

INTERPRETATION OF THE DATA

Huge amount of the information was gathered about ribulose bisphosphate carboxylase large chain 2 of *Thiobacillus ferrooxidans* using Uniprot database. This protein is about 473 amino acids and encoded by the genes CBBL2 and RBCL2. Rubisco catalyzes two reactions: the carboxylation of d-ribulose 1,5-bisphosphate, the primary event in photosynthetic carbon dioxide fixation, as well as the oxidative fragmentation of the pentose substrate in the photorespiration process. Both reactions occur simultaneously and in competition at the same active site. Molecular weight of the protein is 52810 mw and is comprised of 8 domains with active site at 194 th amino acid. Individual entry from where all this information is derived is given in databox 3.1 and explanation to all the abbreviation in the entry are given in Table 3.3. Information retrieved from Uniprot could be further used to explore the secondary structure of the protein. Various hyperlinks i.e. Pfam, prosite, interpro etc. are there in the data file could be used for exploration of protein families.

Table 3.3: Explanation to Various Terms had in *Uniprofile*

Line code	*Content*	*Occurrence in an entry*
ID	Identification	Once; starts the entry
AC	Accession number(s)	Once or more
DT	Date	Three times
DE	Description	Once or more
GN	Gene name(s)	Optinal
OS	Organism species	Once or more
OG	Organelle	Optinal
OC	Organism classification	Once or more
OX	Taxonomy cross-reference(s)	Once or more
RN	Reference number	Once or more
RP	Reference number	Once or more
RC	Reference comment(s)	Optional
RX	Reference cross-reference(s)	Optional
RG	Reference group	Once or more (Optinal if RA line)
RA	Reference authors	Once or more (Optional if RG line)
RT	Reference title	Optional
RL	Reference location	Once or more
CC	Comments or notes	Optional
DR	Database cross-references	Optional
KW	Keywords	Optional
FT	Feature table data	Optional
SQ	Sequence header	Once
(blanks)	Sequence data	Once or more
//	Termination line	Once; ends the entry

Pfam secondary database was selected here for exploring the protein family to which the present protein belongs. Pfam database identifies it to belong to RuBISCO_large family and is made of 2 domains. One domain called Rubisco_large_N range from 14 to 139 amino acid position and the other called as Rubisco_large is spread from 147 to 455. In between are the connecting amino acids. Pfam database is based on a statistical analysis of sequence with same motif using Hidden Markov Models. The result here is a profile of sequences with matches, mismatches and gaps. In Fig. 3.7 results of pfam searches is shown in graphical format depicting the position of matched domains. It also gives a table of with 4 entries of significant hits.

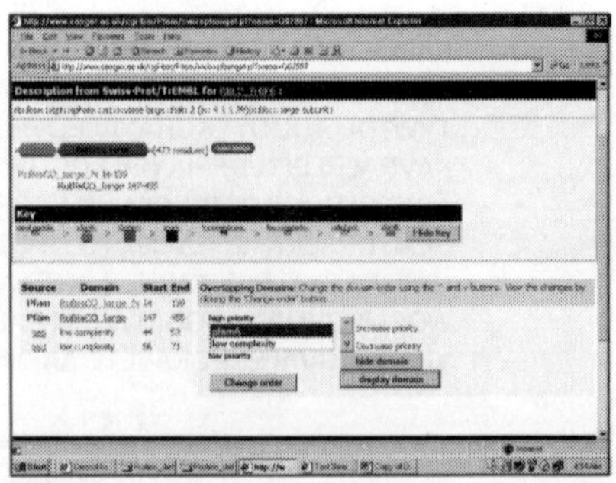

Fig. 3.7: A result of Pfam searches

SUMMARY

There are many different types of protein databases: primary, secondary, structural databases and specialised. Primary databases contain amino acid sequence data UniProt (Universal Protein resource) is the world's most comprehensive and latest catalogues of information of proteins. It is central to repository of protein sequence and function created recently by joining the information contained in Swiss-Prot, TrEMBL and PIR, which existed independently earlier. Secondary databases contain pattern data i.e. diagnostic signature of protein families. Secondary protein databases are classified into three types based on single motif or multiple motif or full domain alignment is used for construction of the database. Three-dimensional structure of a protein is usually determined by making crystals of the proteins and using X-ray diffraction to determine the position of molecules that are fixed within the crystals. The three-dimensional coordinates of each atom in the protein molecule are stored as a pdb file, which can be easily viewed with a molecular viewer for example rasmol. The number of known protein structures is increasing very rapidly and these are available through the Protein Data Bank (PDB).

All possible information regarding sequence and structure of enzyme, Ribulose bisphosphate carboxylase large chain 2 of *Thiobacillus ferrooxidans* was extracted using UNIPROT (the Universal Protein Resource). Information regarding protein family to which it belongs and domain and active site is also be gathered. This not only introduced the students to the latest protein database but also made familiar with internet.

EXERCISE

1. Describe different types of protein databases. How do the classification of protein structure resemble the classification of secondary structure databases?
2. Define the followings:
   (*a*) Motif (*b*) Fold (*c*) Domain (*d*) Module (*e*) Family
3. Describe several criteria used in formulation of secondary structure databases.
4. Log into the web site of PIR, Swiss-prot and trEMBL individually and identify the differences in the information obtained in case of Experiment 3A.
5. Start with the file saved as experiment_3A.doc. Find out sequences similar to this.
6. Retrieve the Swiss-Prot entry for bovine pancreatic trypsin inhibitor and the complete PIR entry for this protein. What information does each have that other does not have?

4

NUCLEIC ACID DATABASES

Nucleic acids are called nucleic because a large proportion of this is present in cell nucleus and relatively much less in other cell organelles like mitochondria and chloroplast. They are acidic in nature when dissolved in an aqueous solution. Nucleic acids are categorized as DNA or RNA based on whether deoxy-ribose or ribose sugar is present. Both DNA and RNA are linear sequences of nucleotides each composed of a nitrogenous base, a sugar and phosphate group; in case of DNA we have A, T, G, C and in RNA instead of T we have U. The bases of DNA molecules carry genetic information, whereas their sugar and phosphate group perform a structure role. Structure of DNA and RNA chain and their structure is discussed in details in chapter 16.

DNA sequencing has really been easy and automated as a result of efforts of several scientists (Box 2.1). This has resulted in accumulation of nucleic acid sequences and in turn the nucleic acid sequences databases. According to DBCAT, as in 2008 there are 87 DNA and 29 RNA databases. This site provide the complete list of these databases. Their effective usage by the scientific community requires a proper organization and management of these databases. Various abbreviations used in DNA and RNA sequences for entry into biological databases are given in Table 4.1.

4.1 TYPES OF NUCLEIC ACID DATABASES

Nucleic acid databases are given different names based on the type of sequence they contain. If they contain a RNA sequence they are called RNA database and if they contain information from a genome they are called genome databases. For example, yeast genome database contain DNA sequences from genome of yeast. Similarly, if nucleic acid database contain a collection of sequences that perform a special function, they are called specialized databases. These sequences are often given special names based on function they are performing. For example, promoter is a nucleic acid sequence present adjacent to the structural gene sequence and has a role in regulation of expression of the structural gene. The eukaryotic promoters contain several sequence elements like CAAT boxes GC boxes and TATA boxes, etc. The prokaryotic promoter gene in case of many *E. coli* contains TTGACA sequence element at -35 site and TATAAT at - 10 sites. Nucleic acid database containing promoter sequences is called promotor sequence database. Similarly structural gene sequence could be classified into exon and intron sequence based on whether they are expressed in the proteins or not. These are classified as exon/intron databases. These specialized databases are very useful in sequence analysis and gene annotation.

The nucleic acid databases in the present book, for the sake of convenience are classified into three categories:

1. Primary nucleic acid databases
2. RNA databases
3. Specialized nucleic acid databases

4.2 PRIMARY NUCLEIC ACID DATABASES

The primary nucleic acid sequence databases are an annotated collection of all publicly available nucleotide sequences. Main sources for DNA and RNA sequences are direct submissions from individual researchers, genome sequencing projects and patent applications. There are three main primary nucleic acid databases viz. EMBL, Genbank and DDJB. EMBL is a prominent bioinformatics web site and is discussed in details in section 1.4. The EMBL Nucleotide Sequence Database (also known as EMBL-Bank) constitutes Europe's primary nucleotide sequence resource. The database is produced in an international collaboration with GenBank (USA) and the DNA Database of Japan (DDBJ). Each of the three groups collects a portion of the total sequence data reported worldwide, and all new and updated database entries are exchanged between the groups on a daily basis. DNA sequence databases were first assembled in Los Alamnos National Laboratory (LANL), New Mexico by Walter Goad and colleagues in 1979 (Historical Profile 7). He not only conceived a prototype of Genbank but also helped to establish a working relationship between the three partners from 1992 to 2000.

Table 4.1: Abbreviations used in Nucleic acid sequence

Symbol	*Meaning*
A	Adenine
G	Guanine
C	Cytosine
T	Thymine
U	Uracil
Y	pYrimidine (C or T)
R	puRine (A or G)
W	Weak (A or T)
S	Strong (C or G)
K	Keto (T or G)
M	aMino (C or A)
B	Not A (C or G or T)
A	C or G or T
D	Not C (A or G or T)
H	Not G (A or C or T)
X, N,?	Unknown
O	Deletion
-	Deletion

1. **The GenBank** sequence database is an annotated collection of all publicly available nucleotide sequences and their protein translations. This database is produced at NCBI as part of an international collaboration with EMBL Data Library from the EBI and the DNA Data Bank of Japan (DDBJ). GenBank and its collaborators receive sequences produced in laboratories throughout the world from more than 100,000 distinct organisms. GenBank continues to grow at an exponential rate, doubling every 10 months. GenBank is built by direct submissions from individual laboratories, as well as from bulk submissions from large-scale sequencing centers. Direct submissions are made to GenBank using BankIt [http://www.ncbi.nlm.nih.gov/BankIt/], which is a Web-based form, or the stand-alone submission program, Sequin [http://www.ncbi.nlm.nih.gov/Sequin/index.html]. Upon receipt of a sequence submission, the GenBank staff assigns an Accession number to the sequence and performs quality assurance checks. The submissions are then released to the public database, where the entries are retrievable by Entrez. Bulk submissions of Expressed Sequence Tag (EST), Sequence Tagged Site (STS), and High-Throughput Genome Sequence (HTGS) data are most often submitted by large-scale sequencing centers. The GenBank direct submissions group also processes complete microbial genome sequences. Both ESTs and STS have applications in gene identification in genomes. They are discussed in details in section III on genomes.

BOX 4.1

Term Commonly Used in Context of Nucleic Acid Sequence Databases

Transposon

A segment of DNA that is capable of independently replicating itself and inserting the copy into a new position within the same or another chromosome or plasmid. Transposons act somewhat similar to viruses and in humans are an underlying cause of hemophilia, certain cancers, and other diseases. In other organisms, they can become a permanent and even beneficial part of the genome, as in maize corn, where transposons account for half the genome, and in certain bacteria, where genes for antibiotic resistance can spread by means of transposons. They are also called *jumping gene*.

Repetitive Sequences

Nucleotide sequences present in multiple copies in the genome. They include direct, inverted, tandem, and terminal repeat sequences and the alu family repeat.

Operon

A segment of DNA containing adjacent genes including structural genes and an operator gene and a regulatory gene.

Regulatory Gene

A gene that produces a repressor molecule that inhibits an operator gene.

Promoter Gene

A region of DNA to which RNA polymerase binds before initiating the transcription of DNA into RNA. The nucleotide at which transcription starts is designated _1 and nucleotides are numbered from this with negative numbers indicating upstream nucleotides and positive downstream nucleotides. Most bacterial promoters contain two consensus sequences that seem to be essential for the binding of the polymerase. The first, the pribnow box, is at about 10 and has the consensus sequence 5' TATAAT 3'. The second, the 35 sequence, is centred about 35 and has the consensus sequence 5' TTGACA 3'. Most factors that regulate gene transcription do so by binding at or near the promoter and affecting the initiation of transcription. Much less is known about eukaryotic promoters. Each of the three RNA polymerases in eukaryotes have a different promoter.

RNA polymerase I recognises a single promoter for the precursor of rRNA.

RNA polymerase II, that transcribes all genes coding for polypeptides, recognises many thousands of promoters. Most have the Goldberg Hogness or TATA box that is centred around position 25 and has the consensus sequence 5' TATAAAA 3'. Several promoters have a CAAT box around 90 with the consensus sequence 5' GGCCAATCT 3'. There is increasing evidence that all promoters for housekeeping genes contain multiple copies of a GC rich element that includes the sequence 5' GGGCGG 3'. Transcription by polymerase II is also affected by more distant elements known as enhancers.

RNA polymerase III synthesises 5s ribosomal RNA, all tRNAs and a number of small RNAs. The promoter for RNA polymerase III is located within the gene either as a single sequence, as in the 5s RNA gene or as two blocks, as in all tRNA genes.

Operator gene

A gene that activates the production of messenger RNA by adjacent structural genes.

Structural gene

A segment of a gene that contains information used in coding for protein synthesis. It includes the regions preceding and following the coding DNA as well as introns between the exons. Genetic information within the structural genes is discontinuous in eukaryotes, split among the exons that encode for messenger RNA and absent from the DNA sequences in between, which are called introns. Genetic splicing, catalyzed by enzymes, results in the final version of messenger RNA, which contains only genetic information from the exons.

Transcription

Synthesis of RNA by RNA polymerases using a DNA template. This is the First step in protein synthesis.

Translation

It is second step of protein synthesis. Information encoded in a mRNA transcript guides the synthesis of a new polypeptide chain from amino acids; occurs at ribosomes.

Splicing

The process by which introns are removed from pre-mRNA to produce mature messenger RNA that contains only exons.

Reverse Transcription

Synthesis of DNA from mRNA with the help of the enzyme reverse transciptase. DNA synthesized by reverse transcription is called cDNA or complementary DNA.

EST When cDNA reads are available in fragments it is called a EST or expressed tags.

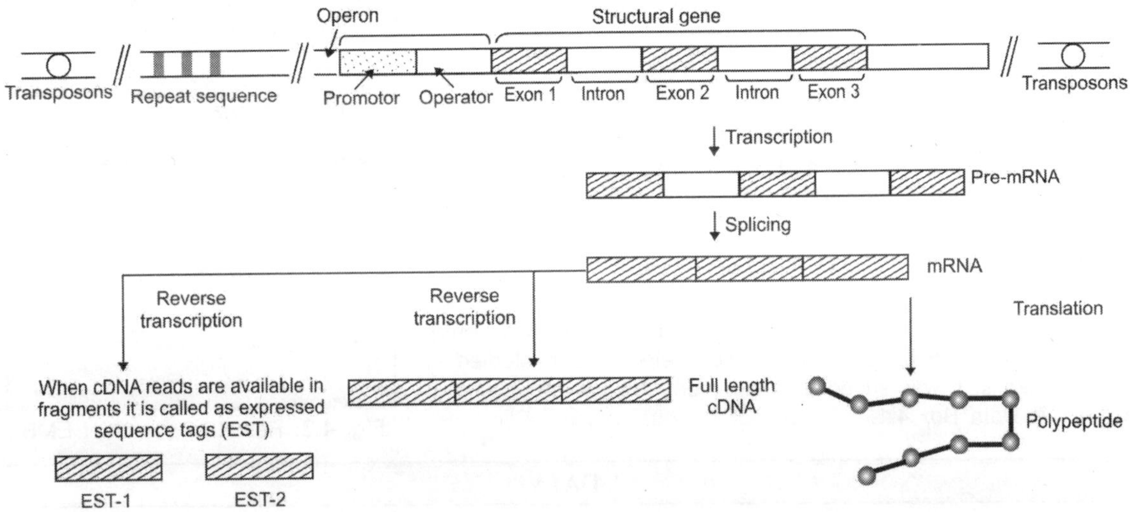

Fig. 4.1: Several type of nucleic acid sequences distinguished on the basis of their location, structure and function

2. EMBL database is European Molecular Biology Laboratory database: It is a database from the Europe Bioinformatics Institute (EBI). The arte of growth of the database has been following an exponential trend with a doubling time of 9-12 months. EMBL database is studied in details in Experiment 4A.
3. DDBJ: As explained earlier DDBJ refers to DNA data Bank of Japan. It is compiled by the National Institute of Genetics (Mishima, Japan) and is the official DNA data bank in Japan, and as such it collects data (mainly for Japanese researchers). This site has background information about the database, newsletters and tools for sequence submission. There is also access to various nucleotide sequence databases: data retrieval from DDJB, homology search using FASTA and BLAST, multiple alignment using malign and clustalw, and searching DDBJ, EMBL, GenBank, PIR and SWISSPROT using SFGATE. Other databases developed at the Institute are also available here: Codon Usage Database (CUTG), Protein mutant DB, *C. elegans* EST database and *Bacillus subtilis* Non-Redundant Database (NRSub).

EXPERIMENT 4A

Study of Primary Nucleic Acid Database EMBL

Hypothesis: It would be possible to extract all possible information regarding nucleotide sequence for gene rbcL coding enzyme, Ribulose bisphosphate carboxylase large chain 2 of *Acaryochloris marina* using EMBL.

TESTING THE HYPOTHESIS

Computer set up: A normal PC with internet connection in windows environment:

1. Log on to website of EMBL at (www.ibi.ac.uk/embl). This takes to the welcome page of the web site (Fig. 4.2).

Fig. 4.2: Web view of EMBL

2. Select text search and enter Ribulose bisphosphate carboxylase in the text box for search. Results obtained are shown in Fig. 4.3.
3. Observe the results obtained and select only the entries for large chain 2 of *Acaryochloris marina.*
4. It gives 20502 entries about the enzyme. Of these entry about *Acaryochloris marina* was selected and studied in details. It was saved as experiment_4A.doc and is shown in Data Box 4.1.

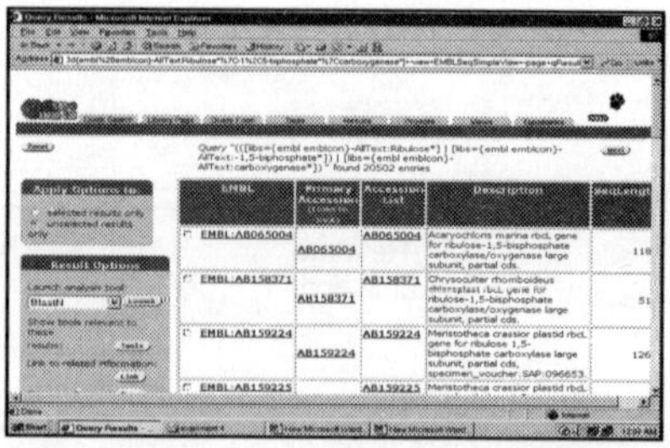

Fig. 4.3: Result of search at EMBL

DATA

Data Box 4.1: EMBL entry for *Acaryochloris marina* large chain 2 for Ribulose biphosphate carboxylase.

```
ID    AB065004   standard; genomic DNA; PRO; 1188 BP.
XX
AC    AB065004;
XX
SV    AB065004.1
XX
DT    08-JUL-2004 (Rel. 80, Created)
DT    15-JUL-2004 (Rel. 80, Last updated, Version 2)
XX
DE    Acaryochloris marina rbcL gene for ribulose-1,5-bisphosphate
DE    carboxylase/oxygenase large subunit, partial cds.
XX
KW    .
XX
OS    Acaryochloris marina
OC    Bacteria; Cyanobacteria; Acaryochloris.
XX
RN    [1]
RP    1-1188
RA    Miyashita H., Shimada A.;
RT    ;
RL    Submitted (06-JUL-2001) to the EMBL/GenBank/DDBJ databases.
```

```
RL      Hideaki Miyashita, Kyoto University, Hall of Global Environmental Studies;
RL      Yoshida-Honmachi, Sakyo-ku, Kyoto, Kyoto 606-8501, Japan
RL      (E-mail:miyashita@hm1.mbox.media.kyoto-u.ac.jp, Tel:81-75-753-7928,
RL      Fax:81-75-753-7928)
XX
RN      [2]
RA      Miyashita H., Shimada A., Miyachi S., Chihara S.;
RT      "Molecular phylogeny of Acaryochloris marina (Cyanophyceae), a novel
RT      photosynthetic prokaryote containing chlorophyll d as a major pigment,
RT      inferred from 16S ribosomal DNA and RbcL sequences";
RL      Unpublished.
XX
FH      Key       Location/Qualifiers
FH
FT      source    1..1188
FT                /db_xref="taxon:155978"
FT                /mol_type="genomic DNA"
FT                /organism="Acaryochloris marina"
FT                /strain="MBIC11017"
FT      CDS       <1..>1188
FT                /codon_start=1
FT                /transl_table=11
FT                /gene="rbcL"
FT                /product="ribulose-1,5-bisphosphate carboxylase/oxygenase
FT                 large subunit"
FT                /protein_id="BAD27388.1"
FT                /translation="WTTVWTDLLTDLDRYKGRCYDIEAVANEDNQYIAYIVYPLDLFEE
FT      GSVVNLLTSLVGNVFGFKALRALRLEDIRIPVAYMKTFQGPPHGITVERDKINKYGRPL
FT      LGCTIKPKLGLSAKNYGRAVYECLRGGLDFTKDDENINSQPFMRWRDRFLFVAEAIHKA
FT      QAETGEIKGHYLNVTAATCEEMLKRAEFAKELEMPIIMHDFITGGFTANTTLSHWCRDN
FT      GVLLHIHRAMHAVIDRQKNHGMHFRVLSKCLRMSGGDHIHTGTVVGKLEGEKGITMGFV
FT      DLLRENYVEKDLSRGIYFTQDWASMGGVMAVASGGIHVWHMPALVEIFGDDSVLQFGGG
FT      TLGHPWGCAPGATANRVALEACVQARNEGRDMAREGGDILREAAKWSPELAVALEV"
XX
SQ      Sequence 1188 BP; 262 A; 275 C; 320 G; 331 T; 0 other;
        tggaccactg tgtggactga cctgttaacc gatttggatc gatacaaagg ccgttgctac     60
        gacatcgaag cagttgccaa cgaagataat cagtacattg cctatattgt ctatcccctc    120
        gatttgtttg aggaagggtc agtcgtcaac ttgttgacct cccttgtggg taacgtgttt 180
        ggtttcaaag cacttcgggc cttacgtctt gaagatattc gtatcccagt tgcctacatg240
        aagaccttcc aaggtcctcc tcacggaatc acggttgagc gtgacaagat caacaagtat    300
        ggccgtcctt tgttgggttg caccattaag cccaagttag gtctatccgc gaagaactac    360
```

Data Box 4.2: Details of abbreviations used in data box 4.1.

ID	—	identification code for the sequence in the database
AC	—	accession number giving origin of sequence
DT	—	dates of entry and modification
KW	—	Key cross-reference words to look up this entry
OS, OC	—	Source organism
		RN, RP, RX, RA, RT, RL, - Literature reference or source
DR	—	id in other databases
CC	—	Description of biological samples
FH, FT	—	Information about sequence by base position or range of positions
SQ	—	count of A, C, G, T and other symbols

// end of sequence.

INTERPRETATION OF THE DATA

Present experiment was conducted with a limited aim of searching the EMBL database and extracting information about specific entry about nucleotide sequence for gene rbcL coding enzyme, Ribulose biphosphate carboxylase large chain 2 of *Acaryochloris marina*. Using Internet resources it was very easy to extract the relevant information and detailed entry is shown in data box 4.1. It also gave the amino acid sequence, which might be synthesized using this sequence as coding sequence (CDS). However complete interpretation of the information is not possible until the abbreviations used are explained. The extracted information about abbreviations as extracted from EMBL website is given in data box 4.2. EMBL sequence format is very similar to Genbank format. The main differences are in use of ORIGIN in Genbank format to indicate start of the sequence.

HISTORICAL PROFILE 7

Dr. Walter Goad—Man to Have First Assembled a DNA Sequence

To molecular biology, Goad brought a suite of quantitative skills and a strong knowledge of computing. As early as the late 1960s, he made efforts to address the problems associated with the storage, retrieval, and analysis of molecular data, conceiving of the computer as a central tool for collaborative and analytical work in molecular biology. Following a 1979 meeting at the Rockefeller University to discuss how these data could be managed and exploited, Goad and his Los Alamos colleagues established the Los Alamos Sequence Data Bank and began to develop software for sequence analysis. Goad conceived the idea of expanding the database nationally and internationally.

With funding from the National Institute of General Medical Sciences (NIGMS), and other agencies, in 1982, the database was rechristened GenBank. The base of its operations remained at Los Alamos, as it was operated successively under the control of BBN (1982-1987), IntelliGenetics (1987-1992), and (since 1992) the National Center for Biotechnology Information. Goad also helped to establish a strong working relationship with parallel efforts in Europe [the European Molecular Biology Laboratory (EMBL) and Japan (DDBJ)]. The success of GenBank in turn laid much of the groundwork for the Human Genome Project which followed in the late 1980s and 1990s. Begun formally in 1990 as an effort to identify, map, and sequence all of the genes in the human genome, the Human Genome Project benefited both conceptually and practically from GenBank, and Goad was an early member of the Human Genome Advisory Panel. For his efforts on behalf of GenBank, Goad received a Los Alamos National Laboratory Distinguished Service Award, and the Laboratory bestowed its highest honor, making him a Laboratory Fellow. He was also a Fellow of the American Physical Society and the American Association for the Advancement of Science. Walter Goad died of a stroke in Santa Fe on November 2, 2000.

4.3 RNA DATABASES

The RNA databases are much less in number as compared to DNA databases but are much more varied in their function. RNA by connection is known to have an important role in protein synthesis. Different type of conventional RNAs which are used in protein synthesis are given in Fig. 4.4. RNA databases are classified into following types and example of each one of them is discussed separately:

1. mRNA databases
2. Noncoding RNA databases
3. RNA Structure database
4. Mitochondrial RNA database

4.3.1 mRNA Databases

mRNA databases contain collection of mRNA sequences and their control elements. They are especially useful in construction of cDNA library. Transterm is an example of database of mRNAs and translational control elements. It contains a curated collection of motifs in mRNAs that control translation, and also biologically relevant mRNA regions extracted from GenBank. In the main, it is organised on a taxonomic basis with files and summaries for each species. Several types of access are provided via a web browser interface. Transterm defined motifs may be matched in a users sequence or in the database. Alternatively, the user to search specific sections of the database or the user's sequence can enter motifs. This web site is made available on the web by the Department of Biochemistry and Centre for Gene Research, University of Otago, Dunedin, New Zealand.

4.3.2 Noncoding RNAs Databases

There are several types of non-conventional RNAs in addition to coventional RNAs given in Fig. 4.4. It includes RNAs acting on the level of chromatin structure, transcriptional and translational

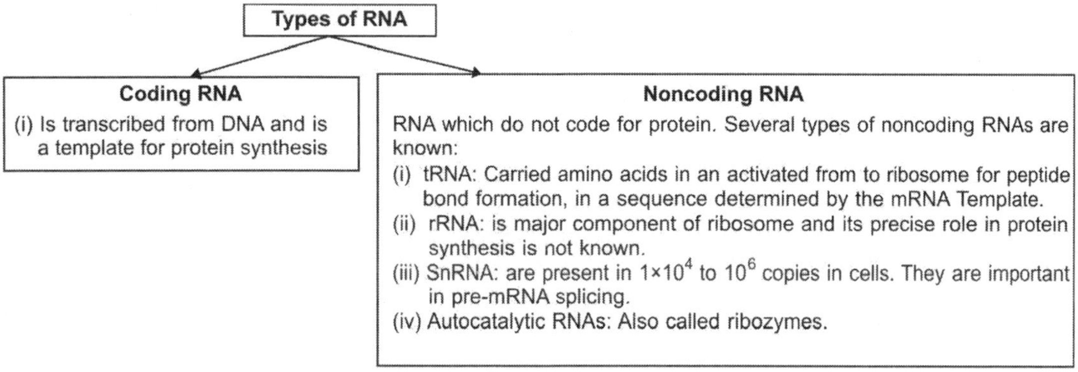

Fig. 4.4: Different types of RNA

regulation of gene expression, modulation of protein function and regulation of subcellular distribution of RNAs and proteins. The noncoding RNA database is a collection of currently available sequence data on RNAs that have no protein-coding capacity and have been implicated in regulation of cellular processes. The RNAs included in the database have been identified in prokaryotic, animal and plant cells, and form a very heterogenous group of molecules that act on different levels of information transmission in the cell. This web site is made available on the web by the Institute of Bioorganic Chemistry, Polish Academy of Sciences, Poznan, Poland.

The Plant ncRNAs database is another example noncoding RNA database and is specially dedicated to plant noncoding RNAs (ncRNAs). These transcripts lack significant open reading frames and encode RNA as their final product. This database also includes summaries with prominent characteristics for each gene, such as gene expression and associated mutant phenotypes.

4.3.3 RNA Structure Database

Structural RNA database provides information about structure of RNA. Rfam is a database of multiple sequence alignments and covariance models representing noncoding RNA families. Concepts of multiple sequence alignment are important in similarity searches and are discussed in details in chapters 10 and 11. Covariation in sequences refers to coincident change at two or more sequence reactions that may influence the secondary structure of RNA. The web site allows the user to search a query sequence against a library of covariance models and view multiple sequence alignments and family annotation.

4.3.4 Mitochondrial RNA Databases

PLMItRNA is a database containing information on the distribution of tRNA molecules and genes in mitochondria of photosynthetic eukaryotes; higher plants and green algae, and Cryptophyta, Rhodophyta and Stramenopiles algae. In addition to background information on tRNAs, a description of the type of information provided by the database, how to use the search facility to retrieve information from PLMItRNA, and a set of references are also provided. PLMItRNA is coordinated by Luigi R. Ceci, and is made available on the Web by the Institute of Biomedical Technology, Universita of Bari, Italy.

4.4 SPECIALISED NUCLEIC ACID DATABASES

These are the databases created with a specific purpose or function. Nucleic acid sequences are given different names based on their location, structure or function. Exons, introns, promoters, repeated sequences, transposons, ESTs, CDNA, STS are all examples of these specialized sequences Fig. 4.3. Genome databases are also examples of specialized nucleic acid databases and are discussed separately in Section 4. Several examples of specialised nucleic acid data bases are discussed below:

4.4.1 Intron/Exon Database

An intron refers to intervening sequence and exon refers to sequence that is expressed. Intron is a segment of DNA that is transcribed but removed from within the transcript by splicing and the exon sequence is translated into a protein. Introns are usually absent in prokaryots. The Exon/Intron Database (ExInt) incorporates information on exon-intron structure of eukaryotic genes. The database accumulates and describes the phases and positions of introns and the exon size for all proteins in GenBank. Data are available through a retrieval system, as flat-files and as a MySQL dump file. The database can be searched by LOCUS name, NID, accession number, or length of the protein. A BLAST service against ExInt is available. This site on the Web is courtesy of BioInformatics Centre, Singapore.

There are several other intron/exon databases created with a more specific organisms like the Ares lab Yeast Intron Database contains information about the location, structure, and function of spliceosomal introns in the nuclear genome of the yeast *Saccharomyces cerevisiae*. Also included are expression data relating to the efficiency of splicing relative to other processes in strains of yeast lacking nonessential splicing factors. Similarly we have intron sequence database for *Drosophila* and humans are discussed in details in Section 3.

4.4.2 Promotor Sequence Database

It is the region of DNA where RNA polymerase binds to initiate transcription. One promoter controls transcription of one polycistronic mRNA. A significant difference between the eukaryotic and prokaryotic promoters is that, in eukaryotes, it involves binding of transcription factors before the three types of RNA polymerases binding to promotor. The eukaryotic promotors contain several sequence elements like CAAT boxs, GC boxes and TATA boxes etc. The prokaryotic promotor gene in case of many *E.coli* contains TTGACA sequence element at -35 site and TATAAT at -10 sites. PlantProm DB, a plant promoter database, is an annotated, non-redundant collection of proximal promoter sequences for RNA polymerase II, with experimentally determined transcription start site (TSS) from various plant species. PlantProm DB contains entries including promoters from monocot, dicot and other plants.

4.4.3 Repeat Sequence Database

Repetitive DNA describes sequences that are present in more than one copy in each genome. Repetitive DNA is often divided into two general types: Moderately repetitive DNA consists of relatively short sequence that are repeated typically 10 to 1000 times in the genome. The sequences are dispersed throughout the genome and are responsible for the high degree of a secondary structure formation in pre-mRNA, when inverted repeats in the introns pair to form duplex regions; Highly repetitive DNA consists of very short sequences (typically <100 bp) that are present many thousands of times in the genome, often organised as long tandem repeats. Neither class represents proteins. A significant part of moderatly repetitive DNA consists of transposons, short sequences of DNA that have the ability to move to new locations in the genome and/or to make additional copies of themselves. Repbase Update (RU) is a database of repetitive DNA sequence elements found in a variety of eukaryotic organisms including mammals, fish, insects, nematodes, and plants. The goal of RepBase is to collect, review and systematically organize representative examples or consensus sequences of repetitive elements from all eukaryotic species. It is updated approximately once a month to include new information. This facility is free to academic users upon registration.

TIGR Plant Repeat Database is a collection of repetitive sequences from the genomes of twelve plant genera and four plant families. This site allows the user to search, download by anonymous FTP, and retrieve sub-sets of the TIGR Plant Repeat Databases. It also provides information on the composition of the TIGR Plant Repeat Databases, codes for plant repetitive sequences related links, help and acknowledgments. Made available on the web by The Institute of Genome Research (TIGR), a not-for-profit organisation funded by the US National Institutes of Health (NIH).

The highly repetitive sequences that are clustered together in a tandem array are called tandem repeats. In a tandem array a very short nucleotide sequence is repeated many times in row. In *Drosoplilla* for example 19 per cent of the chromosomal DNA is highly repetitive DNA found in tandem arrays. Tandem repeats sequences are commonly found in centromeric region of chromosomes and can more than 1,000,000 bp in lengths. The Short Tandem Repeat (STR) DNA Internet Database (Strafe) contains facts and sequence information on each STR system, population data, commonly used multiplex STR systems, PCR primers and conditions, and a review of various technologies for analysis of STR alleles.

4.4.4 MHC Database

MHC refers to major histocompatibility complex. Every individual has a characteristic set of MHC proteins. They are important in graft reactions: a graft of tissue from one individival to another is rejected because of the difference in MHC proteins between donors and recipient, an issue of major medical importance. Major histocompalibility locus refers to set of genes that codes for MHC proteins. The major histocompatibility locus occupies a small segment of a single chromosome in mouse (where it is called as H_2 locus) and in man called as HLA Locus. One of the most intensely studied regions of the human genome is the Major Histocompatibility Complex (MHC), a group of genes that occupies approximately 4-6 megabases on the short arm of chromosome number 6. The MHC genes, known in humans as Human Leukocyte Antigen (HLA) genes, are highly polymorphic and encode molecules involved in the immune response. The MHC database, (http://www.ncbi.nlm.nih.gov/mhc), was designed to provide a neutral platform where the HLA community can submit, edit, view, and exchange MHC data. It currently consists of an interactive Alignment Viewer for HLA and related genes, an MHC microsatellite database (dbMHCms), a sequence interpretation site for Sequencing Based Typing (SBT), and a Primer/Probe database. The MHC database is fully integrated with other NCBI resources, as well as with the International Histocompatibility Working Group (IHWG) Web site, and provides links to the IMmunoGeneTics HLA (IMGT/HLA) database.

4.4.5 EST Sequences Databases

An expressed sequence tag (EST) is a short sequence of DNA taken from a DNA copy of mRNA. The EST is complementary to the mRNA and can be used to identify genes corresponding to mRNA, ESTs are very useful in construction of physical map of genome. Mendel-ESTs is EST database created to apply controlled annotation to plant ESTs. Mendel-ESTs is primarily a database of plant ESTs, which have been compared to Mendel-GFDb, completely sequenced genomes and domain databases. This approach associates ESTs with individual sequences and the controlled annotation of gene families and protein domains. Produced by the Mendel Bioinformatics Group, John Innes Centre, Norwich, and is made available on the Web by UK Cropnet.

The COGEME Phytopathogenic Fungi and Oomycete EST Database contains expressed sequence tags (ESTs) from eleven species of phytopathogenic fungi and two oomycete species. In order to reduce sequence redundancy, ESTs representing the same gene were joined together into a single contig or unisequence. Unisequences have been annotated based on similarity to known sequences and assigned to a functional classification group. The database can be queried by simple text queries against annotation, by functional classification, by similarity to known genes from *Saccharomyces cerevisiae* and using BLAST algorithms. Unisequence datasets can be downloaded.

SUMMARY

The molecular structure of nucleic acids determines their function. Nucleotides which are composed of a sugar, phosphate and nitrogenous base, form the repeating structural unit of nucleic acids. The primary structure is a strand that contains a linear sequence of nucleotides. The formation of secondary structure occurs because strands of DNA (and RNA) can from complementary regions due to hydrogen bonding between adenime and thymine (or uracil), and between guamine and cytosine. Base stacking also stabilizes a double-stranded structure. The most common form of sugar–phosphate linkages, with the bases projecting inward from the backbones and hydrogen bonding with each other.

Nucleic acid databases are classified into primary, RNA and specialised nucleic acid databases. Genbank, EMBL and DDJB are three major primary nucleic acid databases. The RNA databases are much less in number as compared to DNA databases but are much more varied in their function. RNA databases are classified into mRNA databases noncoding RNA databases, structural RNA databases and mitochondrial database. Transterm is an example of mRNA database. Noncoding RNAs database and plant nc RNAs database are examples of non coding RNA databases. Rfam is structural RNA database. PLMItRNA is an example of mitochondrial RNA databases.

The last category of nucleic acid database ie specialised nucleic acid databases are databases created with a specific function or purpose. Intron/Exon databases, promotor sequence database, repeat sequence database, MHC database and EST databases are several types of specialized nucleic acid databases. In experiment 4A all possible information regarding nucleotide sequence for gene rbcl coding enzyme Ribulese bisphosphate carboxylase large chain 2 of *Acaryochlois marina* is searched using primary nucleic acid database EMBL.

EXERCISE

1. A naturally occurring DNA single strand has following sequence:

   '5A AGGAA AAGGGA GGAGAG 3'

   what is the sequence of its complementary strand?
2. Draw the structure of a phosphodiester bond.
3. List differences between structure of DNA and RNA.
4. Describe in details different types of noncoding RNAs.
6. Search for mRNA coded by gene rbcl coding enzyme ribulose biphosphate carboxylase large chain 2 of *Acaryochloris marina* using mRNA database transterm.
7. Log on to the web site of MHC database study the each file entry of the database and compare it with the entry of EMBL.
8. What are the key structural features of DNA molecules that underlie its ability to be faithfully replicated?
9. Some people consider proteins to be active participants in genetics while DNA is considered to be a passive participant. Comment.

10. For each of the following, how would transcription be effected if any of following sequence is missing from a structural gene.

    (*a*) promoter (*b*) terminator (*c*) Ribosme binding site (*d*) start codon (*e*) stop codon (*f*) regulatory sequence.

11. Compare the file retrieved by search in NCBI for human elastase under protein and nucleotide. On photocopies of these two pages mark with a highlighter all items that are common in the two files.
12. Find all structures of sperm whale myoglobin in protein data bank and draw a histogram of their dates of deposition.

5

METABOLIC PATHWAY DATABASES

Pathway refers to an ordered sequence of a series of biochemical reactions of products and substrates in a biological system. From a biochemist's perspective, life at the cellular level can be defined as a network of integrated and carefully regulated metabolic pathways, each contributing the sum of activities a cell must carry out. In a biological cell the pathways, form a complex network. Genes encoded in genomic DNA contribute to four major areas of function in a living cell. These functions are metabolism (metabolic pathways), regulation (regulatory control systems), signaling (signal transduction cascades), and assembly (architectural construction). Metabolic pathways include all the enzyme catalysed reaction that allow a cell to extract chemical energy from the environment for making cell components and for doing other useful work. Regulatory control systems are biochemical mechanisms that turn processes on and off, such as gene expression, or regulate the timing of events. Signal transduction cascades provide internal and external sensory functions of the cell. Signaling can be simple, for example, signal peptides in protein sequences that determine whether a protein is translated into the lumen of the endoplasmic reticulum, or transported through the mitochondrial membrane. Complex networks of interacting proteins are often carry out signaling functions. Architectural construction function includes synthesis of components that form structure of cell and its parts.

Traditional experiments in genetics, biochemistry, and molecular and cellular biology have contributed to the accumulation of a huge body of knowledge on molecular pathways in selected aspects of living systems. Whether the genome is a blueprint or not, it is impossible to make complete sense out of the sequence data, without additional information about molecular pathways of that organism. Application of the existing knowledge of molecular pathways, in gene identification is of limited application because the information existing is not computerised. Besides the complexity of these processes requires specialized database for storage of experimental results, and specialized software to assemble and correlate information. Visualization tools are particularly important in analysis and presentation of these complex data sets. This chapter is designed to introduce key concepts and methods for analysis of pathways, networks and systems.

5.1 METABOLIC PATHWAYS ARE VERY COMPLEX NETWORK

The term metabolism applies to the assembly of biochemical reactions, which are employed by the organism for synthesis of cell materials and for utilization of energy from their environments. In other words, the metabolism of an organism (or a cell) may be defined as sum total of all the enzyme—catalyzed reactions that occur in an organism or a cell. These large numbers of reactions in a cell are organised into a relatively small number of sequences or pathways. It is a highly coordinated and purposeful cell activity in which multi-enzyme system cooperate. Since metabolic processes take place in a series of progressive stepwise individual reactions they can be conveniently represented in the form of a metabolic pathway. In Fig. 5.1, a hypothetical pathway is shown where the reactant A is being converted to P. B, C and D are intermediate compounds and E_1, E_2 and E_3 are the enzymes required and X_1, X_2 and X_3 are cofactors.

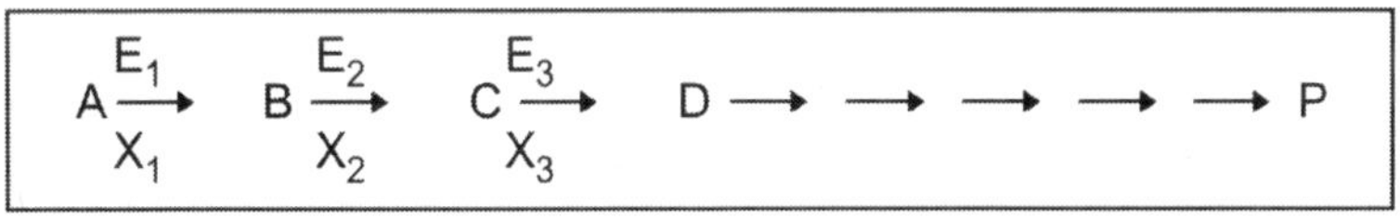

Fig. 5.1: An example of a metabolic pathway

Like, metabolism is classified into catabolism and anabolism the metabolic pathways are also classified as catabolic and anabolic pathways. Catabolism is breaking down large molecules into simple ones and release energy. Anabolism construct biochemical processes forming complex molecules from simple ones and requires energy. At certain steps in catabolic pathway much of the free energy is conserved in the form energy carrying molecule adenosine triphosphate (ATP) and some may be conserved as Nicotinamide adenime dinucleotide (NADH). In anabolism these ATP and NADH are used up synthesize various compounds. In Fig. 5.2 a simplified view of various processes occurring in a living body categorized as anabolism and catabolism is shown.

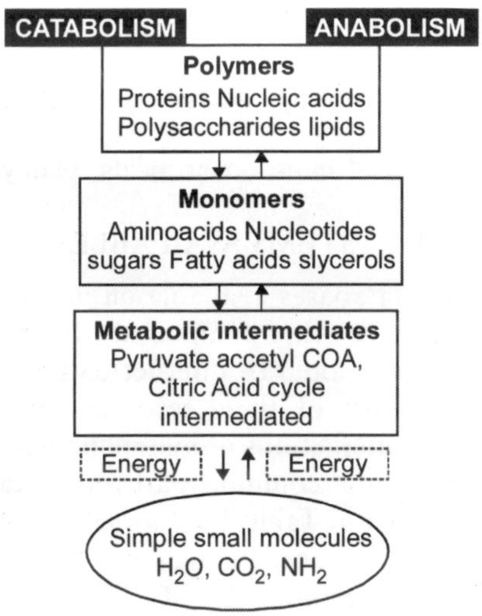

Fig. 5.2: Simplified view of various processes occurring in the body

Catabolic pathways play two roles in cells: They release energy and they give rise to the small organic molecules or metabolites that are building blocks for biosynthesis. The energy is needed in part to drive anabolic pathways in which specific metabolites are used to synthesize the macromolecules and other cellular components needed for cell structure and function.

Metabolic pathway is a complex network of reactions. In a hypothetical example depicted in Fig. 5.3a where we have roughly 10^3 to 10^4 genes present in the cell and we have 10^4 to 10^5 molecules in 10^5 to 10^7 genetic and molecular interaction are occurring. These reactions are different depending upon the location in cell i.e. whether in nucleus or mitochondria. Similarly, the stage of development of a cell or signalling response etc. also makes a difference and leads to addition in number of reactions taking place. So there are expected to be a large number of pathways existing exact number of which is very difficult to determine. Complexity of pathway databases is depicted in Fig. 5.3b where synthesis of only three amino acids, phenylalanine, tyrosine and tryptophan is shown. Like other biological sequence databases, pathway databases are also holding huge amount of data. The idea here, is not to explain the sequences in reactions rather it is to show the complexity of reaction in the body.

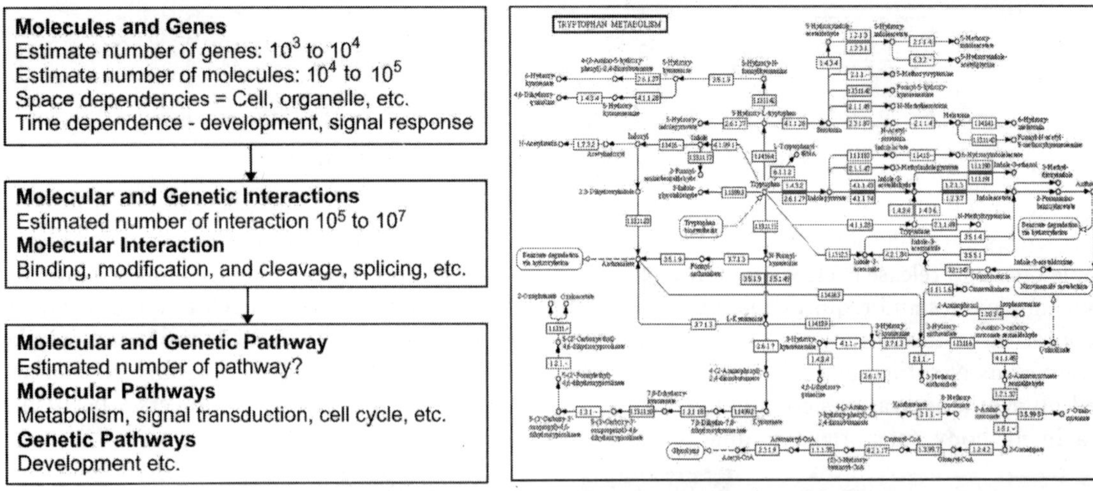

(a) (b)

Fig. 5.3: (a) Complexity of metabolic pathways according to estimated number of genes, molecules and reactions
(b) An example of synthesis of three amino acids, phenylalanine, tyrosine and tryptophan

5.2 PATHWAY DATABASES AND SYSTEMS AVAILABLE

There is large number of pathway databases available on internet to get the information about the metabolic compounds and its link with other compounds. Pathways Resource List (PRL), is a new database site that contains information on 156 internet pathway resources and these are arranged on the basis of web popularity. Most of these resources are databases themselves, containing information such as protein-protein interactions or metabolic reactions. In its present form, PRL only provides links to these resources. Some common pathway databases arranged according to web popularity (as given by PRL) is shown in Table 5.1 short name of one database along with detailed name is given. According to PRL is the most popular pathway database available on net is KEGG the Kyoto encylopedia of genes and genomes. KEGG.

These pathway databases provide detailed information regarding several aspects of metabolic pathways:

1. How many pathways, and how many steps within each pathway, lead from compound A to compound B and give a list of all compounds, enzymes and reactions involved.
2. These highlight metabolic processes and pathways as a networks according to various criteria.
3. Information can easily be retrieved regarding impact of one metabolic process on the other e.g. when one or more gene/proteins are turned off, it would show which paths or pathways may be affected.

4. Biochemical pathways from different organisms and tissues, or at different stages of annotation can easily be compared once we have their computerised database.
5. These databases highlight common features and differences among themselves and predict missing elements ('reconstruction') and represent pathways at different resolution levels.
6. Identification of all positive/negative regulatory cycles in a pathway graph is much easier.

All these features are very easily understood once we study these pathways with the help of specific examples.

Table 5.1: List of metabolic pathway databases according to their popularity

Propality score	*database*	*Details*
765	KEGG	Kyoto Encyclopedia of Genes and Genomes
293	ENZYME	Enzyme nomenclature database
110	BRENDA	Comprehensive Enzyme Information System
104	UM-BBD	Biocatalysis/Biodegradation Database
72	MPB	Metabolic Pathways of Biochemistry
51	EcoCyc	Encyclopedia of *E. coli* Genes and Metabolism
46	BioCyc	BioCyc Knowledge Library
43	LIGAND	Database of Chemical Compounds and Reactions in Biological Pathways
20	PathDB	Pathways Database
17	Reactome	Reactome KnowledgeBase
14	EMP	Enzymes and Metabolic Pathways Database
13	MetaCyc	Metabolic Pathway Database
11	aMAZE	Protein Function and Biochemical Pathways Project

Metabolic pathway databases are classified into four types based on different criterion:

1. **Enzyme function and metabolic pathways:** This includes the database which are used for extracting information about enzyme function and details of metabolic pathway. KEGG, EcoCyc, and BRENDA are examples of common enzyme function and metabolic pathway database. These provide querying facilities with various levels of complex pathways. This ranges from simple browsing and basic queries (string search on the values of selected fields), to pathway analysis.

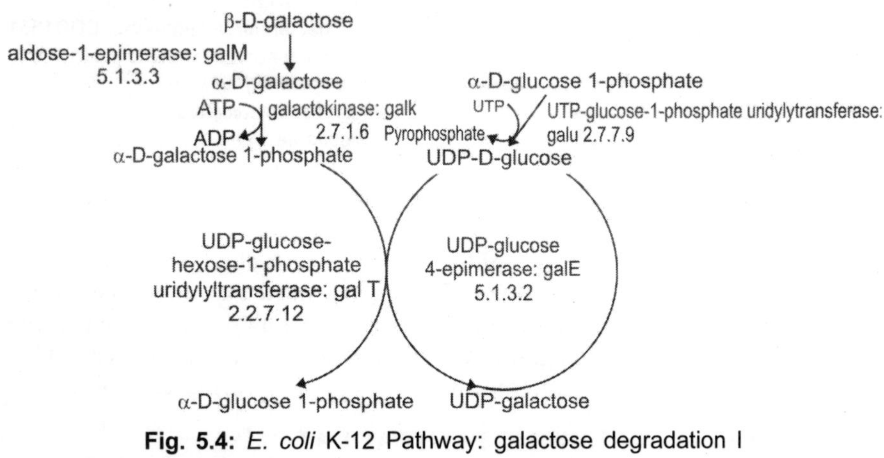

Fig. 5.4: *E. coli* K-12 Pathway: galactose degradation I

2. **Organism specific pathway databases:** These databases usually are a collection of pathway genome databases and each entry describes the genome and metabolic pathways of a single organism, with the exception of the MetaCyc database, which is a reference source on

metabolic pathways from many organisms. A common example typical of this category is EcoCyc which is an organism-specific pathway/genome database that describes the metabolic and signal-transduction pathways of *Escherichia coli*, its enzymes, its transport proteins and its mechanisms of transcriptional control of gene expression. EcoCyc is queried using the pathway tools graphical user interface, which provides a wide variety of query operations and visualization tools. In Fig. 5.4a *E. coli K-12* pathway for galactose degradation I as searched and drawn with EcoCyc is shown.

MetaCyc is a database of non-redundant, experimentally elucidated metabolic pathways from more than 240 different organisms. MetaCyc is curated from the scientific experimental literature. It stores pathways involved in both primary metabolism (including photosynthesis), secondary metabolism, as well as associated compounds, enzymes, and genes. Both EcoCyc and MetaCyc are database based on literature-derived information. However there are several computer based genome pathway databases, which contain organism specific information about pathways, which have been extracted from other databases computationally. In Table 5.2 few selected organism specific computationally derived databases are given.

3. **Biochemical Process Specific Databases:** These are Databases containing information on specific types of biochemical processes. For example CSNbd describes cell signaling networks, and mainly focuses on signal transduction pathways. Information about transcriptional regulation (mainly transcription factor binding sites) is found in Transfac, regulonDB.

Table 5.2: Organism specific pathway databases

Sl. No	*Database*	*Organism*
1.	EcoCyc	*Escherichia coli* K12
2.	AgroCyc	*Agrobacterium tumefaciens*
3.	AnthraCyc	*Bacillus anthracis*
4.	BsubCyc	*Bacillus subtilis*
5.	CtraCyc	*Chlamydia trachomatis*
6.	CauloCyc	*Caulobacter crescentus*
7.	EcoO157Cyc	*Escherichia coli* O157:H7
8.	HinCyc	*Haemophilus influenzae*
9.	HpyCyc	*Helicobacter pylori*
10.	HumanCyc	*Homo sapiens*
11.	MtbCdcCyc	*Mycobacterium tuberculosis* CDC1551
12.	MtbRvCyc	*Mycobacterium tuberculosis* H37Rv
13.	MpneuCyc	*Mycoplasma pneumoniae*
14.	PlasmoCyc	*Plasmodium falciparum*
15.	ShigellaCyc	*Shigella flexneri*
16.	TpalCyc	*Treponema pallidum*
17.	VchoCyc	*Vibrio cholerae*

4. **Path-finding Databases:** KEGG and CSNdb provide path-finding tools, which find all paths between two specified elements, or from a specified element to any other. The results of the path finding query can be displayed by colouring the paths found on pre-drawn static maps (KEGG), or on a dynamically generated diagram (CSNdb). KEGG, EcoCyc and PathDB can also be downloaded and installed locally, offering more extended querying capabilities.

5.3 KEGG AS A MODEL OF METABOLIC PATHWAY DATABASE

KEGG (Kyoto Encyclopedia of Genes and Genomes) is a knowledge base for systematic analysis of gene functions, linking genomic information with higher order functional information. This is the most popular metabolic pathway database available on web. The web site KEGG is shown in Fig. 5.5a. and current statistics is given in part b. The KEGG Project under the Japanese Human Genome Program was initiated in 1995 under supervision of Prof. EM. Kanehisa (Historical Profile 8) in order to fully utilize the biological knowledge on interactions and pathways, as well as the interaction data generated by new systematic experiments. The primary objective of KEGG was to computerize the current knowledge of molecular pathways; namely, metabolic pathways and a number of regulatory pathways including signal transduction, cell cycle, and development pathways. At the same time, KEGG maintained gene catalogs for all the organisms that have been sequenced and linked each gene product to a component on the pathway. KEGG also organised a database of all chemical compounds in living cells and linked each compound to a pathway component. KEGG in real terms is aimed at developing new bioinformatics technologies toward functional reconstruction.

KEGG is organised into following databases:

1. Gene database—maintain gene catalogs of all sequenced organisms and link each gene product to a pathway component.
2. Pathway database—which contain computerized current knowledge of molecular and cellular biology in terms of the pathway of interacting molecules or genes. The pathway database is supplemented by a set of ortholog group tables for information about conserved sub-pathways (pathway motifs), which are often encoded by positionally coupled genes on the chromosomes and which are especially useful in predicting gene function.
3. Ligand database—a database of all chemical compounds in living cells and link each compound to a pathway component.

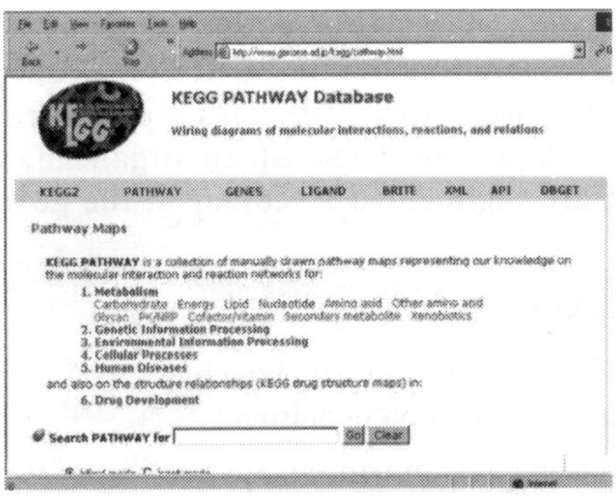

(a)

Number of pathway	18,481
Number of reference pathways	259 (PATHWAY database)
Number of ortholog tables	86 (PATHWAY database)
Number of organisms	231 (GENOME database)
Number of genes	815,240 (GENEOME database)
Number of ortholog clusters	34,387 (SSDB database)
Number of KO assignments	5,605 (KO database)
Number of chemical compounds	11,461 (COMPOUND database)
Number of glycans	10,962 (GLYCAN database)
Number of chemical reactions	6,304 (REACTION database)
Number of reactant pairs	5,991(RPAIR database)

(b)

Fig. 5.5: (a) Web site of KEGG; (b) Current statistics of KEGG

KEGG provides java graphics tools for browsing genome maps; comparing two genome maps and manipulating expression maps, as well as computational tools for sequence comparison graph comparison and path computation. The KEGG databases are daily updated and made freely available. KEGG database can be used freely for academic purposes and commercial evaluation. Besides the databases it also provides several on-line resources: slide shows, tutorials, manuals, references etc.

Important features of a KEGG search and computation are:

1. KEGG pathways are searched by EC numbers (enzymes), compound numbers, and by gene accessions.
2. Combine search with KEGG grouping or the hierarchical classification [e.g. EC numbers from a specific group in the superfamily table (or SCOP table) and searching against pathway diagrams].

HISTORICAL PROFILE 8

Minoru Kanehisa for Developing KEGG Pathway

He is Director and Professor, Bioinformatics Center, Institute for Chemical Research, Kyoto University and Professor, Human Genome Center, Institute of Medical Science, University of Tokyo. After receiving D.Sc. in physics from the University of Tokyo in 1976, Minoru Kanehisa worked in the Johns Hopkins University School of Medicine, in the Los Alamos National Laboratory where he was one of the cofounders of GenBank, and in the National Cancer Institute of the National Institutes of Health. Since 1987 he is Professor in the Institute for Chemical Research, Kyoto University, and since April 2001 he is Director of the newly established Bioinformatics Center of Kyoto University. He and his group engineered a the KEGG database of genetic and molecular pathways. His other activities include: concurrent professorship at the Human Genome Center, Institute of Medical Science, University of Tokyo (1991-1995 and 2002-Present), principal invesitgator of the Japanese Genome Informatics Project (1991-2000), and president of the Japanese Society for Bioinformatics (1999-2003).

3. Search KEGG pathways by sequence similarity (it identifies orthologs and reconstructs pathways from the gene catalog).
4. Given list of enzymes, automatically generate the organism specific pathways by marking the matching enzymes on the diagram. Missing elements imply either gene catalog wrong or known reaction pathway utilizing different enzymes in the catalog.
5. Compute pathways from a given list of enzymes. Deduction from binary relations of substrates and products with optional use of query relaxation for functional hierarchies.

KEGG is also functioning as repository of metabolic pathways for organisms whose genome is completely sequenced. It also provides the regulatory information in the pathway. For many of these organisms, the body of experimental data is very restricted. Protein function inferred from sequence similarity with proteins characterized experimentally in other organisms. Pathways represented as diagrams, manually created and stored as static .gif files. Upon selection of an organism, the reactions for which an enzyme is known in that organism are highlighted in colour in the generic pathway diagrams.

5.4 APPLICATION OF METABOLIC PATHWAY DATABASES

Metabolic pathway databases hold enormous amount of information and are extremely useful as a tool for finding alternative pathways. Several specific example are given below:

1. Genome evolution: In genome evolution pathway databases have enormous application as it can be used to infer the unknown genome compare with known genome and find missing enzymes in the pathways.
2. Biotechnology: Biotechnological applications of pathway databases are identification of alternative enzymes, pathways, substrates and products. This information will really uplift the biotechnological industries.
3. Pharmacology: Research in pharmacology is helped by knowledge of non-homologous gene displacement and species-specific drug targets obtained using metabolic pathway databases.
4. Identification of previously unknown genes.

   Application of metabolic databases is limited because of following reasons:

   (*a*) Discovery of alternative pathways: known pathways have been established on a restricted number of model organisms, alternative pathways remain to be discovered.

   (*b*) Partial information i.e. presence of the missing steps.

   (*c*) Negative results.

   (*d*) Putative interaction resulting from computational analysis.

EXPERIMENT 5A

To Study Galactose Metabolism Pathway with the Help of EMP Pathway Database

About EMP pathways database: Enzymes and Metabolic Pathways database EMP, is a unique and most comprehensive electronic source of biochemical data. It covers all aspects of enzymology and metabolism and represents the whole factual content of original journal publications. The database format has about 300 subject fields. EMP contains about 30,000 records compiled from about 15,000 original experimental journal publications. EMP is being updated with over 8,000 records a year to cope with the essentials published in the literature each year. Each EMP record is a translation of the whole factual content of an original journal publication into a structured, indexed, and easy searchable form. A metabolic part of EMP constitutes a separate database, EMP Pathways (earlier known as MPW). It has over 3,000 metabolic diagrams created by Evgeni Selkov and his coworkers. The information stored in EMP is indispensable for analysis and mathematical simulation of metabolic networks, metabolic design, drug development and bioengineering. EMP can easily be integrated into any academic or industrial bioinformatics environment to play a critical role in the metabolic reconstruction technology. WIT2 Pro system of Integrated Genomics, Inc. is an example of a successful use of EMP for metabolic reconstruction and annotation of over 250 sequenced genomes.

Computer Setup: A normal PC with internet connection:

1. Log on to the web site of EMP pathway at (Fig. 5.6).

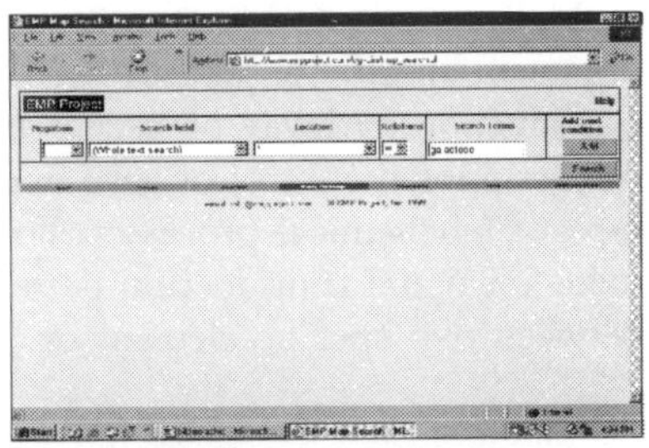

Fig. 5.6: Web view of EMP

2. Enter galactose at search terms text box. Keep search field as whole field. It gives the option for being selective in search like reactant substrate cofactor etc. In location in the cell i.e. chloroplast, cytosol. etc. is to be selected (Fig. 5.7).
3. Enter search and ignore add. Because we are searching one word and the search is not conditional.

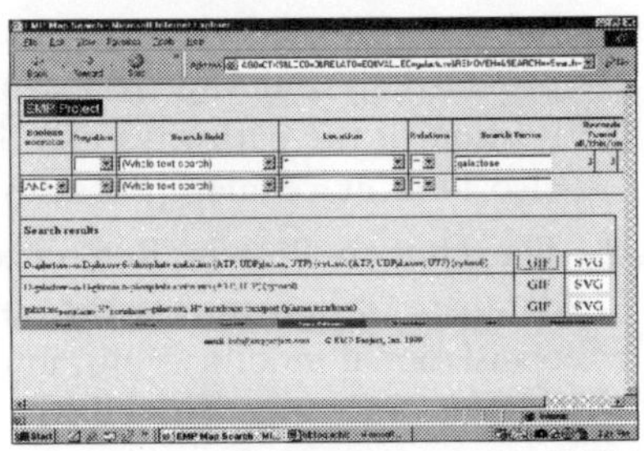

Fig. 5.7: Result obtained with EMP

4. This gives result as shown in Fig. 5.8
5. Select the query of your interest. Like in this case D-galactose-α-D-glucose 6-phosphate anabolism (ATP, UDP glucose, UTP) [cytosol (ATP, UDP glucose, UTP) (cytosol)] is selected.
6. This drew the pathway as shown in Fig. 5.8.

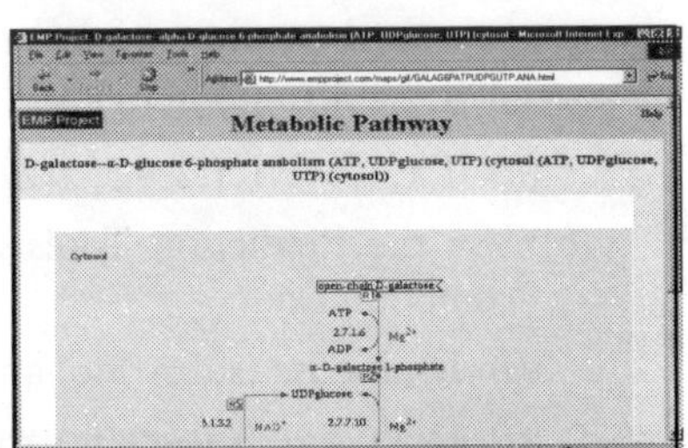

Fig. 5.8: Metabolic pathway drawn by EMP

INTERPRETATION OF DATA

Drawing of metabolic pathway and cycles has become much easier with computation metabolic pathways. Also the links among several pathways could be easily understood. As discussed later in the chapter these pathway databases have a really facilitated the study of genome evolution, gene identification and biotechnology.

SUMMARY

Metabolic pathways in cells are either anabolic (synthetic) or catabolic (degradative). The latter reactions provide the energy to drive former. Metabolic pathways are complex network of reactions. Large number estimated genes and molecules lead to huge number of pathways database. Pathway Resource List (PRL) is a new database site that contains information 156 internet pathway resources and arranged on the basis of web popularity. KEGG at present is the most popular pathway database.

In the present chapter pathway databases are classified into four types based on different criterion. First is enzyme function and metabolic pathway databases such as KEGG, Ecocycl and pathDB contain information about enzyme function and details of metabolic pathway. Second is organism specific databases such an Ecocy is a pathway that describes the metabolic and signal transduction pathway of *Escherchia coli*, its enzymes its transport proteins and its mechanisms of transcriptional control of gene expression. Third type of databases provide information on specific type of biochemical processes. For example CSNbd describes cell signalling networks and mainly focus on signal transduction pathway. Last category is path finding databases which find all paths between two specific element or from a specified elements to any other.

EXERCISE

1. For a impressive view of biochemical pathways go to the biochemical pathways wall chart provided by ExPASy. You can search by enzyme keyword or pull up a thumbnail view of a portion of the pathways map by clicking on the region of interest. In thumbnail view you can also look at cellular processes from a cellular and molecular perspective.
2. The Human Protein Reference Database, at http://www.hprd.org/ is an integrated pathway database of human reference proteins. The HPRD interface displays domain architecture, post-translational modifications, interaction networks, and disease associations for each protein in the human proteome. The HPRD site is self-explanatory. Enjoy!
3. Goto KEGG and enter the Search and Compute with KEGG section.
   (*a*) Search the following EC numbers and find out the name of these pathways:
      (*i*) 1.2.1.12, 1.2.1.51, 2.7.1.11, 2.7.1.40, 2.7.2.3, 4.1.2.13, 4.2.1.11, 5.3.1.9, 5.4.2.1.
      (*ii*) 1.1.1.37, 1.1.1.42, 1.2.7.3, 1.3.99.1, 3.1.2.3, 4.1.3.7, 4.2.1.2, 4.2.1.3.
   (*b*) Use the Prediction Tools of KEGG to generate possible reaction pathways between two compounds:
      (*i*) D-glucose (substrate) and Acetyl-CoA (product)
      (*ii*) D-glucose (substrate) and ethanol (product)
      (*iii*) D-glucose (substrate) and citrate (product)
      (*iv*) D-glucose (substrate) and lactate (product)

6

BIODIVERSITY DATABASES

Biodiversity bioinformatics is a term recently added to the dictionary of scientific terms in early 2000 which mainly deals with data about organisms using informatic resources. Bioinformatics gained importance in 1990s to describe the tools and techniques for storing, handling and communicating the biological data arising principally from genomic research and the term biodiversity came up in 1980 to describe variability of life on earth. Edward O. Wilson, whose guest editorial appeared on page 2279 of Gournal Science coined the term 'biodiversity' in 1980 to encompass the taxonomic and functional diversity of living organisms (Historical Profile 9, Edward O. Wilson). Recently Science in September 2000 released a special issue on biodiversity information (News Update 1) and described it as a quiet revolution. Like genomics, bioinformatics is indispensable to study of biodiversity because: data storage is a key thing in taxonomic research and there exists a massive and ever increasing amount of biological data which is held as card index and taxonomic monographs from a very long time; biodiversity workers are distributed all over the globe; biodiversity priorities are very much interdependent on information from one country and the globe needs to be understood as a whole.

Biodiversity or the biological diversity is defined as the variability among living organisms from all sources including *inter alia*, terrestrial, marine and other aquatic ecosystems and the ecological complexes of which they are part; this includes diversity within species, between species and of ecosystems. The number of species given a scientific name is believed to fall between 1.5 million to 1.8 million. The true, full number including those undescribed can only be guessed at the nearest order of magnitude, is in the vicinity of 10 million according to experts. Reliable biodiversity assessments are limited to relatively well known groups, including the vascular plants and vertebrates. The smaller species like bacteria, mites nematodes, fungi, etc. which are major environmental players certain ground as per counting the species is concerned.

Biodiversity information is critical to a wide range of scientific, educational and governmental uses, and is essential to decision-making in many realms. Initiatives to integrate biodiversity data into viable resources for innovation in science, technology, and decision-making are being developed at local, regional, and global levels. The challenge lying ahead is integration of these initiatives into an organised, well-resourced, global approach to build and manage biodiversity information resources through collaborative efforts. The existence of biodiversity data resources from different fields of knowledge and the strong demand to integrate, synthesize, and visualize this information for different purposes and by different end users is to passing a further challenge. Application of informatics in this field has great potential in diverse realms, with applications ranging from prediction of distributions of known and unknown species prediction of geographic and ecological distribution of infectious disease vectors, prediction of species' invasions, and assessment of impacts of climate change on biodiversity. This potential nonetheless remains largely unexplored, as this field is only now becoming a vibrant area of enquiry and study.

Biodiversity informatics ultimately emerged as a new subdiscipline because of general problems, common to many scientific disciplines. The reason lies in the inherent complexity not only of the problem domain itself but also of its basic data. In biodiversity, we are dealing with concept-related data at the very basic level. To cite but one example: the presence or absence of a certain species at a specific place is not only a question of recording the name, location, and time, but it also relates to the definition of the species—what we find is the organism, what we record is its classification. The matter becomes even more complex in paleontology, where even the location (stratum) where the species is found is the designation of a concept, the basic data are essentially location-referenced numeric data plus information on the methods used to obtain these measurements. As stated above there are an estimated more than 1.5 to 1.8 million known named species. There are somewhere between one and two synonyms for every valid/accepted species (in addition the valid name). Compiling a list of scientific names for a major group takes years of effort. For the resulting list to represent real progress i.e. there is an addition of a valid name. Each original publication must be reviewed (at least briefly) by a taxonomist and the decisions documented with supplemental information. Data gathered along with the name typically include the bibliographic reference, author(s), and date of publication. Additional information may include references to type specimens (institution and catalog number), type locality, and references to subsequent taxonomically significant publications. An example is given in Fig. 6.1.

Scoparia Dulcis L.

Taxonomic Serial No. 34029

Taxonomy and Nomenclature

Kingdom	Plantae
Taxonomic Rank	Species
Synonym(s)	
Common Name(s)	Licorice weed
Taxonomic Status	
Current Standing	Accepted
Taxonomic Credibility Rating	Reviewed

Taxonomy Hierarchy

Kingdom	*Plantae*	Planta, Plantes, Plants, Vegetal
Subkingdom	*Tracheobionta*	Vascular plants
Division	*Magnoliophyta*	Angiospermes, Angiosperms, Flowering plants, phanèrogames, Plantes à fleurs, plantes à fruits
Class	*Magnoliopsida*	Dicots, Dicotylèdones, dicotyledons
Subclass	*Asteridae*	
Order	*Scrophulariales*	
Family	*Scrophlariaceae*	
Genus		
Species Scopria dulcis L.	*Licorice Weed*	

Fig. 6.1: Taxonomic classification *Scoparia dulcis* L.

HISTORICAL PROFILE 9

E.O. Wilson for Giving the Concepts in Biodiversity

Dr. E.O. Wilson

Dr. Edward E.O. Wilson is one of the greatest thinkers in the 20th century. His groundbreaking research, original thinking, and scientific and popular writing have changed the way humans think of nature, and our place in it. Currently he is a research professor and museum curator at Harvard University. He has received many of the world's leading prizes for his research in science, his environmental activism, and his writing. Wilson has been a leader in the Fields of entomology (the study of insects), animal behaviour and evolutionary psychology, island biogeography, biodiversity, environmental ethics, and the philosophy of knowledge. He has written groundbreaking books and articles on all of these subjects. Two of his non-fiction books, *The Ants* (1990, with Bert Hölldobler) and *On Human Nature* (1978) have won Pulitzer Prizes. The Diversity of Life (1992) and Consilience. The Unity of Knowledge (1998), two of his more recent books, have been applauded for their graceful, creative and constructive approaches to challenging subjects. In The Diversity of Life and The Future of Life he conveys his deep concern for humanity's bewildering degradation of our planet's ecosystems. His commitment to protecting our natural heritage has brought him to the forefront of environmental activism.

Wilson grew up in Alabama. In childhood he became fascinated with the natural world, especially insects. In high school he decided to devote his career to the study of ants. He received a BSc from the University of Alabama in 1949 an MS in 1950, and his Ph.D. from Harvard University in 1955. For most of his career Wilson has been a professor at Harvard University. He is currently a Pollogrine University Research Professor and Honorary Curator in Entomology of the Museum of Comparative Zoology at Harvard University. Wilson has won numerous awards, honors, and fellowships for his research, writing, and activism, including two Pulitzer Prizes (1978 and 1990), the 1977 National Medal of Science, the international Prize of Biology from Japan (1993), and the Audubon Medal of the National Audubon Society (1995).

NEWS UPDATE

The Quiet Revolution: Biodiversity Informatics and the Internet

The massive development of biodiversity-related information systems on the Internet has created much that appears exciting but chaotic, a diversity to match biodiversity itself. This richness and the arrays of new sources are counterbalanced by the maddening difficulty in known what is where, or of comparing like with like. But quietly, behind the first waves of exuberance, biologists and computer scientists have started to pull together in a rising tide of coherence and organisation. The fledgling field of biodiversity informatics looks set to deliver major advances that could turn the internet into a giant global biodiversity information system (source Science 29th September 2004 vol. 289).

Biodiversity Informatics is a part of biological informatics sandwiched between—and strongly overlapping with—environmental informatics and molecular bioinformatics. Presently, the objectives of research and other activities in Biodiversity Informatics is to provide a sound information management infrastructure for biodiversity and global change research. The basic objectives of biodiversity informatics research are:

1. Mobilizing existing biodiversity information resources like taxonomic names register and observational biodiversity databases.
2. Increasing research efficiency by timely provision of fundamental data in digital form.
3. Providing the information base and the tools for diverse biodiversity modeling tasks.

Historically roots of Biodiversity Informatics lies in Australia. Australia has been a leading country in biodiversity informatics. Since the mid-1970s, Australian herbaria have been digitizing their data. The Environmental Resources Information Network (ERIN), was established in 1989 to provide geographically-related environmental information for planning and decision-making. In 1993 other countries like Brazil, Mexico, Costa Rica, etc. became fully engaged in biodiversity informatics after the exchange of experience with ERIN experts. In the early 1990s, researchers from diverse fields of expertise held meetings, and began the Biodiversity Information Network—Agenda 21 (BIN21) initiative. In 1998, a research project was launched at the University of Kansas Natural History Museum and Biodiversity Research Center: the Species Analyst (TSA). TSA's main objective was to develop standards and software tools for access to world natural history collection and observation databases. Several global and regional efforts are aiming at organising data stakeholders and making data available for conservation and sustainable development research led to genesis of GBIF (Global Biodiversity Information Facility) in 2001. Although GBIF intends eventually to incorporate all levels of biodiversity data (molecules to ecosystems), its initial phase is focusing on species- and specimen-level information. Its work program emphasizes four priority areas:

1. Digitization of natural history collection data.
2. Data access and database interpretability,
3. Electronic catalogue of names of known organisms, and
4. To outreach and capacity building.

The GBIF data portal, which connects participant nodes and other data providers, came on-line in February 2004, and by October 2004 was serving over 40 million records. Several global and regional efforts in biodiversity informatics are listed in Table 6.1.

Table 6.1: Global and regional efforts in biodiversity informatics

1. ERIN (Australia)	Environmental Resources Information Network
2. CONABIO (Mexico)	Mexico's Commission Nacional Para el Conocimiento y Use de la Biodiversidad
3. Costa (Rica)	Costa Rica's Instituto Nacional de Biodiversidad (INBio)
4. BDT (Brazil)	Brazil's Base de Dados Tropical
5. GBIF	Global Biodiversity Information Facility
6. ENBI	The European Network of Biodiversity Information
7. IABIN	Inter-American Biodiversity Information Network

India is one of the 12 mega biodiversity countries of the world, from about 70 per cent of the total geographical areas surveyed so far 46,000 plant and 81,000 animal species have been described (annual report 2002-03). India, specially the north-eastern region is very well recognized for its richness in biological resources and indigeneous knowledge about economic exploitation of these resources. However, there was a lack of communication both on national and international level, because of which controversies like patents of turmeric, neem, basmati given in US arose. In this context, the Biological biodiversity Bill approved on 11th December 2002 with an intention to protect India's rich biodiversity and associated knowledge against their use by foreign individuals and organisation without sharing the benefits arising out of such as and check biopiracy. In India the biodiversity informatics is at infant stage. Recently, SAMPADA a software is developed for automation of a repository of collections and digitalizing specimens (Software Update).

SOFTWARE UPDATE

Sampada a Software for Digitalizing Museum or Herbarium Specimen

SAMPADA is an Sanskrit word. It means "natural wealth" or "heritage". Biological collections are natural wealth of a nation. SAMPADA, a software developed by us for digitizing your museum or herbarium's specimen data and provides support for biological collections data. Current versions of SAMPADA and technical support for its efficient use are free to nonprofit research collections institutions. Sampada is available at http://www.ncbi.org.in/sampada/index1.htm.

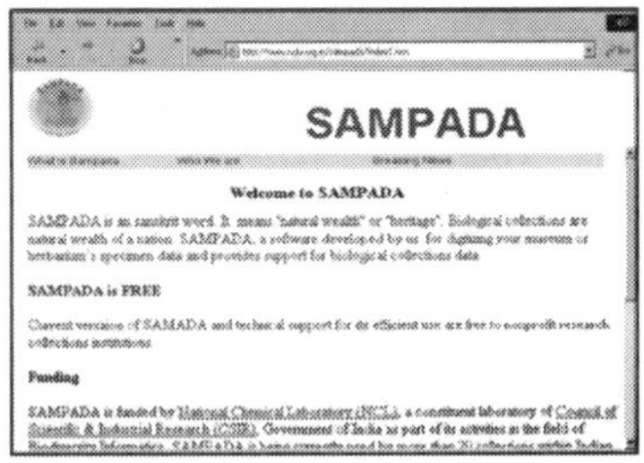

Fig. 6.2: Web view of Sampada

6.1 TAXONOMIC VS OBSERVATIONAL BIODIVERSITY DATABASES

Biological taxonomy—the scientific names of organisms—provides a global (at least internationally recognized) system of designating natural groups of organisms: i.e. species and higher taxa. Classifications assemble smaller groups into larger groups, and provide a way of making statements about or retrieving information about many species at a time. One of the first steps in communicating the discovery of a new kind of organism is to give it a name, and to infer what it is by classifying it, i.e., saying that it is a particular kind of something more general and perhaps already familiar.

Defining a group of organisms (taxon) and naming it are conceptually distinct operations. Determining that a taxon is biologically meaningful or real is a question of scientific judgement and is subject to refutation. Determining what name should be applied to it is a matter of following the rules set forth in the relevant international code of nomenclature. (There are three; a code for microbiology, another for botany, and a third for zoology.) An example of a taxonomic name for licorice weed and its classification is given in Fig. 6.1.

There are mainly two types of biodiversity databases:

1. **Taxonomic databases:** These are used mainly by the taxonomists whose purpose is to create and organize the taxonomic frame work of nomenclature and classification. There are (at least) three further types of these:
   (*a*) Taxon databases: where information contained relates directly to the specimen.
   (*b*) Accession, specimen or collection databases, which contain information on for example individual specimen (in a museum collection), accession (cultivated in a botanical garden), cultures (maintained in a botanical culture collections) or observations (recorded in a mapping project).
   (*c*) Name or nomenclatoral databases which mainly deals with names.
2. **Observational databases:** The databases containing all kind of information about organisms, including their characteristics, chemical constituents, uses, poisons, conservation and management data and so on, which use taxonomic frame work as a reference, is just a

convenient indexing system and not the primary purpose of the database. These databases are mainly indexed in other ways e.g. geography, uses, status, and not by species name, they have wider category of users ranging from ecologists to policy makers. These type of databases which are becoming more and more important as user's demands increase as they offer easy availability of data through internet.

A quick glance at biological and biodiversity web servers reveals a wide range of searchable databases available, most of which however are taxonomic databases, a sure progress towards a complete biological names register. Recently science magazine in 2000 released a quick guide to biodiversity informatics giving detailed information about institutions, networks and software applications (News Update 1).

6.2 MAJOR BIODIVERSITY DATABASES

There are four significant biodiversity databases that aim to provide information about taxonomy and biodiversity of biota on a global scale

1. **Integrated Taxonomic Information System (ITIS):** A checklist. A checklist represents determinations about which taxa are accepted or valid, and for each taxon gives its correct name a list of synonyms, and places all taxa in a single coherent classification. ITIS was established as a collaborative project among US Federal Government agencies. Recently it added international collaborators include Agriculture Agri-food Canada and CONABIO (Mexico). Information given in Fig. 6.1 is derived from ITIS database. Besides providing information about taxonomy and classification of an organism it also provides complete reference about the source of information and name and link to the expert on the said organism (see Experiment 6.1A).
2. **Species 2000:** Species 2000 is also a checklist. Species 2000 is an international project that aims to produce a checklist of all species through a federation of "Global Species Databases" (GSDs). Taxonomists in each discipline have collaborated to build the global species database for that discipline e.g., for birds, flatworms, bromeliads, etc. The project was established in 1996, but progress has been slow, due in part to the difficulty of raising money for international projects. In the Fig. 6.3 website of species 2000 is shown. Species 2000 has the objective of enumerating all known species of organisms on earth (animals, plants, fungi and microbes) as the baseline dataset for studies of global biodiversity. The existing global species databases may presently account for some 40 per cent of the total known species, so substantial investment in new databases will be needed for full coverage of all taxa to be achieved.

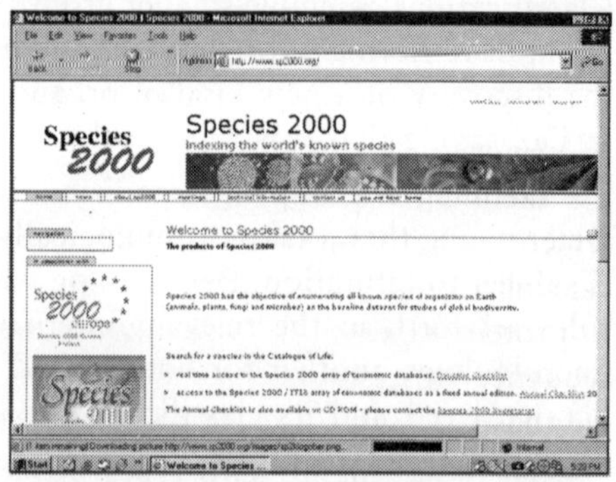

Fig. 6.3: Web view of species 2000

3. **International Plant Name Index (IPNI):** It is a nomenclator i.e., is a compilation of all relevant names, but does not present opinions about which taxa are accepted or valid. IPNI is an international collaboration among three institutions: The Royal Botanic Gardens, Kew (RBGK), the Harvard University Herbaria (HUH), and the Australian National Botanic Garden (ANBG). This project is important because it represents the largest compilation or vascular plant names and because it attempts to support an open contribution and editorial processes.
4. **Index to Organism Names (ION):** It is also a nomenclator database. ION is being developed by BIOSIS (combining their zoological and bacteriological DBs) in collaboration with other

organisations that have developed comprehensive databases for other taxa, such as seaweeds, fungi, and mosses. With more than 1.4 million names, ION represents the largest compilation on biological nomenclature. The only taxa excluded by intensive vascular plants, the domain of IPNI.

The two nomenclator projects (IPNI and ION) are derived from "traditional" literature indexing operations. An indexer reads publications and records the taxonomic names and subjects addressed in each publication; the resulting indexes are published annually.

Table 6.2: Examples of different categories of biodivesity databases

Catagory	*Examples*	*Description*
Global Databases	Species 2000, Species, Species Analyst Global Biodiversity Information Facility	
Geographical, Regional and National Databases	Integrated Taxonomic Information System	North America taxa
	CONABIO	Maxican taxa database
	Nature Serve	North America and Canada Australia
Databases for specific organism	Virus databases online	Virus
	USDA Plants Database	Vascuar plants, mosses, liverworts, hornworts and lichens of U.S.
	International Legume Database Information Service	legume
	Fishbase	Fishes
	Cephbase	Cephalopods

Another category of taxonomic databases is the databases maintained by the sites more known for their molecular data. For example, Taxonomy Browser is a prominent taxonomic database which maintained by the NCBI. This taxonomy is hierarchical and sequence based aiming to centralize the classification of all organisms represented in the databases with at least one nucleotide or protein sequence. The Taxonomy Browser can be used to view the taxonomic position or retrieve sequence data for a particular organism or group. Similarity, NEWT is the UniProt taxonomy database. It integrates taxonomy data compiled in the NCBI database and data specific to the Swiss-Prot protein knowledgebase. NEWT is updated daily.

The strange thing about all the databases is that while there are large overlaps among them, they are collaborators primarily in name and not really in substance. The Web site for each project mentions at least one of the others as partner and the management team for each project has at least one liaison from another project. Yet despite this openness the projects do not share actual data, nor does there appear to be any intention to retionalize the efforts and create data flows between projects. Each has a sufficiently different strategic plant, technology base or funding model that actual collaboration (data sharing) appears politically infeasible, at least for the time being. There are several other biodiversity databases which could be grouped as follows:

1. Global databases.
2. Geographical, regional, and national databases.
3. Databases for specific taxonomic groups.

Examples and description of each of these is given in Table 6.2.

EXPERIMENT 6.1A

To Study the ITIS (Integrated Taxonomic Information System) Database

A brief description of ITIS is given earlier in the chapter. In the present experiment we want to explore all information regarding correct name, synonyms and taxonomic classification of garden pea, *Pisum sativum*.

TESTING THE HYPOTHESIS

Computer set up: Normal PC with internet connection in windows environment:

1. Log on to the website of ITIS shown in Fig. 6.4.
2. Enter pea in text box for search.
3. Search the results obtained for garden pea. The result obtained is shown in Fig. 6.5.
4. Detailed entry for garden pea is shown in Data Box 6.1.

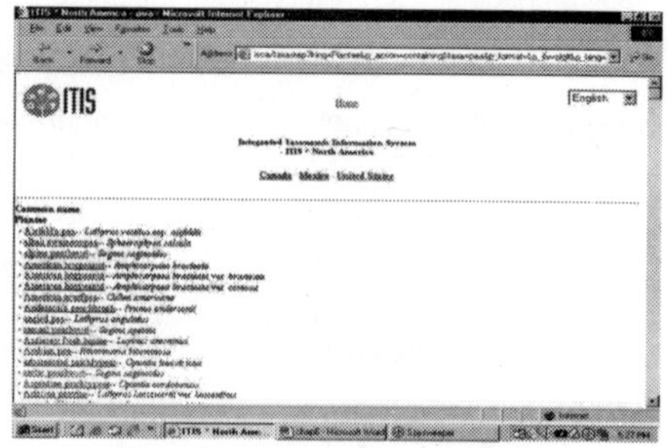

Fig. 6.4: Web view of ITIS

DATA

Data Box 6.1

Scientific Name: Pisum sativum L.
- TSN: 26867
Taxonomy and Nomenclature
Kingdom: Plantae
Rank: Species **Synonym(s):** Pisum sativum var. arvense (L.) Poir.
Pisum sativum var. humile Poir.
Pisum sativum var. macrocarpon Ser.
Pisum humile Boiss & Noe
Pisum arvense L.
Pisum sativum ssp. arvense (L.) Poiret
Vernacular name(s): garden pea [English]
Taxonomic Status:
Current Standing: accepted

Data Quality Indicators
Record Credibility Rating: verified - standards met
Taxonomic Hierarchy

Get the following data in XML format
up to the kingdom
Get the following data in XML format
Kingdom Plantae— *plants*
Subkingdom Tracheobionta—*vascular plants*
Division Magnoliophyta—*flowering plants, angiosperms*
Class Magnoliopsida—*dicots, dicotyledons*
Subclass Rosidae
Order Fabales
Family Fabaceae
Genus Pisum—*pea*
Species Pisum sativum— *garden pea*

References
Expert(s):John Kartesz: Biota of North America Project (BONAP), University of North Carolina.
Other sources: Source: NODC Taxonomic Code database (version 8.0)

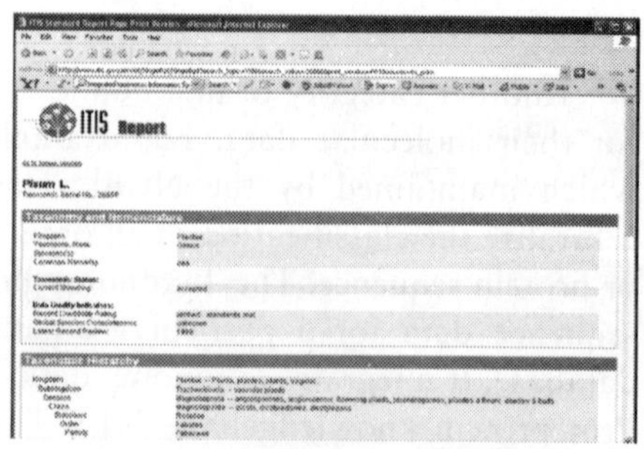

Fig. 6.5: Query result at ITIS

Source: The PLANTS Database database (version 4.0.4); (National Plant Data Center, NRCS, USDA. Baton Rouge, LA 70874-4490 USA. http://plants.usda.gov)
Source: The PLANTS Database database (version 5.1.1); (National Plant Data Center, NRCS, USDA. Baton Rouge, LA 70874-4490 USA. http://plants.usda.gov)
Geographic Information
Geographic Information:North America
Jurisdiction: Continental US, Introduced
Internet Search Portal for: Pisum sativum
Images:
images.google.com
using scientific name and any vernacular names: AltaVista Canada • AltaVista • Excite • Fast Search • Google • HotBot • Lycos
using scientific name and synonym(s):AltaVista Canada • AltaVista • Excite • Fast Search • Google • HotBot • Lycos
searching specialized Internet databases:
GenBank • GoBase • Medline • Scirus • CISTI/CAL • Agricola (Periodicals • Books)
Global Biodiversity Information Facility (GBIF) Data portal

Retrieve biological records and collection information:
- 5339 record(s); 495 record(s) with coordinates
- Dynamic map
- Please read the requirements
- Download this GBIF occurrence data for Google Earth

SUMMARY

Recent developments in information and communication technology are allowing new experiences in the integration, analysis and visualization of biodiversity information, and are leading to development of a new field of research, biodiversity informatics. Although this field has great potential in diverse realms, including basic biology, human economics, and public health, much of this potential remains to be explored.

The success of several concerted international efforts depends largely on broad deployment of biodiversity informatics information and products. Several global and regional efforts are organizing and providing data for conservation and sustainable development research, including the Global Biodiversity Information Facility, the European Biodiversity Information Network, and the Inter-American Biodiversity Information Network. Critical to development of this field is building a biodiversity information infrastructure, making primary biodiversity data freely and openly available over the Internet. In addition to specimen and taxonomic data, access to non-biological environmental data is critical to spatial analysis and modeling of biodiversity.

EXCERCISE

1. Use the data base species 2000 and find out the entry for *Scoparia dulcis* L. Compare the data entry with that given in Fig. 6.1.

7

BIBLIOGRAPHIC DATABASES

Services that produced abstracts of scientific literature began to make their data available in machine-readable form in the early 1960's. The tools that enable scientifically coherent and efficient use of the resources described below are an important activity in the field of bioinformatics called text mining.

7.1 COMMON BIBLIOGRAPHIC DATABASES

Bibliographic indexes include information that will help you find an article, but not the article text itself. These indexes contain the following elements: Title, author, magazine or journal title in which it was published, date, page numbers and may include abstracts (summaries of the article's contents). Commonly used bibliographic databases of interest to biologists are given in Table 7.1. First in the list is medline database. Medline is the NLM's (National Library of Medicine) premier bibliographic database covering the fields of medicine, nursing, dentistry, veterinary medicine, the health care system, and the preclinical sciences. Medline contains bibliographic citations and author abstracts from more than 4,000 biomedical journals published in the United States and 70 other countries. The file contains over 12 million citations dating back to the mid-1960's. Coverage is worldwide, but most records are from English language sources or have English abstracts.

Bibliographic databases are quick way to get into the bit lite literature of a topic combined with use of a general search engine for web sites that do not correspond to articles published in journals, fairly comprehansive information is readily available about most subject. If you are trying to start to learn about an unfamiliar subject try adding the keyword tutorial to your search in a general search engine or keyword review to your search in bibliographic databases.

Major scientific journals have their web site offering on line web articles are important part of bibliographic databases. They also offer link to related articles. US National Institutes of Health have established a central web based library of scientific library called as Pub Med Central. In collaboration with scientific journals, the NCBI is organising the electronic distribution of full texts of published articles. Medline integrates the Medical literature, including many papers dealing with subjects on molecular biology.

IndMED database covers prominent peer reviewed Indian biomedical journals. Databases is designed to provide medical professionals/researchers/students and the medical library professional quick and easy access to Indian literature. Indian contribution in the areas of biomedical research and health care has been significant and conforming to international standards. However, only a small fraction of it is available for reference through international bibliographic databases. The ICMR-NIC Centre for Biomedical Information (Indian MEDLARS Centre or IMC) has designed and developed a bibliographic database of peer reviewed Indian biomedical literature. This database covers prominent Indian journals. These have been selected from more than 200 journals. More journals would be added to the list as their quality improves in coming years. It is proposed to cover the journals from 1985 onwards in this database.

Table 7.1: Important bibliographic databases

Name of database	*Comments*
MEDLINE	is medical database accessible through EBI's SRS.
IndMED	Database covering prominent peer reviewed Indian biomedical journals. Database designed to provide medical professionals, researchers/students and the medical library professional quick and easy access to Indian literature.
PUBMED	is biological database accessible through NCBI's ENTREZ.
EMBASE	is a commercial product for the medical
BIOSIS	The inherior of the old Biological Abstracs, covers a broad biological field; the Zoological Record indexes he zoological literature.
CAB International	maintains absract databases in the fields of agriculture and parasitic diseases.
AGRICOLA	is for the agricultural field what MEDLINE is for the medical field. The bibliographical databases are with the exception of MEDLINE/PUBMED only available through commercial database vendors.

Biosis database is the most complete resource for finding life sciences information quickly and efficiently. We select documents from thousands of sources worldwide, index and abstract them into citations which describe their content, and maintain databases for searching citations — adding more than 600,000 new entries each year. BIOSIS Previews®, Biological Abstracts® and Zoological Record are available on several leading information platforms, the powerful web-based platform that delivers a powerful combination of content, tools and technology.

Important features of biosis database are given below:

1. Interdisciplinary: BIOSIS databases integrate important subjects from many different fields all across biology.
2. International: More than 90 nations are represented in BIOSIS databases.
3. Easy to Search: BIOSIS databases use a unique indexing system that allows flexible, efficient searching.
4. Informed by Diverse Sources: You can find citations for articles, meetings, patents, book chapters, and more in BIOSIS databases.

7.2 PUBMED: A MODEL BIBLIOGRAPHIC DATABASE

PubMed (web view Fig. 7.1) is a bibliographic database which is integrated with other information retrieval tools of the National Center for biotechnology information within the National Library of Medicine. It offers the abstracts of scientific articles.

PubMed is a Web-based retrieval system developed by the NCBI at the NLM. It is part of NCBI's vast retrieval system, known as Entrez. PubMed is a database of bibliographic information drawn primarily from the life sciences literature. PubMed contains links to full-text articles at participating publishers' web sites as well as links to other third party sites such as libraries and sequencing centers. PubMed provides access and links to the integrated molecular biology databases maintained by NCBI.

There are four major PubMed Services:

1. the MeSH Database
2. the Journals Database
3. the Single Citation Matcher
4. Clinical Queries

MeSH Database includes MeSH terms. MeSH is the acronym for "Medical Subject Hearings". MeSH is the authority list of the vocabulary terms used for subject analysis of biomedical literature at NLM. MeSH vocabulary is used for indexing journal articles for MEDLINE and is also used for cataloging books and audiovisuals. The MeSH controlled vocabulary is a distinctive feature of MEDLINE. It imposes uniformity and consistency to the indexing of biomedical literature. MeSH terms are arranged in a hierarchical categorized manner called MeSH Tree. The MeSH Database allows you to: Locate and select MeSH terms (Headings, Subheadings, and Publication Types); Supplementary Concept terms (Substance Names) and Pharmacological Action terms.

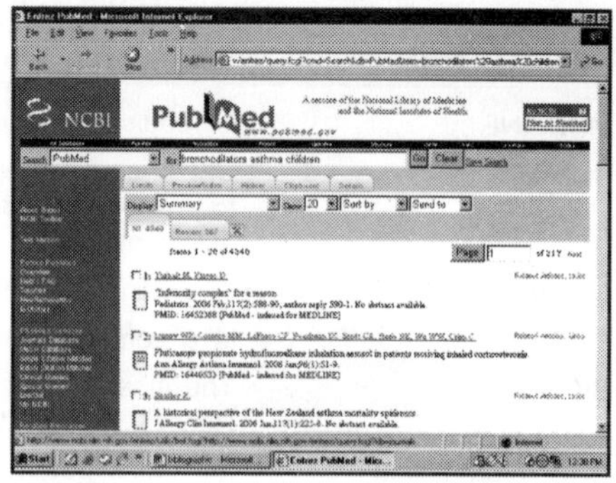

Fig. 7.1: Web view of PubMed

The Journals Database allows you to look up information about a PubMed journal and search for that title. The database includes journals in all Entrez databases (e.g., PubMed, Nucleotide, Protein). You can search for a journal using: journal title, Medline/PubMed journal title abbreviation, NLM ID (NLM's unique journal identifier, abbreviation. For a list of full-text journals available on the web to which PubMed currently links, click on links to "full-text web sites" at the bottom of the Journal Browser screen. New journals are added regularly. As noted before, some journals may require that you register, subscribe, or pay a fee in order to view the full-text of an article.

The Clinical Queries feature is available from PubMed's. There are 3 search filters available from this page:

1. Search by clinical study category
2. Find systematic reviews
3. Medical genetics searches

The Single Citation Matcher allows you to find a single citation with an easy, fill-in-the-blank format. Enter as much information as you have to locate a single citation or items from a particular volume or issue of journal. For example to search for Gene Therapy, 1999, and the title words, mitochondrial gene therapy, enter the information in the appropriate boxes and click Go. Notice the journal name autocomplete feature. This feature will suggest titles as you enter a title in abbreviated form or full title. Titles displayed by the auto-complete menu are in a ranked order based on the number of citations in PubMed.

1. Identify the key concepts in your research question. For example to find citations about bronchodilators for treating asthma in children, the key concepts are bronchodilators, asthma and children.
2. Enter the significant terms into the search box.
3. Press the Enter key or click Go.
4. Result of such query is shown in data box 7.1.

DATA

Data Box 7.1

1. Turkalj M, Plavec D.
"Inferiority complex" for a reason.
Pediatrics. 2006 Feb; 117 (2): 588-90; author reply 590-1. No abstract available.
PMID: 16452388 [PubMed - indexed for MEDLINE]
2. Lumry WR, Conway MM, LaForce CF, Pearlman DS, Scott CA, Herje NE, Wu WW, Crim C. Related Articles, Links
Fluticasone propionate hydrofluoroalkane inhalation aerosol in patients receiving inhaled corticosteroids.
Ann Allergy Asthma Immunol. 2006 Jan; 96(1): 51-9.
PMID: 16440533 [PubMed-indexed for MEDLINE]
3. Beasley R. Related Articles, Links
A historical perspective of the New Zealand asthma mortality epidemics.
J Allergy Clin Immunol. 2006 Jan; 117 (1): 225-8. No abstract available.
PMID: 16429618 [PubMed-indexed for MEDLINE]
4. Kieckhefer GM, Lentz MJ. Related Articles, Links
Nocturnal asthma in children.
Adv Nurse Pract. 2006 Jan; 14 (1): 53-6. Review. No abstract available.
PMID: 16425517 [PubMed-indexed for MEDLINE]
5. Wolfe J, Laforce C, Friedman B, Sokol W, Till D, Della Cioppa G, vanAS A. Related Articles, Links
Formoterol, 24 microg bid, and serious asthma exacerbations: similar rates compared with formoterol, 12 microg bid, with and without extra doses taken on demand, and placebo.
Chest. 2006 Jan; 129(1): 27-38.
PMID: 16424410 [PubMed-indexed for MEDLINE]
6. Nelson HS, Weiss ST, Bleecker ER, Yancey SW, Dorinsky PM;SMART Study Group. Related Articles, Links
The Salmeterol Multicenter Asthma Research Trial: a comparison of usual pharmacotherapy for asthma or usual pharmacotherapy plus salmeterol.
Chest. 2006 Jan; 129(1): 15-26.
PMID: 16424409 [PubMed-indexed for MEDLINE]
7. Frischer M, Heatlie H, Chapman S, Bashford J, Norwood J, Millson D. Related Articles, Links
Ratio of inhaled corticosteroid to bronchodilator treatment and asthma hospitalisation.
Curr Med Res Opin. 20000; 16 (1): 8-13.
PMID: 16422029 [PubMed-indexed for MEDLINE]
8. Aogagi M, Kojima H, Sato K, Watanabe H, Sekine K, Nishimuta T. Related Articles, Links
[Additive effect of intravenous prednisolone for acute severe asthma in infants and toddlers: a randomized controlled clinicaltrial]
Arerugi. 2005 Oct; 54 (10): 1190-6. Japanese.
PMID: 16407665 [PubMed-indexed for MEDLINE]
9. Silverman M, Sheffer A, Diaz PV, Lindmark B, Radner F, Broddene M, de Verdier MG, Pedersen S, Pauwels RA; START Investigators Group. Related Articles, Links
Outcome of pregnancy in a randomized controlled study of patients with asthma exposed to budesonide.
Ann Allergy Asthma Immunol. 2005 Dec; 95(6): 566-70.
PMID: 16400897 [PubMed-indexed for MEDLINE]
10. Roddick LG, Clements B, Wales S, Henry RL; Childrens's Hospitals Australasia. Related Articles, Links
Guidelines for weaning of bronchodilator therapy.
J Paediatr Child Health. 2005 Dec; 41 (12): 696-7. No abstract available.
PMID: 16398884 [PubMed-indexed for MEDLINE]

SUMMARY

Bibliographic databases are another type of bioinformatics database. They are a quick way to get into literature on a topic. Bibliographic indexes include information that will help you to find an article and not the article itself. PubMed is a web-based bibliographic database developed by NCBI. It contains full articles. Ind MED database covers peer reviewied Indian biomedical journals. Biosis is digital form of biological abstracts.

EXERCISE

Conduct exercises 1 to 3 within PubMed web site.

1. Use the MeSH Database to build a strategy that will find citations to references discussing the economics of community-acquired pneumonia.
2. Find references using the following information and PubMed's Single Citation Matcher:
   (*a*) Arthritis Rheum, 1982, page 1271.
   (*b*) V. Lee, Biochemical Pharmacology, vol. 29, issue 14.
   (*c*) "A powder formulation of measles vaccine for aerosol delivery.
3. Use the Journals Database to see it PubMed includes the journal, *Molecular Microbiology.* If so, search for this journal from the Journals Database.

8

SEQUENCE FORMATS IN BIOLOGICAL DATABASES

Different databases use different formats for submitting sequence and retrieving information. The reason for that is the best format for a human being to read does not mean that it is necessarily the best and most efficient for a computer. That is why every program introduces its native format optimized for a particular program needs, and some programs include converters from one format to each other. To recognize formats and to be able to inter-convert files in different formats is very important for successful data analysis. In absence of knowledge about sequence formats in biological databases the experiments conducted throughout he book could not be done.

All the formats are standard ASCII files and may differ in presence of certain characters and words that indicate where different type of information and sequence itself are to be found. Most of the sequence formats are same for nucleic acids and proteins. Few things are common in all the databases, one of which is that amino acids and nucleic acids are written by one letter code given in Tables 3.1 and 4.1 respectively. Sequence file is usually created and edited using a text editor. Computers store sequence information as simple rows of sequence characters called as string. Each character is stored in binary code in the smallest unit of memory called as byte. Each byte comprised of 8 bits having a possible value of 0 or 1 producing 255 combinations. By convention many of these combinations have a specific definition called their ASCI equivalent. Some ASCI values are defined as keyboard characters others as special control characters such as signaling the end of line character. A file with only ASCI character is called as ASCI file. For convenience all binary codes may be written in a hexadecimal format, which corresponds to our decimal format 0, 1, 2 ....9 plus the letters A, B ....F. A DNA sequence is usually stored and read in computer as a series of 8-bit words in this binary format.

The file formats are classified as input and output/retrieval sequence format and studying both of them is very important for effective studies with data. The input file format is very important because if the sequence in not in the right format the program does not accept the sequence. Out put sequence format is usually the result obtained as a result of similarity searches or sequence analysis and is equally important. Each program uses its own abbreviations for input and output files and tutorials of that particular program need to be consulted to understand the explanation of that. Selected popular formats are discussed in details here.

The commonly used sequences formats along with a brief description about them are given below:

8.1 FASTA FORMAT

The simplest and commonest format for DNA/protein sequences is a Fasta format or Pearson format because it was developed by Pearson (Historical Profile 7) who along with Lipman in 1996 developed FASTA program for similarity searches. This format was originally given for FASTA but now it is used in a variety of molecular biology software. It has three characteristic features.

1. Every sequence in the file starts with "greater than" character (>). This character is followed by an identifier of a sequence (e.g. name, description, gi number).
2. The sequence in one letter symbols, and is continuous without any spaces.
3. An optional '*' which indicates end of the sequence

The example of a sequence in FASTA format is shown below:

```
>gi|2978501|gb|AAC06133.1| vacuolar ATPase proteolipid subunit [Giardia intestinalis]
MSSIDSPVAVEKCPAGASFWSMLGQVVAVVFSSIGAAYGTAKAGSGLGV
AGLINPAPVTKLTLPVIMAGILSIYGLITSLLINSRVRSYTNGMPLYVS
YAHFGAGLCCGLAALAAGLAIGVSGSAAVKAVAKQPSLFVVMLIVLIFS
EALALYGLIIALILSTKSADSNFCVNNVNQ
```

8.2 PHYLIP FORMAT

Phylip is another format, which is commonly used. This came from PHYLIP package a package for phylogenetic analysis. An example of PHYLIP format is given below:

```
6    13
Archaeopt CGATGCTTAC CGC
HesperorniCGTTACTCGT TGT
BaluchitheTAATGTTAAT TGT
B. virginiTAATGTTCGT TGT
BrontosaurCAAAACCCAT CAT
B.subtilisGGCAGCCAAT CAC
```

The first line of the input file contains the number of species and the number of characters, in free format, separated by blanks (not by commas). The information for each species follows, starting with a ten-character species name (which can include punctuation marks and blanks), and continuing with the characters for that species.

The sequences can continue over multiple lines; as a consequence, there are two types of the format:

1. Interleaved
2. Sequential

In sequential format all of one sequence is given, possibly on multiple lines, before the next starts. In interleaved format the first part of the file should contain the first part of each of the sequences, then possibly a line containing nothing but a carriage-return character, then the second part of each sequence, and so on. Only the first parts of the sequences should be preceded by names. Here is a hypothetical example of interleaved and sequential format:

Interleaved PHYLLIP Format

```
5    42
Turkey    AAGCTNGGGC ATTTCAGGGT
Salmo gairAAGCCTTGGC AGTGCAGGGT
H. SapiensACCGGTTGGC CGTTCAGGGT
Chimp     AAACCCTTGC CGTTACGCTT
Gorilla   AAACCCTTGC CGGTACGCTT
GAGCCCGGGC AATACAGGGT AT
GAGCCGTGGC CGGGCACGGT AT
ACAGGTTGGC CGTTCAGGGT AA
AAACCGAGGC CGGGACACTC AT
AAACCATTGC CGGTACGCTT AA
```

Sequential format

```
5    42
Turkey    AAGCTNGGGC ATTTCAGGGT
GAGCCCGGGC AATACAGGGT AT
Salmo gairAAGCCTTGGC AGTGCAGGGT
GAGCCGTGGC CGGGCACGGT AT
H. SapiensACCGGTTGGC CGTTCAGGGT
ACAGGTTGGC CGTTCAGGGT AA
Chimp     AAACCCTTGC CGTTACGCTT
AAACCGAGGC CGGGACACTC AT
Gorilla   AAACCCTTGC CGGTACGCTT
AAACCATTGC CGGTACGCTT AA
```

SOFTWARE UPDATE

SEQIO: A Package for Reading and Writing Sequence Files

The SEQIO package is a C/C++library that has been designed to reduce that effort, both for people with little or no programming experience and for more experienced software developers.
The main advantages for non-programmers come when programs use the SEQIO package to perform their reading and writing. The SEQIO package can read and write any of the 15 following file formats, eliminating, for programs that use the package, much of the worry about converting sequences from one file format to another in order to use the various programs: GenBank, PIR (CODATA), EMBL, Swiss-Prot, FASTA, NBRF, IG/Stanford, ASN.1 text, GCG, MSF, PHYLIP, Clustalw, FASTA-output, BLAST-output where FASTA-output and BLAST-output are the formats of the output produced by the FASTA and BLAST suites of programs.

For those with a little programming experience, the SEQIO package code has been designed to make reading and writing sequence files as simple to use as the C stdio package. For example, the complete program below scans a database's sequences and outputs the entry text for every entry whose sequence has a match to the given keyword. (Note: This is a simple version of the grepseq program.) The package is ideal for writing those "quick and dirty" programs to extract entries from a database or perform a new analysis on the sequences of a file or database.

8.3 CLUSTAL (.ALN) FORMAT

This was originated in the alignment program CLUSTAL. The file starts with word CLUSTAL. The alignment is written in blocks of 60 residues. Every block starts with sequence names. The example of the alignment in CLUSTAL format is shown below:

```
CLUSTAL X (1.8) multiple sequence alignment
R.sodomens   ——————————————————————CAACCUGA-GAGUU-U-GA-U-CCU-G
R.rubrum4    ——————————————————————UUCCCUGAA-GAGUU-U-GA-U—U-G
Ag.tumefac   ——————————————————————CUCAACUUGA-GAGUU-U-GA-U-CCU-G
Ag.rhizog2   ——————————————————————UUCCCUGAA-GAGUU-U-GA-U-CCU-G
Rhb.legum4   ——————————————————————UUCCCUGAA-GAGUU-U-GA-U-CCU-G
Rhb.legum6   ——————————————————————UUCCCUGAA-GAGUU-U-GA-U-CCU-G
Bdr.japon8   ——————————————————————UUCCCUGAA-GAGUU-U-GA-U—U-G
                                   * * * * *
R.sodomens   GCUC-A-G-AAC-GAAC-GC—U-GGC-GGC-A-GG-C-CU—AACACA-UGCAA—G-
R.rubrum4    GCUC-A-G-GAC-GAAC-GC—U-GGC-GGC-A-GG-C-CU—AACACA-UGCAA—G-
Ag.tumefac   GCUC-A-G-AAC-GAAC-GC—U-GGC-GGC-A-GG-C-UU—AACACA-UGCAA—G-
Ag.rhizog2   GCUC-A-G-AAC-GAAC-GC—U-GGC-GGC-A-GG-C-UU—AACACA-UGCAA—G-
Rhb.legum4   GCUC-A-G-AAC-GAAC-GC—U-GGC-GGC-A-GG-C-UU—AACACA-UGCAA—G-
Rhb.legum6   GCUC-A-G-AAC-GAAC-GC—U-GGC-GGC-A-GG-C-UU—AACACA-UGCAA—G-
Bdr.japon8   GCUC-A-G-AGC-GAAC-GC—U-GGC-GGC-A-GG-C-UU—AACACA-UGCAA—G-
                 **** * *   * **** **  * *** *** * ** *  *  ****** *****
```

The characteristic features of CLUSTAL format are given below:

1. The word CLUSTAL should be on the first non-space line of the file.
2. The alignment is displayed in blocks of a fixed length.
3. Each line in the block corresponds to one sequence.
4. The line starts with the sequence name (of any length), followed by at least one space character.
5. Then the sequence is displayed: upper or lowercase, '-' denotes gaps, residue number is displayed optionally at the end.
6. In between blocks a line shows conservation information.

8.4 GCG FILE FORMAT

A sequence file in GCG format contains exactly one sequence, begins with annotation lines and the start of the sequence is marked by a line ending with two dot ("..") characters. This line also contains the sequence identifier, the sequence length and a checksum.

This format should only be used if the file was created with the GCG package.

```
Example:
ID   AA03518     standard; DNA; FUN; 237 BP.
XX
AC   U03518;
XX
DE   Aspergillus awamori internal transcribed spacer 1 (ITS1) and 18S
DE   rRNA and 5.8S rRNA genes, partial sequence.
XX
SQ   Sequence 237 BP; 41 A; 77 C; 67 G; 52 T; 0 other;
AA03518  Length: 237  Check: 4514  ..
1      aacctgcgga aggatcatta ccgagtgcgg gtcctttggg cccaacctcc catccgtgtc
61     tattgtaccc tgttgcttcg gcgggcccgc cgcttgtcgg ccgccggggg ggcgcctctg
121    ccccccgggc ccgtgcccgc cggagacccc aacacgaaca ctgtctgaaa gcgtgcagtc
181    tgagttgatt gaatgcaatc agttaaaact ttcaacaatg gatctcttgg ttccggc
```

8.5 MOLGEN/STANFORD FILE FORMAT

The first line in a sequence file is a comment line. This line begins with a semi-colon in the first space. This line need not be present. If it is present it holds descriptive text. There may be as many comment lines as desired at the first of sequence file. The second line must be present and contains an identifier or name for the sequence in the first ten spaces.

The sequence begins on the third line and occupies up to eighty spaces. Spaces may be included in the sequence for ease of reading. The sequence continues for as many line as needed and is terminated with a 1 or 2. 1 indicates a linear sequence while 2 marks a circular sequence.

```
Example:
LINE 1      :; Describe the sequence any way you want
LINE 2      : ECTRNAGLY2
LINE 3      : ACGCACGTAC ACGTACGTAC   A C G T C C G T ACG TAC GTA CGT
LINE 4      :  GCTTA   GG G C T A1
```

8.6 NBRF/PIR FILE FORMAT

The PIR format is similar to FASTA format, each sequence starts with a sequence type code (described in the table below) on the next line the sequence name and a description appears the sequence is on the following line and is ended with an asterisk (*)

```
Example:
>RL;Homo sapiens
Homo sapiens RNA sequence
AGUCGAGUC—GCAGAAACGCAUGAC-GACCACAUUUU-CCUUGCAAAG*
>RL;Pan paniscus
Pan paniscus RNA sequence
AGUCGCGUCG—GCAGAAACGCAUGACGGACCACAUCAU-CCUUGCAAAG*
>RL;Gorilla gorilla
Gorilla gorilla RNA sequence
AGUCGCGUCG—GCAGAUACGCAUCACGGAC-ACAUCAUCCCUCGCAGAG*
>RL;Pongo pigmaeus
Pongo pigmaeus RNA sequence
AGUCGCGUCGAAGCAGA—CGCAUGACGGACCACAUCAUCCCUUGCAGAG*
```

8.7 PAUP/NEXUS FILE FORMAT

Several programs use NEXUS format: PAUP, Spectrum, MacClade

Only the alignment information (not comments) can be read by Webin-Align Webin-Align accepts interleaved NEXUS formats. Nexus non-interleaved formats should be converted to FASTA format or nexus interleaved format.

```
Example:
:
#NEXUS
[TITLE: Four Anthropoidea]
begin data;
dimensions ntax=4 nchar=50;
format interleave datatype=RNA missing=N gap=-;
matrix
Homo_sapiens       AGUCGAGUC—GCAGAAACGCAUGAC-GAC
Pan_paniscus       AGUCGCGUCG—GCAGAAACGCAUGACGGAC
Gorilla_gorilla    AGUCGCGUCG—GCAGAUACGCAUCACGGAC
Pongo_pigmaeus     AGUCGCGUCGAAGCAGA—CGCAUGACGGAC
Homo_sapiens       CACAUUUU-CCUUGCAAAG
Pan_paniscus       CACAUCAU-CCUUGCAAAG
Gorilla_gorilla    -ACAUCAUCCCUCGCAGAG
Pongo_pigmaeus     CACAUCAUCCCUUGCAGAG
;
endblock;
begin assumptions;
options deftype=unord;
```

8.8 GENBANK FORMAT

It is also called GenBankFlatFile format. It is used to store the information in the GenBank database. Every record in GenBank consists of 3 parts: the header, the features that describe the annotations on the record, and the sequence itself. Very detailed description of every field in a GenBank record is given here:

```
The example of GenBank record is shown below:
LOCUS AF286477      2292 bp     mRNA              BCT        27-JUL-2000
DEFINITION      Thermoplasma acidophilum A-ATPase A-subunit mRNA, complete cds.
ACCESSION       AF286477
VERSION         AF286477.1  GI:9502269
KEYWORDS
SOURCE          Thermoplasma acidophilum.
ORGANISM        Thermoplasma acidophilum
                Archaea; Euryarchaeota; Thermoplasmales; Thermoplasmaceae;
                Thermoplasma.
REFERENCE 1     (bases 1 to 2292)
AUTHORS         Senejani, A.G. and Gogarten, J.P.
TITLE           Thermoplasma acidophilum A-ATPase A-subunit and Intein
JOURNAL         Unpublished
REFERENCE 2     (bases 1 to 2292)
AUTHORS         Senejani, A.G. and Gogarten, J.P.
TITLE           Direct Submission
JOURNAL         Submitted (12-JUL-2000) Molecular and Cell Biology, University of
                Connecticut, 75 North Eagleville Road, U-44, Storrs, CT 06269-3044, USA
FEATURES        Location/Qualifiers
source          1..2292
                /organism="Thermoplasma acidophilum"
                /db_xref="taxon:2303"
CDS             1..2292
                /note="intein"
```

```
                    /codon_start=1
/transl_table=11
/product="A-ATPase A-subunit"
/protein_id="AAF88065.1"
/db_xref="GI:9502270"
/translation="MGKIIRISGPVVVAEDVEDAKMYDVVKVGEMGLIGEIIKIEGNR
                    STIQVYEDTAGIRPDEKVENTRRPLSVELGPGILKSIYDGIQRPLDVIKITSGDFIAR
                    GLNPPALDRQKKWEFVPAVKKGETVFPGQILGTVQETSLITHRIMVPEGISGKVTMIA
DGEHRVEDVIATVSGNGKSYDIQMMTTWPVRKARRVQRKLLSRDPLVTAQSGNRCAFP
                    VAEAANCRVPGPFGSGKCVSGDTPVLLDAGERRIGDLFMEAIRPKERGEIGQNEEIVR
                    LHDSWRIYSMVGSEIVETVSHAIYHGKSNAIVNVRTENGREVRVTPVHKLFVKIGNSV
IERPASEVNEGDEIAWPSVSENGDSQTVTTTLVLTFDRVVSKEMHSGVFDVYDLMVPD
                    YGYNFIGGNGLIVLHNTVIQHQLAKWSDANIVVYIGCGERGNEMTEILTTFPELKDPN
TGQPLMTGLSFIANTSNMPVAAREASIYTGITIAEYYRDMGYDVALMADSTSRWAEAL
                    REISGRLEEMPGEEGYPAYLGRRVSEFYERSGRARLVSPDERYGSITVIGAVSPPGGD
                    ISEPVSQNTLRVTRVFWALDAALANRRHFPSINWLNSYSLYTEDLRSWYDKNVSSEWS
                    ALRERAMEILQRESELQEVAQLVGYDAMPEKEKSILDVARIIREDFLQQSAFDEIDAY
CSLKKQYLMLKAIMEIDTYQNKALDSGATMDNLASLAVREKLSRMKIVPEAQVESYYN
                    DLVEEIHKEYGNFIGEKNAEASL"
BASE COUNT     661 a    500 c    672 g    459 t
ORIGIN
        1      atgggaaaga taatcagaat ttcaggtcca gtagtcgtgg ctgaagatgt tgaagacgcc
       61      aagatgtacg atgttgtcaa ggtcggagag atgggcctca tcggtgagat aataaagatt
      121      gaggggaaca gatcgaccat acaggtctat gaggatactg caggcataag gcctgacgaa
      181      aaggttgaga acaccaggag gccgctgtcg gtggagctcg gcccaggcat actcaaatcg
      241      atatacgatg gaatacagag gccact
```

8.9 UNIPROT FORMAT

This format is studied and explained in details in chapter 3.

8.10 SEQUENCE FORMATS CONVERSIONS

Sequence format conversion sometimes is mandatory for similarity searches and analysis of a sequence. Very often an automatic format conversion is performed when needed by the program by many online soft wares used for sequences analysis and similarity searches. Besides there exist many tools (Table 8.1) to process and convert the format of a sequence. These tools are made available one's own needs or when the automatic conversion fails. In Experiment 8A one of the most common software for sequence conversion is used to make the students familiar with the software and the different formats.

Table 8.1: List of sequence format conversion software

READSEQ	Basic tool for sequence conversion performs additional functions: reverse sequence, gap removal and pretty output. This program is selected for sequence conversion in experiment 8A.
FMTSEQ	a more elaborate tool, which encompasses readseq functions, as well as: Clustalw conversions (to and from) and input format option Input formats include FASTA, Genbank, EMBL, SWISS-PROT, PIR, GCG, SELEX, Clustal, and GCG MSF. Available output format include fasta, EMBL/SWISS-PROT, Genbank, GCG, MacStrider, Zuker MFOLD, ig, PIR/CODATA, ELEX/HMMER/Pfam, GCG MSF, a2m.
MVIEW	MVIEW not only helps to visualize an alignment, but also allows to convert it into different formats (MSF, clustalw, Pearson Fasta, Pir, ...); extracting some sequences or some positions is also possible. This loads several sequences in a file. This is discussed in Chaper 11 with referenece to multiple sequence alignment.
EMBOSS	cutseq: Removes a specified section from a sequence pasteseq: Insert one sequence into another. nthseq: Writes one sequence from a multiple set of sequences. extractseq: Extract regions from a sequence. EMPOSS is duscussed in deails in chapter

EXPERIMENT 8A

Study of Primary Different Sequence Formats in a DNA or Protein Sequence Using the Program READSEQ

About WWW READSEQ Sequence Conversion Program: It converts input DNA/AA sequence to specified format. Input format is determined automatically. Some of these formats are useful for special type of analysis such as multiple sequence alignment and phylogenetic tree analysis. Also it converts any badly written sequence to plain text sequence format. Sequence Conversion uses the READSEQ program developed by Dr. Don Gilbert at Indiana University. Information on READSEQ is maintained at the IUbio Archive site at University of Indiana (http://iubio.bio.Indiana.edu). The input sequence is piped through the Readseq program to perform the format conversion. The program is available from the IUbio archive site http:// bimas.dcrt.nih.gov/molbio/readseq/ pretty_form.html)

Output sequence formats supported are listed below:

IG/Stanford	Phylip3.2
GenBank/GB	Phylip
NBRF	Plain/Raw
EMBL	PIR/CODATA
GCG	MSF
DNAStrider	ASN.1
Fitch	PAUP/NEXUS
Pearson/Fasta	Pretty

Hypothesis: It would be possible to study the different formats of sequences using READSEQ program in the given protein sequence.

TESTING THE HYPOTHESIS

Computer set up: A Normal PC with internet connection at windows environment

1. Log in to the site for READSEQ (http://bimas.dcrt.nih.gov/molbio/readseq/pretty_form.html) as shown in Fig. 8.1.
2. Copy sequence from Experiment - 8A. doc' paste to the form in the text box for pasting the sequence.
3. Select the format as FASTA.
4. Save the result as 'fasta.doc'
5. Repeat the same for IG/Stanford, Genbank/GB, NBRF, EMBL, GCG, DNAstrider, FITCH, PHYLIP, Plain/row, PIR/CODATA, MSF, ASN.1, and PAUP/nexus and their files are saved with their respective names.
6. Underline the characteristic feature of each format.

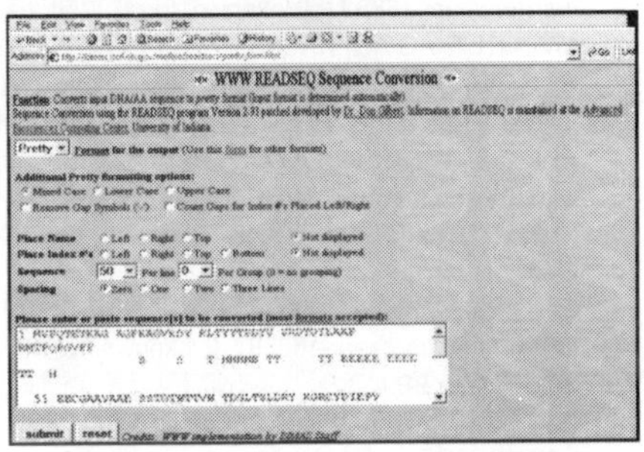

Fig. 8.1: Web view of READSEQ

DATA

Data Box 8.1: You are provided with a protein sequence in a separate file saved as Experiment_6A.doc is given below:

```
1 MVPQTETKAG AGFKAGVKDY RLTYYTPDYV VRDTDILAAF RMTPQPGVPP
S    S   T HHHHB TT      TT EEEEE EEEE TT   H
51 EECGAAVAAE SSTGTWTTVW TDGLTSLDRY KGRCYDIEPV PGEDNQYIAY
HHHHHHHHHH TTT  SS  G GGGGTTHHHH    EE EEEE   TT SS EEEE
```

```
101 VAYXIDLFEE GSVTNMFTSI VGNVFGFKAL RALRLEDLRI PPAYVKTFVG
EEE GGGTTT T HHHHHHHH SSSGGGTTTB SEEEE EEE   HHHHTT
```

Data Box 8.2

FASTA

```
>/usr/tmp/aaaa18802  [Unknown form], 256 bases, 1361 checksum.
MVPQTETKAGAGFKAGVKDYRLTYYTPDYVVRDTDILAAFRMTPQPGVPP
SSTHHHHBTTTTEEEEEEEEEETTHEECGAAVAAESSTGTWTTVWTDGLTS
LDRYKGRCYDIEPVPGEDNQYIAYHHHHHHHHHHTTTSSGGGGGTTHHHH
EEEEEETTSSEEEEVAYXIDLFEEGSVTNMFTSIVGNVFGFKALRALRLE
DLRIPPAYVKTFVGEEEGGGTTTTHHHHHHHHSSSGGGTTTBSEEEEEEE
HHHHTT
```

IG/Stanford

```
;/usr/tmp/aaaa18811  [Unknown form], 256 bases, 1361 checksum.
/usr/tmp/aaaa18811
MVPQTETKAGAGFKAGVKDYRLTYYTPDYVVRDTDILAAFRMTPQPGVPP
SSTHHHHBTTTTEEEEEEEEEETTHEECGAAVAAESSTGTWTTVWTDGLTS
LDRYKGRCYDIEPVPGEDNQYIAYHHHHHHHHHHTTTSSGGGGGTTHHHH
EEEEEETTSSEEEEVAYXIDLFEEGSVTNMFTSIVGNVFGFKALRALRLE
DLRIPPAYVKTFVGEEEGGGTTTTHHHHHHHHSSSGGGTTTBSEEEEEEE
HHHHTT1
```

Genbank/GB

```
LOCUS /usr/tmp/aaaa18828   256 bp
DEFINITION      /usr/tmp/aaaa18828  [Unknown form], 256 bases, 1361 checksum.
ORIGIN
       1        MVPQTETKAG AGFKAGVKDY RLTYYTPDYV VRDTDILAAF RMTPQPGVPP
      51        SSTHHHHBTT TTEEEEEEEE ETTHEECGAA VAAESSTGTW TTVWTDGLTS
     101        LDRYKGRCYD IEPVPGEDNQ YIAYHHHHHH HHHHTTTSSG GGGGTTHHHH
     151        EEEEEETTSS EEEEVAYXID LFEEGSVTNM FTSIVGNVFG FKALRALRLE
     201        DLRIPPAYVK TFVGEEEGGG TTTTHHHHHH HHSSSGGGTT TBSEEEEEEE
     251        HHHHTT
//
```

NBRF

```
>P1;/usr/tmp/aaaa18833
/usr/tmp/aaaa18833  [Unknown form], 256 bases, 1361 checksum.
MVPQTETKAG AGFKAGVKDY RLTYYTPDYV VRDTDILAAF RMTPQPGVPP
SSTHHHHBTT TTEEEEEEEE ETTHEECGAA VAAESSTGTW TTVWTDGLTS
LDRYKGRCYD IEPVPGEDNQ YIAYHHHHHH HHHHTTTSSG GGGGTTHHHH
EEEEEETTSS EEEEVAYXID LFEEGSVTNM FTSIVGNVFG FKALRALRLE
DLRIPPAYVK TFVGEEEGGG TTTTHHHHHH HHSSSGGGTT TBSEEEEEEE
HHHHTT*
```

EMBL

```
ID     /usr/tmp/aaaa18838
DE     /usr/tmp/aaaa18838  [Unknown form], 256 bases, 1361 checksum.
SQ     256 BP
MVPQTETKAG AGFKAGVKDY RLTYYTPDYV VRDTDILAAF RMTPQPGVPP SSTHHHHBTT TTEEEEEEEE ETTHEECGAA
VAAESSTGTW TTVWTDGLTS LDRYKGRCYD IEPVPGEDNQ YIAYHHHHHH HHHHTTTSSG GGGGTTHHHH EEEEEETTSS
EEEEVAYXID LFEEGSVTNM FTSIVGNVFG FKALRALRLE DLRIPPAYVK TFVGEEEGGG TTTTHHHHHH HHSSSGGGTT TBSEEEEEEE
HHHHTT
//
```

GCG

```
/usr/tmp/aaaa18845  [Unknown form]
/usr/tmp/aaaa18845  Length: 256  (today)  Check: 4961  ..
       1        MVPQTETKAG AGFKAGVKDY RLTYYTPDYV VRDTDILAAF RMTPQPGVPP
      51        SSTHHHHBTT TTEEEEEEEE ETTHEECGAA VAAESSTGTW TTVWTDGLTS
     101        LDRYKGRCYD IEPVPGEDNQ YIAYHHHHHH HHHHTTTSSG GGGGTTHHHH
     151        EEEEEETTSS EEEEVAYXID LFEEGSVTNM FTSIVGNVFG FKALRALRLE
     201        DLRIPPAYVK TFVGEEEGGG TTTTHHHHHH HHSSSGGGTT TBSEEEEEEE
     251        HHHHTT
```

DNAStrider

```
; ### from DNA Strider ;-)
; DNA sequence  /usr/tmp/aaaa18855  [Unknown form], 256 bases, 1361 checksum.
;
MVPQTETKAGAGFKAGVKDYRLTYYTPDYVVRDTDILAAFRMTPQPGVPP
SSTHHHHBTTTTEEEEEEEEETTHEECGAAVAAESSTGTWTTVWTDGLTS
LDRYKGRCYDIEPVPGEDNQYIAYHHHHHHHHHHTTTSSGGGGGTTHHHH
EEEEEETTSSEEEEVAYXIDLFEEGSVTNMFTSIVGNVFGFKALRALRLE
DLRIPPAYVKTFVGEEEGGGTTTTHHHHHHHHSSSGGGTTTBSEEEEEEE
HHHHTT
//
```

FITCH

```
/usr/tmp/aaaa18871  [Unknown form], 256 bases, 1361 checksum.
MVP QTE TKA GAG FKA GVK DYR LTY YTP DYV VRD TDI LAA FRM TPQ PGV PPS STH
        HHH BTT
TTE EEE EEE EET THE ECG AAV AAE SST GTW TTV WTD GLT SLD RYK GRC YDI EPV
        PGE DNQ
YIA YHH HHH HHH HHT TTS SGG GGG TTH HHH EEE EEE TTS SEE EEV AYX IDL FEE
        GSV TNM
FTS IVG NVF GFK ALR ALR LED LRI PPA YVK TFV GEE EGG GTT TTH HHH HHH HSS
        SGG GTT
TBS EEE EEE EHH HHT T
```

Phylip3.2

```
1 256 YF
/usr/tmp/a          MVPQTETKAG AGFKAGVKDY RLTYYTPDYV VRDTDILAAF RMTPQPGVPP
          SSTHHHHBTT TTEEEEEEEE ETTHEECGAA VAAESSTGTW TTVWTDGLTS
          LDRYKGRCYD IEPVPGEDNQ YIAYHHHHHH HHHHTTTSSG GGGGTTHHHH
          EEEEEETTSS EEEEVAYXID LFEEGSVTNM FTSIVGNVFG FKALRALRLE
          DLRIPPAYVK TFVGEEEGGG TTTTHHHHHH HHSSSGGGTT TBSEEEEEEE
          HHHHTT
```

Phylip

```
1 256
/usr/tmp/a          MVPQTETKAG AGFKAGVKDY RLTYYTPDYV VRDTDILAAF RMTPQPGVPP
          SSTHHHHBTT TTEEEEEEEE ETTHEECGAA VAAESSTGTW TTVWTDGLTS
          LDRYKGRCYD IEPVPGEDNQ YIAYHHHHHH HHHHTTTSSG GGGGTTHHHH
          EEEEEETTSS EEEEVAYXID LFEEGSVTNM FTSIVGNVFG FKALRALRLE
          DLRIPPAYVK TFVGEEEGGG TTTTHHHHHH HHSSSGGGTT TBSEEEEEEE
          HHHHTT
```

Plain/row

```
          MVPQTETKAGAGFKAGVKDYRLTYYTPDYVVRDTDILAAFRMTPQPGVPP
          SSTHHHHBTTTTEEEEEEEEETTHEECGAAVAAESSTGTWTTVWTDGLTS
          LDRYKGRCYDIEPVPGEDNQYIAYHHHHHHHHHHTTTSSGGGGGTTHHHH
          EEEEEETTSSEEEEVAYXIDLFEEGSVTNMFTSIVGNVFGFKALRALRLE
          DLRIPPAYVKTFVGEEEGGGTTTTHHHHHHHHSSSGGGTTTBSEEEEEEE
          HHHHTT
```

PIR/CODATA

```
\\\
ENTRY /usr/tmp/aaab18915
TITLE          /usr/tmp/aaab18915  [Unknown form], 256 bases, 1361 checksum.
SEQUENCE
                5         10        15        20        25        30
      1        M V P Q T E T K A G A G F K A G V K D Y R L T Y Y T P D Y V
     31        V R D T D I L A A F R M T P Q P G V P P S S T H H H H B T T
     61        T T E E E E E E E E E T T H E E C G A A V A A E S S T G T W
     91        T T V W T D G L T S L D R Y K G R C Y D I E P V P G E D N Q
    121        Y I A Y H H H H H H H H H H T T T S S G G G G G T T H H H H
    151        E E E E E E T T S S E E E E V A Y X I D L F E E G S V T N M
    181        F T S I V G N V F G F K A L R A L R L E D L R I P P A Y V K
    211        T F V G E E E G G G T T T T H H H H H H H H S S S G G G T T
    241        T B S E E E E E E E H H H H T T
///
```

MSF

```
/usr/tmp/aaaa18933  MSF: 256  Type: N  January 01, 1776  12:00  Check: 4961 ..
 Name: /usr/tmp/aaaa18933 Len:   256  Check:  4961  Weight:  1.00
//
/usr/tmp/aaaa18933  MVPQTETKAG AGFKAGVKDY RLTYYTPDYV VRDTDILAAF RMTPQPGVPP
/usr/tmp/aaaa18933  SSTHHHHBTT TTEEEEEEEE ETTHEECGAA VAAESSTGTW TTVWTDGLTS
/usr/tmp/aaaa18933  LDRYKGRCYD IEPVPGEDNQ YIAYHHHHHH HHHHTTTSSG GGGGTTHHHH
/usr/tmp/aaaa18933  EEEEEETTSS EEEEVAYXID LFEEGSVTNM FTSIVGNVFG FKALRALRLE
/usr/tmp/aaaa18933  DLRIPPAYVK TFVGEEEGGG TTTTHHHHHH HHSSSGGGTT TBSEEEEEEE
/usr/tmp/aaaa18933  HHHHTT
```

ASN.1

```
Bioseq-set ::= {
seq-set {
  seq {
    id { local id 1 },
    descr { title "/usr/tmp/aaaa18947  [Unknown form]" },
    inst {
      repr raw, mol aa, length 256, topology linear,
      seq-data
       iupacaa
      "MVPQTETKAGAGFKAGVKDYRLTYYTPDYVVRDTDILAAFRMTPQPGVPPSSTHHHHBTTT
TEEEEEEEEETTHEECGAAVAAESSTGTWTTVWTDGLTSLDRYKGRCYDIEPVPGEDNQYIAYHHHHHHHHHHTTTSS
GGGGGTTHHHHEEEEEETTSSEEEEVAYXIDLFEEGSVTNMFTSIVGNVFGFKALRALRLEDLRIPPAYVKTFVGEEE
GGGTTTTHHHHHHHHSSSGGGTTTBSEEEEEEEHHHHTT"
      } } ,
} }
```

Paup/nexus

```
>/usr/tmp/aaaa18961  [Unknown form], 256 bases, 1361 checksum.
MVPQTETKAGAGFKAGVKDYRLTYYTPDYVVRDTDILAAFRMTPQPGVPP
SSTHHHHBTTTTEEEEEEEEETTHEECGAAVAAESSTGTWTTVWTDGLTS
LDRYKGRCYDIEPVPGEDNQYIAYHHHHHHHHHHTTTSSGGGGGTTHHHH
EEEEEETTSSEEEEVAYXIDLFEEGSVTNMFTSIVGNVFGFKALRALRLE
DLRIPPAYVKTFVGEEEGGGTTTTHHHHHHHHSSSGGGTTTBSEEEEEEE
HHHHTT
```

SUMMARY

Different databases use different formats for submitting sequence and retrieving information. All the formats are standard ASCII files and may differ in presence of certain characters and words that indicate where different type of information and sequence itself are to be found. Most of the sequence formats are same for nucleic acids and proteins. Few things are common in all the databases, one of which is that amino acids and nucleic acids are written by same one letter code in different formats. The file formats are classified as input and output/retrieval sequence format and studying both of them is very important for effective studies with data. FASTA, PHYLIP, CLUSTAL, GCG, MOLGEN/STANFORD, NBRF/PIR, PAUP/NEXUS, GENBANK and UNIPROT are the main type of formats discussed in details in the chapter. Several different softwares available for sequence format conversion are discussed and study of different sequence formats in a DNA or protein sequence using the program READSEQ is the experiment done.

EXERCISE

1. What are the characteristic features of FASTA format?
2. Explain sequential phylip format.
3. Use web resources to find programs performing the same function as READSEQ.

4. You are given a nucleic acid sequence given below. Convert it to format Genbank/GB, NBRF, EMBL, GCG, DNAstrider, FITCH, PHYLIP and Plain/row using readseq.
5. Convert the following sequence into complete amino acid sequence using one letter code. Write it in FASTA format manually.

   Ala-ala-glu- unknown-val-phe-glu-nonsense
6. Write a PERL script to extract the amino acid sequence from an entry in the PIR protein sequence database and convert it to FASTA format.

SECTION - 2

METHODS IN BIOINFORMATICS

9

SEQUENCE ANALYSIS BY COMPUTER PROGRAMMING

Geneticists use computers to collect, store, manipulate and analyse data. Molecular genetic data which comes in the form of a DNA, RNA or amino acid sequence is particularly amenable to computer analysis. In recent years a merger of genetics, genomics and biocomputing has yielded the science of bioinformatics. Several scientific journals are devoted largely to this topic. Genetic sequence data in a computer file, already discussed and listed in section one, can be investigated in many ways corresponding to myriad questions a researcher might ask about the genetic sequence and it's functional significance. These include: Does the given a sequence contain a gene? Where are the functional sequences, such as promoters, regulatory sites and splice sites located within the sequence? Does the sequence encode a polypeptide? If so, what is the amino acid sequence of the polypeptide? Does this sequence predict certain structural features for DNA, RNA or protein? To answer these and many other questions computer programs are written to analyze the sequences in a particular way.

A computer program is a defined series of operations that can analyses data in a desired way. For example a computer program might be designed to take a DNA sequence and translate it in an amino acid sequence. The computer programs are devised by theoreticians who understand the basic genetics and genomics principles and can design computational strategies for analysing genetics sequences. Most practicing molecular biologists and other related scientists now finds it difficult to work without computer programming. Very much like human language, there is a language which computer compiler understands, is called programming language. Programming languages are used to convey instructions to computers. Study of programming language is important for molecular biologists and there are many reasons for them to train themselves in programming some of which are given below:

1. By learning about programming language we can produce programs, which apply correct molecular and other biology concepts which are difficult to be understood by other professional programmers thus are more efficient and more reliable.
2. A programming conversant biologist can more efficiently tap the present bioinformatics resources.

3. By learning about advances in programming languages we equip ourselves for learning the programming languages of tomorrow. Computing technology is a fast-moving field. It is essential to keep learning new skills and new technologies.

The present chapter provides a brief introduction to computer programming and programming languages with reference to DNA and protein sequence analysis. Only single sequence analysis with computer programming is discussed here. Sequence alignment and similarity searches are discussed separately in next few chapters.

9.1 COMPUTER PROGRAMS CAN BE DESIGNED TO SCAN VERY LONG SEQUENCES AND LOCATE MEANINGFUL FEATURES WITHIN THE SEQUENCE

Sequence elements refer to genetic sequences with specific functions. At molecular level, the function of genetic material is based largely on a specific genetic sequence that play distinct roles. For example codons are three base sequences that specify a particular amino acid in a sequence, it is called start codons when it is AUG and stop codons, when it is UAA, UAG or UGA. Similarily promoters are the sequences that provide the binding site for RNA polymerase to initiate transcription and in many *E.coli* it contains TTGACA at -35 location and TATAAT at -10 site. The minus sign here means it is a left-hand side of the structural gene. Start of structural gene is designated as 0. In eukaryotes on the other hand the promoters contain CAAT boxes, GC boxes, TATA boxes usually. Several other types of sequence elements and their characteristic features are given in Table 9.1. The most commonly occurring sequence of bases within a sequence element is called consensus sequence. Occurrence of TTGACA at – 35 location and TATAAT at –10 location in *E.coli* is an example of a consensus sequence that is found in bacterial promoters. It can differ in one or two nucleotides and still functions as promoter. The –35 site is primarily for recognition by sigma factor subunit of RNA polymer use. The –10 site is also known as pribnow box, is the site where DNA will begin to unwind to allow transcription to occur.

Table 9.1: Several types of sequence elements and their characteristic feature

Promoters	Many *E.coli* promoters contain TTGACA (–35 site) and TATAAT (-10 site). Eukaryotic core promoters may contain CAAT boxes, GC boxes, TATA boxes, etc.
Response Elements	Glucocorticoid response elements (AGRACA), cAMP response elements (GTGACGTRA). R here refers to a purine base (A or G)
Start Codon	AUG
Stop Codon	UAA, UAG, UGA
Splice Site	
Polyadenylation Signal	GURAGU———————ΨNΨURAC(Ψ)x, AG. Ψ here refers to $(\Psi)_n$ to a pyramidinyebase (Trc). R refers to a purine-base AAUAAA
Highly Repetitive Sequences	Relatively short sequences that are repeated several times in a genome.
Transposable Elements	Are usually characterized by a pattern in which direct repeat flank inverted repeats.

Sequence elements are very easy to identify with programming and lots of programs are already written for this purpose. In the present discussion we limit ourself to programs specially designed for specific sequence feature like ORF, introns, promoters, motifs, etc. In Table 9.2 example programs for each category is given. ORF refers to open reading frame. Every DNA sequence can be read in six ways i.e. 5′ first nucleotide, 5′ second nucleotide third nucleotide, 3' first, second and third nucleotides. Program NCBI ORF finder will detect presence of ORFs on the basis of occurrence of start and stop codons. It is a very popular program for gene identification.

Table 9.2: Common programs designed for identification of sequence elements

ORF Finder	Identifies all possible ORFs in a DNA sequence by locating the standard and alternative stop and start codons.
DNA Sequence Translation	A program for identification of ORF in DNA sequences and its translation in to amino acid sequence.
SplicePredictor	A method to identify potential splice sites in (plant) pre-mRNA by sequence inspection.
RBS Finder	RBSfinder is a Perl script that implements an algorithm to find ribosome binding sites for genes in bacterial and archaeal genomes. It is normally run as a post-processor to the Glimmer gene finder or to other prokaryotic gene finders.

1. Structural genes encode for mRNA or a protein. Prokaryotic structural genes usually lack introns and consequently have a much simpler structure (Fig. 9.4 and box 9). All types of structural genes in eukaryotes however, contain interrupted genes i.e. are split into exons and introns. Exons are the sequences represented in the mature mRNA. By definition a gene starts and ends with an exon corresponding to 5′ and 3′ end of an mRNA. Introns are the intervening sequences that are removed when primary transcript pre-mRNA is processed to give mature mRNA. Comparison of related genes in different species show that sequences of corresponding exons are usually conserved but the sequences of introns are much less related. Study of exons and their interrelationships in different organism have been very often used to solve evolutionary mysteries. Identification of structural genes and finding their evolutionary relations in a given sequence has been very handy with computer programs. Similarly it has been possible to predict potential splice site in a pre - mRNA sequence with program splicePredictor.
2. Other common terms associated with analysis of DNA sequences by bioinformatics tools are UTR, CDS, cDNA, EST, etc. UTRs are the untranslated regions that occur both in DNA and RNA; they are the portions flanking the CDS (the coding sequence or the exons) and are not translated into proteins. UTRs particularly at the 3′ end are highly specific both to the gene and to the species from which the sequence is derived. Given a piece of DNA sequence it is possible to translate the DNA to a protein by looking up successive codons in a genetic code table; this is called a conceptual translation. The program DNAsequence translation does the conceptual translation of any given DNA sequences. For carrying out conceptual translation in any arbitrary DNA sequence it is not known whether the first, second or third base marks the start of CDS. There are six reading frames possible for any DNA sequences where we want to carry out the conceptual translation. Three frames in forward direction starting from first, second and third bases and three reverse frames determined by reversing the DNA sequences and beginning on first, second and third base. The correct reading frame normally is deemed as longest and is normally uninterrupted by a stop codon. Study of these features is useful in gene identification (chapter 13).

9.2 THE TECHNIQUE OF COMPUTER PROGRAMMING

All computer programs do three basic things:

1. They manipulate data, which involves storing, retrieving and modifying data.
2. They perform operations (such as adding and subtracting) that modify the data structure created to receive results of an operation.
3. They jump around through branching statements. Branching statements in high-level languages are called as loop and conditional statements.
4. Output is given as results.

Several steps involved in designing a program are briefly outlined in Fig. 9.1. Let us discuss this in details with the help of a hypothetical situation where the scientist has a DNA sequence file that he or she may want to translate into an amino acid sequence. Using this example we would not only familiarize with several terms associated with computer programming but also understand basic functioning of a program. Based on knowledge of genetic code a theoretician would device computation procedures or algorithms that would relate a DNA sequence to an amino acid sequence through conceptual translation. This becomes the goal of the program. A thorough work must be done by geneticists, theoreticians and designers before the goal of the program is finalised. They would decide about the name of the program the codon table to be followed and about the programming language to be used. Let us assume that it is decided that name of this program is etranslate.

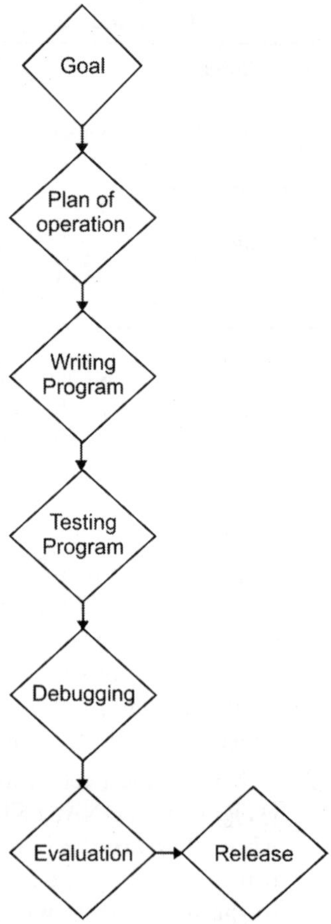

Fig. 9.1: Steps involved in designing a program

In a computer program the procedures or algorithms are executed as a stepwise plan of operations that manipulates the data in a sequence file. In our program etranslate, the steps in plan of operation are broadly:

(*a*) Taking input as a DNA sequence

(*b*) Reading it alphabetic manner

(*c*) Taking input from genetic code i.e. which amino acid stands for a particular codon

(*d*) Output as amino acid sequence

For a computer to perform the plan of operation, the program instructions must be written in a programming language that computers can decipher. A short description on programming languages is given in next subsection. A similar program is written in language HTML and JavaScript in Experiment 9A. After this is accomplished the program is tested under different situations to see if it works. Usually errors reported, also called as bugs are overcome in a process called debugging. After debugging the program is evaluated by several users, it is ready for use. An ideal the program is that which accurately manipulates the data and at the same time is easy to use.

9.3 PROGRAMMING LANGUAGES

Computer programs must be expressed in a formal language for machine execution called as programming language. Programming languages are indispensable tools in computer science. Programming language like other languages are different ways to convey the same things. There are various programming languages, like C, Java, C++ Pascal etc. Programming language C was designed by Brian Kernghan and Dennis M. Ritchie (Historical Profile 10) and is most popular language among scientist and molecular biologist. Programming languages occur in families with collections of languages having similar features and properties (C-like languages, algol-like languages, ML-like languages, Java-like languages and such). A list of common programming languages is given in Table 9.3. Java and C++ are the examples of object-oriented programming (OOP). At its core OOP is simply a way of thinking about problems and their solutions. Instead of tackling programs in a top down, linear fashion as with the traditional languages such as C or pascal, OOP attempts to break a problem into component parts called objects. The solution focuses on these independent objects and their interrelationship. Further discussion on OOP is beyond the scope of the book.

Table 9.3: Common Programming Language

Algol	An early procedural programming language
BASIC	A Programming Language originally designed to teach introductory programming concepts
C Language	The most common modern systems programming language to date, by Brian Kernighan and Dennis M. Ritchie
C++	C++ is an attempt to turn the C language into a high level, object-oriented programming language
Cobol	An acronym for Common Business-Oriented Language, a very old programming language
Fortran	One of the first programming languages. It is still very popular among scientists and engineers.
HTML	Hypertext mark up Language for Internet. It is used for programming in Experiment 9A.
Java	Java is a strongly- and statically-typed procedural, garbage-collected object-oriented programming language with syntax like that of the C language released by Sun Microsystems
Perl	As the name "Practical Extraction and Report Language" coins out, Perl was originally created to extract information from text files and use it to prepare reports. Used with CGI for back end programming. Very popular among scientists and engineers.
Visual BASIC	A Microsoft Programming Language favored by developers for its ease of use.
Javascript	Is a scripting language created by Netscafe. It is widely for developing interactive website.

In any case, data objects are central to programs and in whatever language the program is. The data types can be separated into two groups, primitive and compound. The primitive values are usually numbers, boolean values, and characters. The composite values are usually arrays, records, and recursively defined values. Strings may occur as either primitive or composite values. Lists, stacks, trees, and queues are examples of recursively-defined values. Now we try to clarify what is boolean, number and character value and what are strings, lists, stacks, trees, arrays, records with the example of a DNA sequence given below:

ATATAATTAATATATAGAGATACGGGGGGGACAGAGATTTTTTTTTTTTTGA
CAGGGGGGGGTAGATAGACAGTAGGCCAAAAAAAAAATATAATATAGACA
GTCGATAT

The length of this sequence is a number and would be declared as *int*. Whether DNA or not the option would be declared as *boolean*. Each nucleotide in the sequence i.e. a collection of alphabets would be declared as string. Boolean values are implemented using a single bit of storage. It has a logical values: False or true.

It always makes sense to take care of the computer space needed to store the declared variable and use the minimum possible space while making declaration. In Table 9.4, range of data types and space in bytes used in C++ is given. Integer values are most often implemented using a hardware defined integer storage representation, often 32-bits or four bytes with one bit for the sign. In case of C++ compiler it uses 2 bytes of space. We write *int* to declare when the value expected is integer. Floating point declaration is a value that may contain decimal places. In the DNA sequence given above it makes it sense to store the length of the sequence as unsigned *int* in ANSI C, because the sequence length will neither be a negative value nor a decimal value. Where strings are treated as a primitive type, they are usually of fixed length and their operations are implemented in hardware. Thus the present sequence could best be stored as a string whose syntax in c++ is:

```
var sequence = {ATATAATTAATATATAGAGATACGGGGGGGACAGAGATTTTTTTTTTTTTGA
CAGGGGGGGGTAGATAGACAGTAGGCCAAAAAAAAAATATAATATAGACA
GTCGATAT};
```

HISTORICAL PROFILE 10

Dennis M. Ritchie for Giving C Programming Language

Professor Dennis M. Ritchie

Dennis MacAlistair Ritchie (born September 9, 1941) is an American computer scientist notable for his influence on ALTRAN, B, BCPL, C, Multics, and Unix. He received the Turing Award in 1983 and the National Medal of Technology in 1998. Ritchie is currently the head of Lucent Technologies' System Software Research Department. Ritchie is best known as the creator of the C programming language and a key developer of the Unix operating system, and as co-author of the definitive book on C, *The C Programming Language*, commonly referred to as 'K/R' or *K&R* (in reference to the authors Kernighan and Ritchie). Ritchie's invention of C and his role in the development of Unix along side Ken Thompson, has placed him as an important pioneer of modern computing. The C language is still widely used today in application and operating system development and its influence is seen in most modern programming languages. Unix has also been influential, establishing concepts and principles that are now well-established precepts of computing. The popular Linux operating system and its tools are descendants of Ritchie's work and the Windows operating systems include Unix compatibility tools and C compilers for developers.

The description of the properly formed programs (or sentences) of a language is called as syntax. An example of syntax in programming language javascript is

var sequence = fi. elements [0]. value;

Here var in the syntax is meaning that we have asked computer to keep space for a string, fi is an object and elements [0] refers to an array. Arrays are defined for effective usage of computer memory space.

Table 9.4: Size and range of C++ basic data types

Type	*Bytes*	*Range*
char	1	128 to 127
unsigned char	1	0 to 255
signed char	1	128 to 127
int	2	32768 to 32767
unsigned int	2	0 to 65535
signed int	2	32768 to 32767
short int	2	32768 to 32767
unsigned short int	2	0 to 65535
signed short int	2	32768 to 32767
long int	4	2147483648 to 2147483647
signed long int	4	2147483648 to 2147483647
unsigned long int	4	0 to 4294967295
float	4	3.4E38 to 3.4E + 38
double	8	1.7E308 to 1.7E + 308
long double	10	3.4E4932 to 1.1E + 4932

There are three basic computational models—in the syntax of a programming language functional, logic, and imperative. In addition to the set of values and associated operations, each of these computational models has a set of operations, which are used to define computation. The functional model uses function application, the logic model uses logical inference and the imperative model uses sequences of state changes. The functional, logic and imperative models of computation are equivalent in the sense that any problem that has a solution in one model is solvable (in principle) each of the other models.

1. **The Functional Model:** A function is an example of object, which returns a value, called the output when it is applied to its inputs. The inputs are also known as its parameters or arguments. An example of declaration of a function in language javascript is given below:

```
function check(fi)
{
var sequence=fi.elements[0].value;
slength=sequence.length;
if (sequence.length==0){alert ('Please enter a sequence');
exit;
}
```

   Function named check is declared here. It takes input as if which is an array. The function returns the value of sequence length stored as slength.

2. **The Logic Model:** The logic model of computation consists of a set of values, definitions of relations and logical inference. Programs consist of definitions of relations and a computation is a proof (a sequence of inferences). In the example given above the logical statement of it is asking the program to exit after giving alert "Please enter a sequence" in case the sequence length in zero. The imperative model is often called the procedural model because groups of operations are abstracted into procedures.
3. **The Imperative-Model:** The imperative-procedural model is important because it models change and changes are an integral part of our environment. It is the model of computation that is closest to the hardware on which programs are executed. Its closeness to hardware makes it the easiest to implement and imperative programs tend to make the least demands for system resources (time and space).

EXPERIMENT 9A

To Write a Small Program for Identification of Promoter Region in a *E. coli* Sequence

About Promoters: A detailed description about promoters in eukaryotes and prokaryotes is given in chapter 4 (section 4.1. 3.2).

About HTML: HTML refers to Hypertext Markup language. HTML tags are used in writing web pages. The web browser interprets the HTML and displays the page. HTML is used to specify the hyperlinks, images and all other elements seen on a web page. Web pages are text on ASCII files in which HTML tags are written and are saved with extension *.html.

About Javascript: Javascript is a scripting language created by Netscafe, however it works equally well with internet explorer. It is used for developing interactive web sites. Client side Javascript is traditional embedded into a standard HTML program. Javascript is embedded between <SCRIPT> .... </SCRIPT> HTML tags (Data Box 9.2).

Hypothesis: You are given that many *E.coli* promotor region contain TTGACA (–35 site) and TATAAT (–10 site). Write a program using HTML and javascript on Windows platform to identify the presence of this type of promoter region in the given sequence.

TESTING THE HYPOTHESIS

Computer set up: A normal PC with Windows operating system.

Starting Material: You are given 2 DNA sequences as shown in Data Box 9.1. Save these sequence on a notepad as Experiment - 9A doc.

1. Open a new notepad and copy the coding given in Data Box 9.2 on this and save this as Experiment 9A. html (Fig. 9.2)

Fig. 9.2: A text editor

2. Use internet explorer and open the file Experiment - 9A.html. It shows a view like in Fig. 9.3. Copy sequence 1 in the texy box to search for the presence of promoter. Press search.
3. Result obtained is shown in Fig. 9.4. The sequence contains no promoter region. Repeat the same for sequence 2. Result obtained is shown in Fig. 9.5.

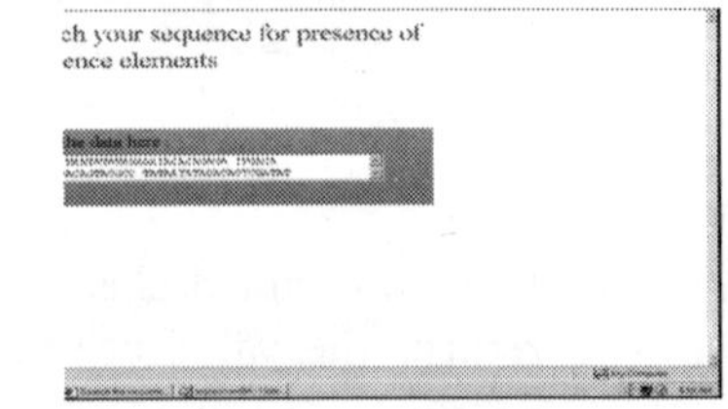

Fig. 9.3: Internet explorer opening the experimental file

DATA

Data Box 9.1: The given sequence

Sequence 1
ATATAATTAATATATAGAGATACGGGGGGGGACAGAGATTTTTTTTTTTTTGACAGGGGGGGGGTAGATAGACAGTAGGCCAAAAAAAAAAATATAATATAGACAGTCGATAT

Sequence 2
ATATAATTAATATATAGAGATACACAGAGATTGACATAGATAGACAGTAGGCC TATAATATAGACAGTCGATAT

Data Box 9.2: Coding for front end with HTML and Javascript

```
<html>
<head>
<title>Search the sequence element</title>
<script language = "javascript">
function check(fi)
{
var sequence=fi.elements[0].value;
slength=sequence.length;
if (sequence.length==0){alert ('Please enter a sequence');
exit;
}
if (slength<35){alert) 'There is no promoter region expected in this sequence because the length of the sequence is shorter than 35 nucleotides');
exit;}
var j=1;
for (var n=0;n<=slength; n++)
{
var se_35 = sequence.substring(n,n+6);
if (se_35=='TTGACA')}
var se_10=sequence.substring(n+25,n+31);
if (se_10=='TATAAT');
document.write ('<HTML><font color="red">');
document.write ('Your sequence is likely to contain');
document.write (j);
```

```
document.write (‘ promoter region(s). The promoter region is underlined below :’);
document.write (‘</font><br><br><font color=blue>’);
document.write (j);
document.write (‘.’);
document.write (sequence.substring(0,n));
document.write (‘<u>’);
document.write (sequence.substring(n,n+35));
document.write (‘</u>’);
document.write (sequence.substring(n+35,slength));
document.write (‘<br></font></HTML>’);
j=j+1;}
}}
if (j==1){
alert (‘There is no promoter region expected in this sequence’);
}}
</script>
</head>
<body bgcolor=#FFFFFF>
<form onSubmit="check(this)">
<table border=0 width=600 bgcolor=#9C9C9C>
<tr><font size=+2 color=blue>Search your sequence for presence of sequence elements</font></th></tr>
<br><br><br>
<tr><td font color="red">Enter the data here :</td></tr></font>
<tr><td><textarea rows 50 cols=50 scroll=yes></textarea> </td></tr>
<tr><td><input type=submit value=search></td></tr>
</table>
</form>
</center>
</body>
</html>
```

INTERPRETATION OF DATA

The present experiment or computer program design was a very easy and clear example of introducing the concepts of programming language and the algorithm in a program. As might be clearly seen from a glance at the program given in Data Box 9.2 that the algorithm for writing this program were to find whether the given sequence contain TTGACA (–35 site) an TATAAT (–10 site) sub sequence at the respective sites. If it did not find the above subsequences it gave an alert box stating that the given sequence probably does not contain promoter region. This was the case with sequence 1 in Data Box 9.1. If the given sequence contained the desired subsequences it underlined the proposed promoter region and gives the message for probable presence of promoter region in the sequence. This was the case with sequence 2. The algorithm for the above experiment is given in Fig. 9.5, which is to be matched with.

Javascript and HTML were ideal of the above design because both of them are very simple to understand and apply. Same thing could have been done with CGI—perl also. Thus, programs could easily be written covering aspects of sequences:

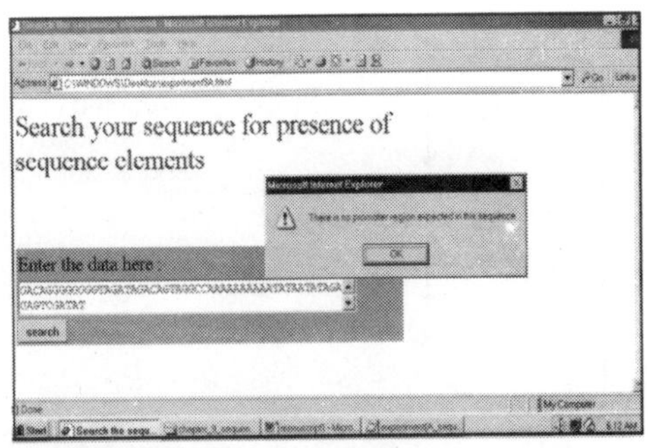

Fig. 9.4: Result obtained in case-1

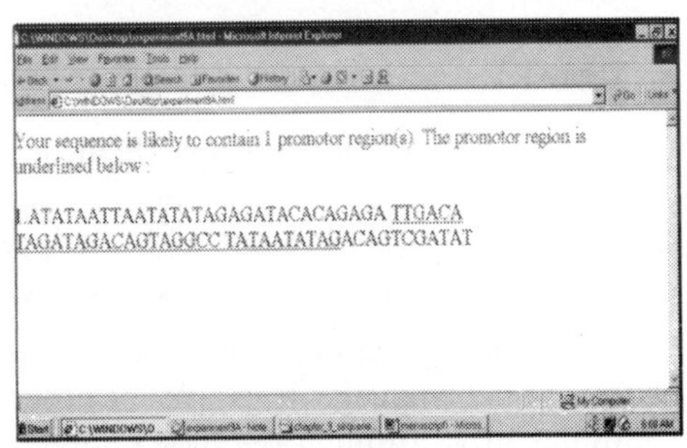

Fig. 9.5: Result obtained in case-2

1. Locate specialised sequences within a very long sequence. A specialized sequence is called sequence elements or a sequence motif. There are several types of sequential elements: promoters, start codons, stop codons, repetitive sequences, transposable elements, etc. Each one of them has their characteristic features and a computer program is written on the basis of these features.
2. Study of organisation of sequences or motifs. For example, occurrence of promoters usually occur upstream of the structural gene. Similarly occurrence of start codons in sequence along with a promoter is signifying a start of a gene.
3. Does a sequence encode for a polypeptide. If so what is the sequence of amino acids encoded by this DNA or RNA sequence. Programs are written to perform simple operations with the sequences like transcription, translation and reverse translation.
4. Does a sequence predict certain structural feature for RNA, DNA or proteins? For example, a DNA sequence is likely to be in B-DNA conformation? What is the secondary structure of a RNA sequence or a polypeptide?
5. Is the sequence homologous to any other known sequence?
6. What is the evolutionary relationship between two or more genetic sequences?

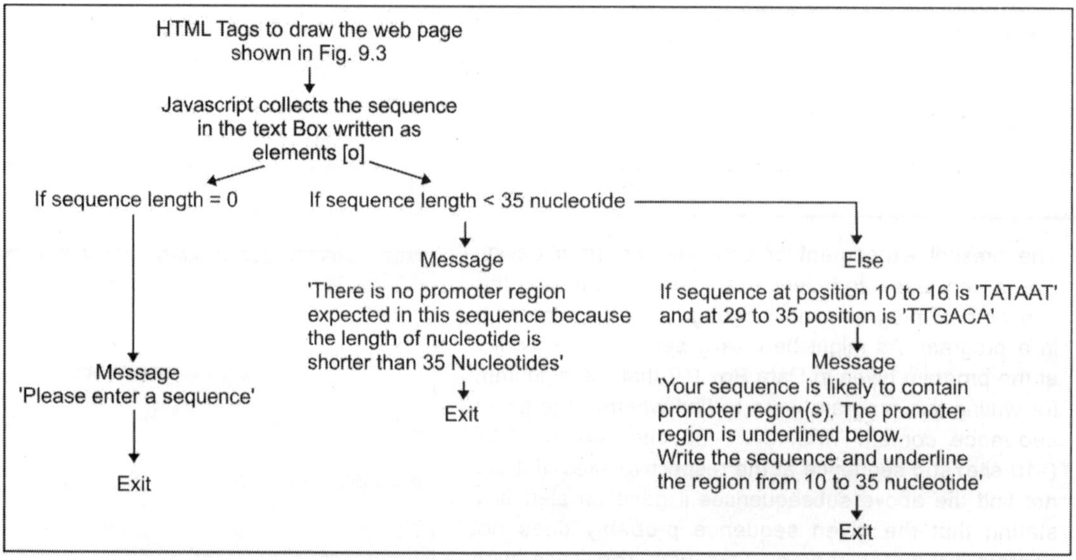

Fig. 9.6: Algorithm for working of Experiment - 9A

SUMMARY

A computer program is a specification of computation and it is extremely useful in biological sequence analysis. They are useful in identification of sequence elements such as promoters, regulatory sites and splice sites located within a particular gene. Similarly whether this sequence codes for a protein or not could easily be known with program.

Computer programe must be written in a formal language for machine execution is called programming language. Examples of programming languages are C++, Java, Pascal etc., Understanding computer programming requires knowledge of terms like variables, constants, scope of definitions, algorithms and function etc. Historical Profile - 10 gives the contributions of Dennis M. Ritchie for giving C programming language. In Experiment 9A a small program for identification of promoter region in *E. coli* sequence is written using HTML and Javascript.

EXCERCISE

1. What are the contents of a computer file?
2. What is the general goal of writing a computer program in molecular genetics? Give five examples.
3. What is a user friendly program?
4. Write a program using HTML and Javascript to identify all the start codons in the following DNA sequence:

   5′ – AGGGGTATACCGATCACTTTATTTA TGGGCTAAACTGTAATATAGGGG—3′
5. Discuss how computer technology is useful in identifying functional sequences?
6. What is the consensus sequence of the following six DNA molecules:

   G G C A T T G A C T
   G C C A T T G T C A
   C G C A T A G T C A
   G G A A A A T G G A
   G G C T T T G T C A
   G G C A T A G T C A

10

SEQUENCE ALIGNMENT CONCEPTS AND ALGORITHMS

Sequence alignment is the identification of residue-residue correspondence in two or more sequences. Sequence alignment is a procedure for comparing two or more sequences by searching for a series of individual characters or character patterns that are in the same order in the given sequences. Consider two sequences one of which has evolved from the other. What kind of differences do they have? Two sequences are aligned, by writing them across a page in two rows. Identical or similar characters are placed in the same column and the non-identical are in the same column as a mismatch or opposite a gap in the other sequence.

Alignment of two hypothetical very small amino acid sequences is depicted in Fig. 10.1. The vertical line indicates the presence of identical amino acid in the protein and '–' indicates a gap. Based on presence of gaps and every identical amino acid an alignment is given a score. One generally assigns 1 to a match, 0 to a mismatch and a penalty for a gap. By adding these values along an alignment one can obtain a score for this alignment. The score indicates a qualitative measure of significance of the alignment. Higher the score better is the sequence alignment. A distance function for two sequences is defined by looking for the alignment that yields the maximum score.

Consider Two Sequences

```
Sequence 1  H H G Q C I G H Y G P T V C A S G C Q V S Y Y
            |   | | |         |       |                 |
Sequence 2  H W G Q C T U V I G Y S G C K T C T S G C Q Y S D Y Y
```

After Gap Insertion

```
Sequence 1  H H G Q C I G H Y G P T V C A - - - S G C Q V S - Y Y
            |   | | |         |       |         | | | | | |   | |
Sequence 2  H W G Q C T U V I G Y S G C K T C T S G C Q Y S D Y Y
```

Fig. 10.1: Alignment of two hypothetical amino acid sequences, sequence 1 and sequence 2

When two sequences are being compared, insertions in one sequence is equivalent to deletions from the other. Hence insertions and deletions are equivalent operations and are termed indels. Indels and substitution are together termed as edit operations. Minimal number of edit operations that can convert one sequence to another is called edit distances. When comparing two sequences each edit operation is given a score (weight). The alignment algorithm search for the set of edit operation with maximum of total score.

For example in the sequence given below three edit operations are needed: insertion of gaps in first sequence; (ii) substitution of C with T or *vice versa* (iii) insertion of gaps in second sequence.

A _ C C T G A

A G C T T – A

A gap is the maximal continuous run of spaces in a single sequence in a given alignment. The length of the gap is the number of indel operations in it. Gap penalty function is a function that measures the cost of the gap as (non-linear) function of its length.

Sequence alignment is a basic tool of bioinformatics, which provides information about two sequences whether the two sequences are similar, what is the level of similarity, where the two sequences are similar, and where they differ. Sequence alignment is very useful for discovering functional, structural and evolutionary information in biological sequences. Given two or more sequences, sequence alignment studies would throw light on the following aspects of these sequences:

1. **Measure their similarity:** Statistical measure of similarity could be provided by sequence alignment. Multiple sequence alignment not only depicts residue-residue correspondence, but also help in search for weak but significant similarities in sequence databases. Similarity, search methods like profile searches are examples of weak but significant similarity.
2. **Demonstration of homology between sequences:** This aspect is discussed in details in chapter 12.
3. **Molecular phylogeny:** Inferring an evolutionary relationship based on multiple sequence alignment is studied in molecular phylogenetics. As discussed above determination of edit operation that convert one sequence to another is given score. Phylogenetic trees are drawn on the basis of this edit operations (see chapter 14).
4. **Identification of functionally important sites:** Multiple sequence alignment allows the identification of highly conserved and variable residues. The conserved sites are usually very important for structure or function of protein. Thus, they may be useful for designing mutagenesis experiments.
5. **Structure prediction:** Multiple alignments increase the efficiency of protein secondary structure prediction.
6. **Function prediction:** Multiple alignments are also used in prediction of function of a protein. A detailed example of this for cellulose binding domain of cellobio hydrolase is discussed in chapter 11, section 11.3.
7. Multiple sequence alignment is also useful for design of primers for PCR (polymerace chain reaction) for identification of related genes.

10.1 SEQUENCE ALIGNMENT CAN BE LOCAL OR GLOBAL

The alignment that is the best, given a defined set of rules and parameter values for comparing different sequences is called as optimal alignment. In optimal alignment non-identical characters and gaps are placed in such a way so as to bring as maximum identical or matching characters as possible into vertical register. There is no such thing as the single best alignment, since optimality always depends on the assumptions one bases the alignment on. There are two major types of alignment viz., local and global alignment and both of them is discussed in details one by one:

10.1.1 Local Alignment

An alignment that searches for segments of the two sequences that match well is called local alignment. There is no attempt to force entire sequences into an alignment in local alignment. Using another set of hypothetical sequence in Fig. 10.2, when we compare only a localized region the alignment would be presented as shown in Fig. 10.2a. Local similarity is typically used in contextual investigations and in determining whether unannotated sequences might be homologous to annotated

A. Local Alignment	
A A A G C G G A A G T C A C A G | | | | | | | A A G G C T G A A G T A T A G	If considers only the segment that matches well
B. Global Alignment A A A G C G G A A G T C A C A G | | | | | | | | | | A A G G C T G A A G T - A T A G	If matches end to end even though its not best alignment

Fig. 10.2: Local and global alignment of given hyhothetical sequence

sequences. Local alignment methods are very useful for scanning databases or other circumstances when you wish to find matches between small regions of sequences, for example between protein domains or cDNA and genome alignment. In Fig. 10.19 local alignment in sequences from a Human alpha actinin2 mRNA with mouse alpha actinin2 mRNA is shown.

10.1.2 Global Alignment

In global alignment two sequence are matched from end to end, even though we do not get the best alignment. It is an alignment that assumes that the two sequences are basically similar over the entire length of one another and attempts to match them to each other from end to end, even though parts of the alignment are not very convincingly matching is called global alignment. A tiny example is given in Fig. 10.2b. Sequences that are quite similar in composition and are of approximately same length are suitable candidates for global alignment. For example global alignment of two sequences from a Human alpha actinin2 mRNA with mouse alpha actinin2 mRNA using EMBOSS-Align software is depicted in Fig. 10.19.

10.2 CONCEPT OF SCORING MATRIX

Each alignment is given a score. Scoring can be as simple as giving a score of 1 to every match and 0 to a mismatch. There is a gap penalty also. The individual score for matches mismatches and gap penalties however, is different depending upon the criteria in the type of scoring scheme used. At its simplest, by criteria we mean the decision of whether two residues in our comparison match, and what value to assign to that match. For example, in a DNA-to-DNA comparison we might simply say that identity is our criterion. Alternatively, we might say that purines match other purines and pyrimidines match other pyrimidines or we might assign values like given in Table 10.1. In such a case the alignment with maximum score is considered to be best aligned alignment.

Scores assess the likelihood that the alignment's columns are drawn from the population of correct columns, as compared to the likelihood they are generated by chance. There exist independent and reliable means of deciding if a particular scoring matrix gives good results. The frequency that the column x-over-y appears in a correct alignment depends on the evolutionary distance.

For protein-protein alignment there are 20 possible matches and, for each amino acid, 19 possible mismatches in case of protein-protein alignment. Also in case of alignment of related protein sequences one amino acid can be substituted with the other based on similarity in the amino acids. Tables called as substitution matrices are used in protein similarity comparisons.

Table 10.1: Simplest criterion for scoring a sequence alignment

Sl.No.	*Nucleotide type*	*Score*
1.	N = N	1 "point"
2.	N not = N, but R = R or Y = Y	0.5 "points"
3.	R versus Y	0 "points"

Where N = Nucleotide; R = Purine (A or G); Y = Pyrimidine (C or T)

A substitution matrix gives a score that reflects the ease by which a particular amino acid in a protein is substituted by another in an evolutionarily related sequence. Different types of substitution matrices are discussed below:

10.2.1 PAM/MDM/Dayhoff Matrix

The late Margaret Dayhoff (Historical Profile 1) was a pioneer in protein databasing and comparison. She and her coworkers developed a model of protein evolution, which resulted in the development of a set of widely used substitution matrices. Her observations were based on 1572 changes in 71 groups of protein sequences that are 85 per cent similar. These are frequently called Dayhoff, MDM (Mutation Data Matrix), or PAM (Percent Accepted Mutation) matrices. By default it looks like as shown in Fig. 10.3.

X = 0

	C	S	T	P	A	G	N	D	E	Q	H	R	K	M	I	L	V	F	W	Y
C	12																			
S	0	2																		
T	-2	1	3																	
P	-3	1	0	6																
A	-2	1	1	1	2															
G	-3	1	0	-1	1	5														
N	-4	1	0	-1	0	0	2													
D	-5	0	0	-1	0	1	2	4												
E	-5	0	0	-1	0	0	1	3	4											
Q	-5	-1	-1	0	0	-1	1	2	2	4										
H	-3	-1	-1	0	-1	-2	2	1	1	3	6									
R	-4	0	-1	0	-2	-3	0	-1	-1	1	2	6								
K	-5	0	0	-1	-1	-2	1	0	0	1	0	3	5							
M	-5	-2	-1	-2	-1	-3	-2	-3	-2	-1	-2	0	0	6						
I	-2	-1	0	-2	-1	-3	-2	-2	-2	-2	-2	-2	-2	2	5					
L	-6	-3	-2	-3	-2	-4	-3	-4	-3	-2	-2	-3	-3	4	2	6				
V	-2	-1	0	-1	0	-1	-2	-2	-2	-2	-2	-2	-2	2	4	2	4			
F	-4	-3	-3	-5	-4	-5	-4	-6	-5	-5	-2	-4	-5	0	1	2	-1	9		
W	0	-3	-3	-5	-3	-5	-2	-4	-4	-4	0	-4	-4	-2	-1	-1	-2	7	10	
Y	-8	-2	-5	-6	-6	-7	-4	-7	-7	-5	-3	2	-3	-4	-5	-2	-6	0	0	17

Fig. 10.3: PAM matrix

This is just a score matrix where values range from –8 to 17. For example, if the pair of amino acids to be compared is leucine and arginine, (L and R respectively in Fig. 10.3), the score matrix gives a score of –3. An identity generally gives a large positive score (tyrosine-tyrosine gives a score of 17 referred to Y in Fig. 10.3) with the largest scores for the rare amino acids. The matrix is not calculated according to physico-chemical properties of amino acids, it is statistically derived from comparison of many related proteins.

The characteristic features of PAM matrix are summarised as follows:

1. The number with the matrix (PAM1, PAM40, PAM100) refers to the evolutionary distance; greater numbers are greater distances. The PAM1 is the matrix calculated from comparisons of sequences with no more than 1% divergence. Other PAM matrices are extrapolated from PAM1.
2. This type of matrix is derived from global alignments of closely related sequences and is not based on physico-chemical properties of amino acids.
3. Matrices for greater evolutionary distances are extrapolated from those for lesser ones.

10.2.2 BLOSUM Substitution Scoring Matrix

Also called as Blocks Amino acid substitution matrix. Steve Henikoff (Histrocial Profile 6) and colleagues in 1992 created the BLOSUM series of matrices. Matrix values are based on observed amino acid substitution in a large set of ~2000 conserved amino acids called blocks. These blocks were derived from a database of protein sequence representing more than 500 families of related proteins and act as a signature of protein families.

Blosum 62

	A	C	D	E	F
A	4	0	–2	–1	–2
C	0	9	–3	–4	–2
D	–2	–3	6	2	–3
E	–1	–2	2	5	–2
F	–2	–2	–3	–3	6

Fig. 10.4: A portion of blosum matrix

A portion of BLOSUM 62 matrix is shown in Fig. 10.4. The number after the matrix (BLOSUM 62) refers to the minimum percent identity of the blocks used to construct the matrix; greater numbers are lesser distances. BLOSUM 62 is a matrix calculated from comparisons of sequences with no less than 62 per cent divergence. It is a 20 × 20 matrix in which every possible identity and substitution is assigned a score based on the observed frequencies of such occurrences in alignments of related proteins. Identities are assigned the most positive scores. Frequently observed substitutions also receive positive scores and seldom-observed substitutions are given negative scores.

Important features of BLOSUM matrix are as follows:

1. It is derived from local, ungapped alignments of distantly related sequences.
2. The BLOSUM series of matrices generally perform better than PAM matrices for local similarity searches.
3. All BLOSUM matrices are based on observed alignments; they are not extrapolated from comparisons of closely related proteins.
4. BLOSUM 62 is the default matrix in BLAST 2.0. Though it is tailored for comparisons of moderately distant proteins, it performs well in detecting closer relationships. A search for distant relatives may be more sensitive with a different matrix.
5. The relationship between BLOSUM and PAM substitution matrices is shown in Fig. 10.5. BLOSUM matrices with higher numbers and PAM matrices with low numbers are both designed for comparisons of closely related sequences. BLOSUM matrices with low numbers and PAM matrices with high numbers are designed for comparisons of distantly related proteins. If distant relatives of the query sequence are specifically being sought, the matrix can be tailored to that type of search.

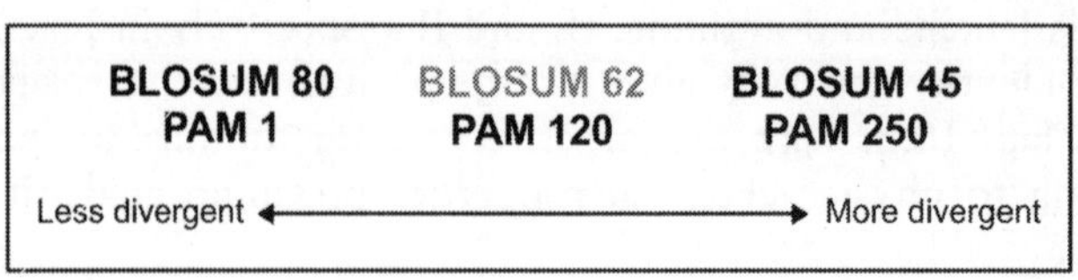

Fig. 10.5: Relationship between PAM and BLOSUM matrices

10.2.3 Specialized Matrices

Specialised matrices are matrices developed for a specific purpose. For example a set of substitution matrices is designed explicitly for finding possible frameshifts in protein sequences.

There is no such thing as a perfect substitution matrix; each matrix has its own limitations. In some cases it may possible to use multiple matrices so that each one complements the limits of the others.

10.3 ALIGNING MORE THAN TWO SEQUENCES

Sequence alignment could further be classified based on number of sequences to be aligned at one time. The simplest of these is sequence alignment in two sequences called pairwise sequence alignment. Sequence alignment when done in more than two sequences at a time is called as multiple sequence alignment. Multiple sequence alignment is little complicated and needs a more systematic approach, because there is large number of long sequences to be compared simultaneously. Examples for both sequence alignment in pairs of sequences and multiple sequence alignment are discussed in details in chapter 11.

10.4 METHODS OF SEQUENCE ALIGNMENTS

Methods of sequence analysis are classified into four types:

1. Dot matrix plots
2. Needleman and Wunsch Method (Global Alignment)
3. Smith and Waterman Method (Local Alignment)
4. Approximate Methods

10.5 DOT MATRIX METHOD PROVIDES AN EASY COMPARISON BETWEEN TWO SEQUENCES

Dot matrix or a dot plot is a table or matrix, where rows correspond to the residue of one sequence and columns to the residue of the other sequence. In its simplest form the positions in dot plot are left blank if the residue are different and filled if they match. Stretches of similar residues show up as diagonal in upper left-lower right direction. An example of dot plot between two similar words SEQUENCER and SEQUENCESEARCH is depicted in Fig. 10.6. It shows two important features of matrix methods. First is that regions of homology are recognized by a series of dots that lie along the diagonal line. Second a gap can be inserted to align sequences that are of unequal length but still have some similarity. Dot plot gives a quick pictorial statement of relationship between similar sequences.

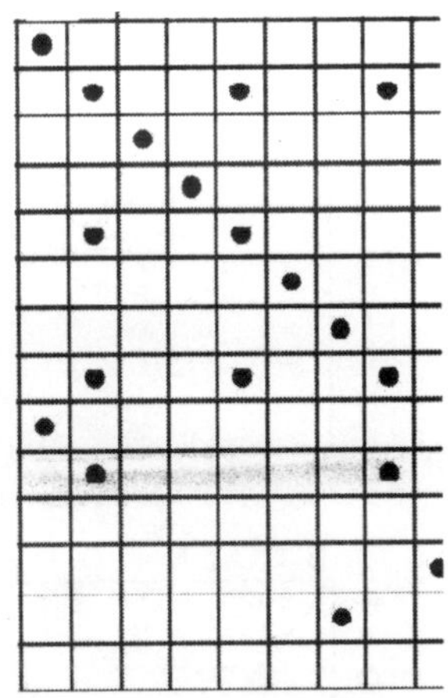

Fig. 10.6: Dotplot between words SEQUENCER and SEQUENCESEARCH

Dot matrix is the simplest and oldest method to visualize the similarity between two protein sequences. Fig. 10.7 depicts another example of dot plot used to evaluate the similarity between sequence, BIOINFORMATICS and several other similar sequences. Figure 10.7a depicts simple dot matrix diagram when the sequence is compared with itself. Each point on the grid corresponds to one position of each sequence. The matrix shows all such pairs to be compared simultaneously. Dots are placed where the same letter occurs at corresponding positions. The key observation is that similarity is distinguished by occurrence of many dots along a diagonal line within the matrix. In contrast Fig. 10.7b compared two unrelated sequences. In this comparison no diagonal lines are seen. In some cases two examples may related to each other but differ in length. In Fig. 10.7c and 10.7d two such examples are shown. In this example the second sequence is four letters longer than first. In dot matrix two diagonal lines occur. To align these two lines a gap must be created in the first sequence. Fig. 10.7d shows that insertion of gap could align the two sequences better. Now the two diagonal lines fall along the same line. Fig. 10.7e shows comparison of two sequences which have a deletion or addition occurred and in Fig. 10.7f dot plot of a sequence with itself to show the inverted repeats.

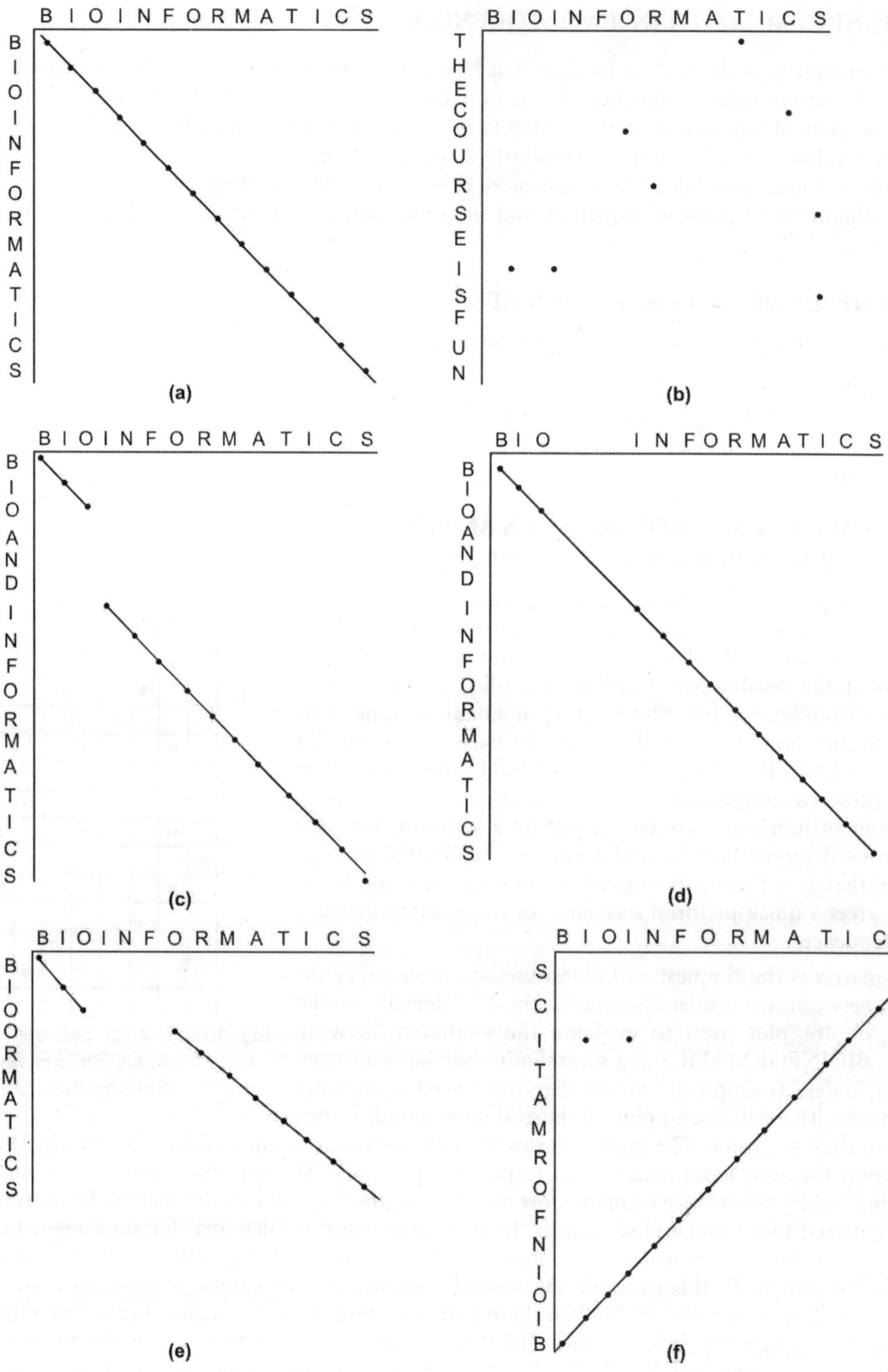

Fig. 10.7: (a) to (f) Use of dotplot to evaluate similarity between two sequences

In 1970 A.J. Gibbs and G.A. McIntyre described this method of dot matrix for comparing the amino acid and nucleotide sequences. Since then several programs have been developed using dot matrix as a method for comparing the two sequences. Few examples of softwares and web sites are given in Table 10.2. Dotter is a dot matrix based program with interactive greyscale rendering for genomic DNA and protein sequence analysis. This was a very popular program for dot matrix analysis and has been used in several studies. Another program Dotplot is equally popular and works very well on windows platform and is used in experiment 10A for alignment of two sets of sequences.

Table 10.2: Program for dot matrix plot

Dotter	Is a program given by Karolinska institute, Center for Genomics Research. Is a very good program for comparison of 2 sequences.
Dotlet	Is a program given by Swiss Institute of Bioinformatics
Dotplot	Is another commonly used program for Dotplot is discussed in details in experiment 10A

Dot matrix programs have extensively been employed in analysis of sequences. Dot matrix method quite readily reveals the presence of insertions or deletions between sequences because they shift the diagonal horizontally or vertically by the amount of change. Comparing a single sequence by itself reveals the presence a repeat of the same (direct repeats) sequences or reverse (inverted repeat or palindromes) orientation. In Experiment 10A dotplot program is used to reveal the presence of repeat sequences in human epidermal growth factor receptor protein sequence. The method of self-comparison can reveal several other features like similarity between the chromosomes tandem genes, repeat domains in a protein sequence regions of low sequence complexity where same characters are often repeated or self-complementary sequences in RNA that can potentially pair with each other to give a double stranded structure.

Dotlet program was written by Marco Pagni and Thomas Junier in 1998. It is fully available to academic users. Dotlet program is also commonly used for sequence comparisons and demonstrate several kinds of features of the sequences using dot plot method. These includes: repeated domains, conserved domains, exons and introns, terminators, frameshifts and low-complexity regions. When comparing nucleic acid to nucleic acid in addition to comparing sequences the usual way, dotlet will reverse complement one of the sequences and perform a second comparison. This enables to see structures like stem-loops for example. It is also possible to compare a protein to a nucleic acid sequence. In this case nucleic acid sequence is translated in three forward frames and the comparison is performed with each frame.

Chief merits of using dot matrix programs are as follows:

1. Unless the sequences are known to be very much similar dot matrix method if used first displays any possible alignment as diagonal on the matrix. Thus it provides investigator a choice of identifying the most significant alignment.
2. It can be easily be used to reveal the presence of insertion deletions, inverted repeated, introns exons that are more difficult to find with the other more automatic methods.
3. The self-comparison by this method can reveal several features such as similarity between chromosomes, tandem genes, repeated domains in protein sequences, regions of low sequence complexity where characters are often repeated or self complementary sequence in RNA that can potentially base pair to give a double-stranded structure.

EXPERIMENT 10A

Draw a Dotplot for Given Sequences using DOTPLOT Software

About DOTPLOT Version 1.0 for Windows: This program fills a niche in the PC molecular biology freeware/shareware world. This program was selected for the present study dotplots are easily implemented on a PC, not being too CPU intensive (unless the sequences to be compared are large) and being fun to play around with if made interactive. The windows version was fairly easy write because windows tend itself to graphics-oriented programs. The operating system does a lot of the work for you, such as the mouse movement and the menus. There's just a lot more setting up to do than there is for MSDOS

DOTPLOT can read the files in following formats: IG/Stanford, Pearson/Fasta, GenBank/GB, NBRF/PIR, EMBL, Phylip3.4/Phylip, GCG, Phylip3.3/Interleaved, DNAStrider and Plain/Raw. (see chapter 7 for details of these formats). The calculation of scores are made even faster by calculating scores diagonal-by-diagonal rather than point-by-point. CHARGE.MAT, DNA53.MAT and DNA35.MAT are the score files. The CHARGE.MAT file scores identities as +5 if both amino acids are charged (D, E, R, K). Non-charged residues are scored as 0. Opposite charges are scored as -3 and identical charges with non-identical residues are scored as +3. The DNA53.MAT gives a score of +5 for an identity and –3 for different nucleotides.

Concept of Window Size and Stringency: The program will prompt you for a window size and stringency. For the simplest case, where the program puts a dot on the screen for every identity, the window size is one and the stringency is one. This will be very NOISY or confusing, as can be seen in the dotplot of two well-known sequences in Fig. 10.8. Dots enclosed in a small square creates lots of confusion. The same plot (shown on left Hand Side in Fig. 10.8) when drawn with window size equals 2 and stringency equal to 2 will direct the plot to ignore all small diagonals where size is less than 2 alphabets. This removes all single points and thus clarifies the picture. The noise decreases for increasing window size and stringency, but eventually the signal decreases too, if it is too large.

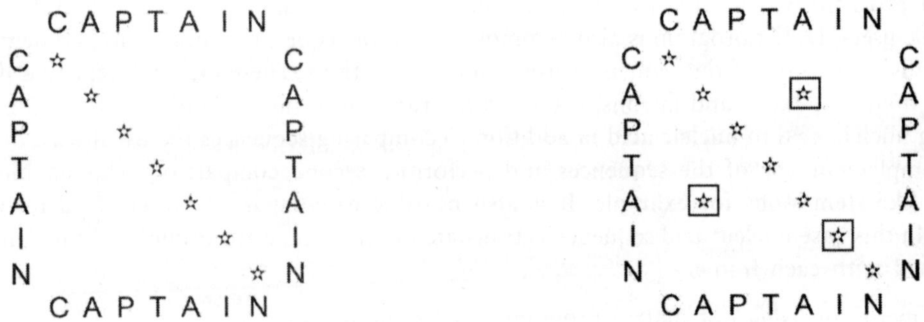

(a) Window size equal to two and stringency

(b) Window size equal to one and stringency equal to one

Fig. 10.8: Concept of window size

About the Sequences from Swissprot

Sequence 1: After binding acetylcholine, the achr Neuronal nicotinic acetylcholine receptor subunit responds by an extensive change in conformation that affects all subunits and leads to opening of an ion-conducting channel across the plasma membrane. Neuronal achr is composed of two different type of subunits: alpha and non-alpha (beta). Alpha subunits two and three from rat were selected for drawing the dotplot.

Sequence 2: The epidermal growth factor (egfr) receptor, mediates the biological signal of epidermal growth factor (egf) and also of transforming growth factor alpha and vaccinia virus growth factor. Binding of egf to the efgr leads to internalization of the egf-receptor complex, induction of the tyrosine kinase activity stimulation of cell DNA synthesis, and cell proliferation. Human efgr is known to have presence of internal repeats.

Hypothesis: It would be possible to draw a plot for two neuronal nicotine acetylcholine receptor subunits and to find out inverted repeats in epidermal growth factors.

TESTING THE HYPOTHESIS

Computer set up:

1. Try comparing two neuronal nicotinic acetylcholine receptor subunit sequences from **swissprot** saved under the filenames a2 and a3 is given in Data Boxes 10.1 and 10.2 respectively. Log into the website **Dotplot**. The interface is like any other Windows program Fig. 10.9 and 10.10. Use the File pop up menu to load the horizontal sequence. You get a dialog box listing all the files in the current directory. Select a2, then the program gives a SeqInfo dialog box, which tells you a little bit about the sequence you just loaded i.e. its name, length, sequence format and whether it's DNA, RNA or protein. Click OK. Then repeat for the vertical sequence, choosing a3.

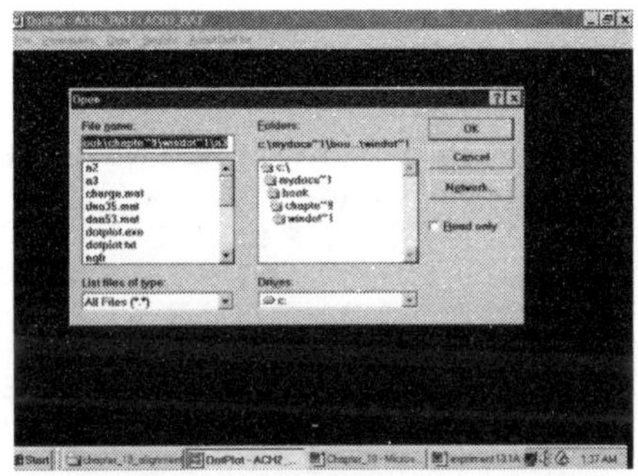

Fig. 10.9

2. For a quick fix, use the Identities option on the Draw pop up menu. The default parameters are a window size of 10 and a threshold of 6. This gives a diagonal line across your screen. Try tweaking the dotplot by clicking on parameters and setting the threshold to 1.

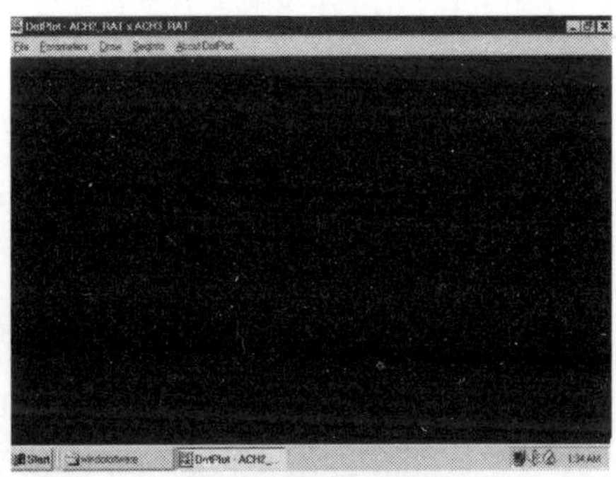

Fig. 10.10

3. Same program is used to find the internal repeats in a human epidermal growth factor receptor Fig.10.11 and 10.12. To do this, load the sequence called egfr.doc is given in Data Box 10.3, as both the horizontal and the vertical sequence. Then set the parameters as window = 20 and threshold = 1. Load the PAM250.MAT score matrix using the File pop up menu "Open Matrix" option. The feature table gives:

FT REPEAT 75 300 APPROXIMATE.
FT REPEAT 390 600 APPROXIMATE.

DATA

Data Box 10.1: The given sequence as contained in file a2. The sequence is extracted from swissprot, however the complete entry is not shown.

>P1;ACH2_RAT

NEURONAL ACETYLCHOLINE RECEPTOR PROTEIN, ALPHA-2 CHAIN PRECURSOR. - RATTUS NORVEGICUS (RAT).

M T L S H S A L Q F W T H L Y L W C L L L V P A V L T Q Q G S H T H A E D R L F K H L F
G G Y N R W A R P V P N T S D V V I V R F G L S I A Q L I D V D E K N Q M M T T N V W L

K Q E W N D Y N V R W D P A E F G N V T S L R V P S E M I W I P D I V L Y N N A D G E F
A V T H M T K A H L F F T G T V H W V P P A I Y K S S C S I D V T F F P F D Q Q N C K M K
F G S W T Y D K A K I D L E Q M E R T V D L K D Y W E S G E W A I I N A T G T Y N S K K Y
D C C A E I Y P D V T Y Y F V I R R L P L F Y T I N L I I P C L L I S C L T V L V F Y L P S E C
G E K I T L C I S V L L S L T V F L L L I T E I I P S T S L V I P L I G E Y L L F T M I F V T L S
I V I T V F V L N V H H R S P S T H N M P N W V R V A L L G R V P R W L M M N R P L P P M
E L H G S P D L K L S P S Y H W L E T N M D A G E RE E T E E E E E E E D E N I C V C A G
L P D S S M G V L Y G H G G L H L R A M E P E T K T P S Q A S E I L L S P Q I Q KA L E G
V H Y I A D R L R S E D A D S S V K E D W K Y V A M V V D R I F L W L F I I V C F L G T I G
L F L P P F L A G MI*
C;Species: RATTUS NORVEGICUS (RAT).

Data Box 10.2: The Given Sequence as Contained in File a3. The Sequence is Extracted from Swissprot, However the Complete Entry is not Shown.

>P1;ACH3_RAT

NEURONAL ACETYLCHOLINE RECEPTOR PROTEIN, ALPHA-3 CHAIN PRECURSOR.—RATTUS NORVEGICUS (RAT).

M G V V L L P P P L S M L M L V L M L L P A A S A S E A E H R L F Q Y L F E D Y N E I I R P
V A N V S H P V I I Q F E V S M S Q L V K V D E V N Q I M E T N L W L K Q I W N D Y K LK W
K P S D Y Q G V E F M R V P A E K I W K P D I V L Y N N A D G D F Q V D D K T K A L L K Y
T G E V T W I P P A I F KS S C K I D V T Y F P F D Y Q N C T M K F G S W S Y D K A K I D L
V L I G S S M N L K D Y W E S G E W A I I K A P G Y KH E I K Y N C C E E I Y Q D I T Y S L
Y I R R L P L F Y T I N L I I P C L L I S F L T V L V F Y L P S D C G E K V T L CI S V L L S L
T V F L L V I T E T I P S T S L V I P L I G E Y L L F T M I F V T L S I V I T V F V L N V H Y R
T P T T H T M P T W V K A V F L N L L P R V M F M T R P T S G E G D T P K T R T F Y G A
E L S N L N C F S R A D S K S C K E G Y PC Q D G T C G Y C H H R R V K I S N F S A N L T
R S S S S E S V N A V L S L S A L S P E I K E A I Q S V K Y I A E N M KA Q N V A K E I Q D D
W K Y V A M V I D R I F L W V F I L V C I L G T A G L F L Q P L M A R D D T *
C;Species: RATTUS NORVEGICUS (RAT).

Data Box 10.3: Contents of Sequence in egfr. doc file

```
ID     EGFR_HUMAN
AC     P00533;P06268'
SQ     SEQUENCE   1210 AA;  134291 MW;  7507159 CN;
       MRPSGTAGAA LLALLAALCP ASRALEEKKV CQGTSNKLTQ LGTFEDHFLS LQRMFNNCEV
       VLGNLEITYV QRNYDLSFLK TIQEVAGYVL IALNTVERIP LENLQIIRGN MYYENSYALA
       VLSNYDANKT GLKELPMRNL QEILHGAVRF SNNPALCNVE SIQWRDIVSS DFLSNMSMDF
       QNHLGSCQKC DPSCPNGSCW GAGEENCQKL TKIICAQQCS GRCRGKSPSD CCHNQCAAGC
       TGPRESDCLV CRKFRDEATC KDTCPPLMLY NPTTYQMDVN PEGKYSFGAT CVKKCPRNYV
       VTDHGSCVRA CGADSYEMEE DGVRKCKKCE GPCRKVCNGI GIGEFKDSLS INATNIKHFK
       NCTSISGDLH ILPVAFRGDS FTHTPPLDPQ ELDILKTVKE ITGFLLIQAW PENRTDLHAF
       ENLEIIRGRT KQHGQFSLAV VSLNITSLGL RSLKEISDGD VIISGNKNLC YANTINWKKL
       FGTSGQKTKI ISNRGENSCK ATGQVCHALC SPEGCWGPEP RDCVSCRNVS RGRECVDKCK
       LLEGEPREFV ENSECIQCHP ECLPQAMNIT CTGRGPDNCI QCAHYIDGPH CVKTCPAGVM
       GENNTLVWKY ADAGHVCHLC HPNCTYGCTG PGLEGCPTNG PKIPSIATGM VGALLLLLVV
       ALGIGLFMRR RHIVRKRTLR RLLQERELVE PLTPSGEAPN QALLRILKET EFKKIKVLGS
       GAFGTVYKGL WIPEGEKVKI PVAIKELREA TSPKANKEIL DEAYVMASVD NPHVCRLLGI
       CLTSTVQLIT QLMPFGCLLD YVREHKDNIG SQYLLNWCVQ IAKGMNYLED RRLVHRDLAA
       RNVLVKTPQH VKITDFGLAK LLGAEEKEYH AEGGKVPIKW MALESILHRI YTHQSDVWSY
       GVTVWELMTF GSKPYDGIPA SEISSILEKG ERLPQPPICT IDVYMIMVKC WMIDADSRPK
       FRELIIEFSK MARDPQRYLV IQGDERMHLP SPTDSNFYRA LMDEEDMDDV VDADEYLIPQ
       QGFFSSPSTS RTPLLSSLSA TSNNSTVACI DRNGLQSCPI KEDSFLQRYS SDPTGALTED
       SIDDTFLPVP EYINQSVPKR PAGSVQNPVY HNQPLNPAPS RDPHYQDPHS TAVGNPEYLN
       TVQPTCVNST FDSPAHWAQK GSHQISLDNP DYQQDFFPKE AKPNGIFKGS TAENAEYLRV
       APQSSEFIGA
```

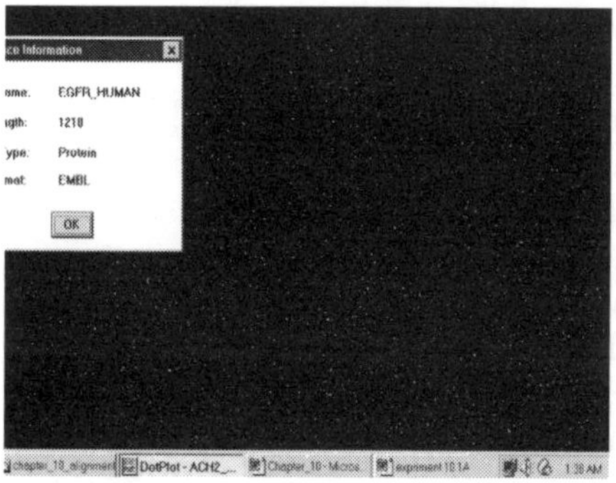

Fig. 10.11

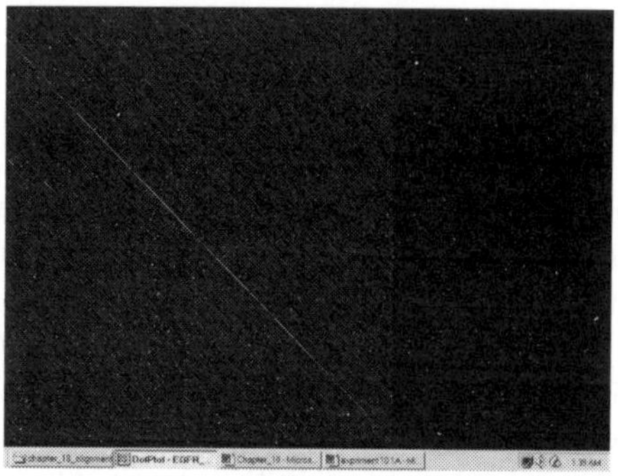

Fig. 10.12

INTERPRETATION OF DATA

Drawing a dotplot for similarity search in two sequences achr a2 and achr a3 using software DOTPLOT gave results similar that would have been obtained if such a plot had been drawn manually as shown in Fig. 10.7. The pixels colour depends on how similar the two sequence at these 2 positions, each pixel represents a score with high, a high score meaning good matches. Light and dark regions of the line in Fig. 10.12 indicate that similarity is very high in certain regions of the sequence. You can see three transmembrance regions in the middle of the protein and a final transmembrance region at C-terminus. The same program is very effectively used to find the Inverted repeats present in the sequence egfr. doc (Fig. 10.12).

The chief limitations with dot matrix method of sequence alignment are:

1. Although the dot matrix method can be used to detect sequence similarity, it does not readily resolve similarity that is interrupted by regions that do not match very well or that is present only in one of the sequences.
2. Dot matrix method is a visual method and not automatic. It requires human intelligence for interpretation.
3. It is difficult to compare more than 2 sequence at a time with dot matrix.
4. Dot matrix method is not a suitable method for optimal alignment i.e. the very best possible method. Such an alignment if tried with dot matrix method can be presented by writing the sequences on successive lines with matching characters places in the same column and unmatching characters placed in the same columns as mismatched or next to gap as an insertion or deletion in the other sequence. To find an optimal alignment in which all possible matches, insertions and deletions have been considered is difficult for a protein of length of 300 AA, because of number of comparisons done is roughly proportional to the square of average sequence length and if the alignment is to include gaps of any length at any portion in either sequence, the total number of comparisons to be done become astronomical and this is not possible to do this using dot matrix.

Considering the limitations of dot matrix methods, other methods were tried. Three variations in alignments were tried i.e. entire sequence against the entire sequence, regions of one sequence against entire other sequence and region of one sequence against region of other sequence. The global alignment algorithm was first applied to biological sequence alignment by S.B. Needlemen and Wunsch method was adopted for local alignments by T. Smith and M. Waterman.

Computational approach for finding the optimal path simply follow drawing a chart indicating containing all possible alignments which can be easily constructed as a matrix similar to that used for drawing the dot plot. Any path through the dot plot from upper left to lower right passes through a succession of cells, each of which picks out a pair of positions, one from the row and one from the column, that corresponds to an alignment or that indicates a gap in one of the sequences. The path need not pass through filled in points only. However more filled in points on diagonal segments of the path, more matching residues in the alignment. The Fig. 10.7c is redrawn in Fig. 10.13 to explain method of scoring in Dot Matrix alignment.

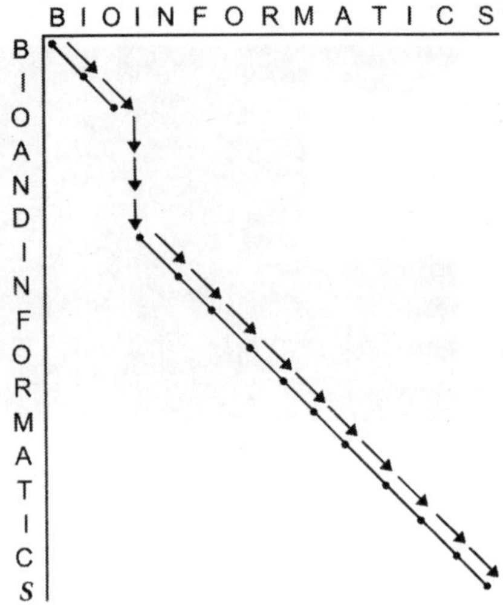

Fig. 10.13: Method of scoring in dotmatrix

The task is to find the path that has lowest cost and the difficulty is that there are a large number of paths to consider. This is much like finding shortest route to drive from Pitampura to Delhi University. There are many choices to produce the complete route, we need to find the one which best suitable in terms of fuel consumption, time taken and comfort in terms of quality of roads and sceneries as for sight seeing. An algorithm will be developed which will explore all possible combinations to find the optimal overall route. Dynamic programming methods ensure the optimal alignment by exploring all possible alignments and choosing the best. It does this by reading in a scoring matrix that contains values for every possible residue or nucleotide match.

In dynamic programming a best possible or an optimal alignment with highest possible score is generated for a given set of user-defined variables including scoring matrices and gap penalty. The procedure for generating all possible alignments is to move sequentially through all matched positions within a matrix much like a dot matrix graph. At each position in the matrix the highest possible score that can be achieved up to that point is placed at that position, allowing all possible starting points in either sequence and any combination of matches mismatches insertions and deletions. The best alignment is the highest scoring position in the graph and then tracking through the path generated the high scoring positions. The sequences are then aligned so that the sequence characters corresponding to this path are matched. The alignment with highest score is defined as optimal alignment.

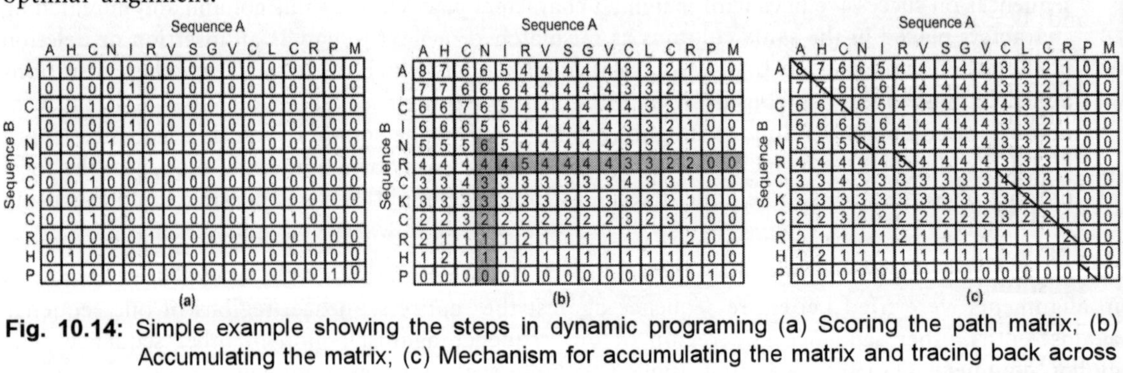

Fig. 10.14: Simple example showing the steps in dynamic programing (a) Scoring the path matrix; (b) Accumulating the matrix; (c) Mechanism for accumulating the matrix and tracing back across

In Fig. 10.14 the step in dynamic programming methods is depicted diagramatically and mathematically. In the first step scoring of the path matrices is done. 0 is given for every mismatch and 1 for match. In the second step accumulation of matrix is done. After this tracing back through the matrix to obtain the highest scoring path giving the alignment of the two sequences is done.

Due to large number of computer steps and memory required time required, to align two sequences of length n and m is proportional to n × m because this size of the edit matrix that must be filled in. This means dynamic programming method is not convenient to use for searching entire sequence databases for a match and even is less convenient for all-against-all alignments.

Both global and local types of alignments may be made by simple changes in basic dynamic programming algorithm. Dynamic programming methods have been used for global sequence alignment by Needleman and Wunsch algorithm (1970) and for local alignment by Smith and waterman algorithms (1981).

10.6 NEEDLEMAN AND WUNSCH METHOD FOR GLOBAL ALIGNMENTS

In 1970 Needleman and Wunsch developed an algorithm by dynamic programming for comparing two amino acids at a time. They started at the end of each sequences and then moved ahead one amino acid pair one at a time allowing all possible combinations of matched pairs, mismatched pairs or an extra amino acid in one sequence. Their simplified approach is given in example in Fig. 10.15. For every position a score is given and the objective is to determine the position with highest score.

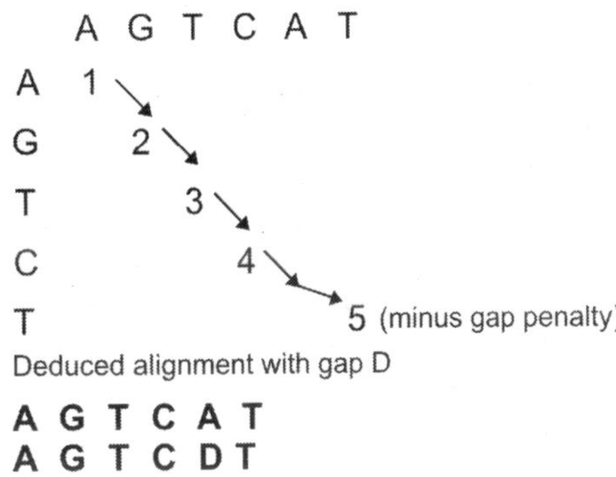

Fig. 10.15

First, all matches in the two sequences are given a score of 1, and mismatches a score of 0 (not shown), chosen arbitrarily for this example. Second, the diagonals are added sequentially, in this case to a total score of 4. At this point the row cannot be extended by another match of 1 to a total score of 5. However, an extension is possible if a gap is placed in GATCA to produce GATCDA, where D is the gap. To add the gap, a penalty score is subtracted from the total match score of 5 now appearing in the last row and column. The best alignment is found starting with the sequence characters that correspond to the highest number and tracing back through the positions that contributed to this highest score.

Let us consider alignment of two strings S = a c g c t g and T = c a t g t are aligned according to simple scoring scheme: match = 0; mismatch = –1; insertion or deletion = 1. The alignment according Needleman and Wunsch algorithm is deplicted in Fig. 10.16. The top row and left most column have been initialized and element in the second row and second column has been entered as –1 mismatch. The value of –1 was chosen as minimum of –1 + –1 (vertical move or insert a gap into string catgt), 0 + –1 (diagonal move or substitution C ⇆ a) and –1 + –1 (horizontal move or insert a gap into string acgctg). Both vertical move and horizontal moves provided the minimum value. The matrix after completion of values is shown in Fig. 10.16. It includes trace

i \ j		0	1	2	3	4	5
			c	a	t	g	t ←T
0	0	0	–1	–2	–3	–4	–5
1	a	–1	1	1	0	–1	–2
2	c	–2	1	0	0	–1	–2
3	g	–3	0	0	–1	2	1
4	c	–4	–1	–1	–1	1	1
5	t	–5	–2	–2	1	0	3
6	g	–6	–3	–3	0	3	2
	↑ S						

Fig. 10.16: Needleman Wunsch scoring method

```
-a c g c t g     a c g c t g      a c g c t g
 |   |   |       |     | |        |     | |
c a t g - t      - c a - t g t    - c - a t g t
```

back information in the form of arrows pointing from each cell to its predecessor(s). This gives a total of three optimal alignments:

This algorithm determines optimal global alignment of two sequences. It is inappropriate for detection of local regions of high similarity within two sequences or for aligning a long sequence with a short fragment because it imposes gap penalty outside the similar region. The method of Smith and Waterman solves this problem.

10.7 SMITH-WATERMAN ALGORITHM FOR LOCAL SEQUENCE ALIGNMENT

The Smith-Waterman evaluation finds an optimal, local alignment of nucleotide or peptide sequences and is typically used when the sequences to be compared are not of nearly the same length. These evaluations were first used for protein sequence comparison by Temple Smith and Mike Waterman in 1981 as a modification of the Needleman and Wunsch algorithm. This algorithm not only aligned the sequence for optimal local alignment but also recognized insertions or deletions of any size likely to be found due to in evolutionary change.

i \ j	0	1 x	2 x	3 x	4 c	5 d	6 e	T
0	0	0	0	0	0	0	0	
1 a	0	0	0	0	0	0	0	
2 b	0	0	0	(0)	0	0	0	
3 c	0	0	(0)	0	(2)	1	0	
4 x	0	2	2	(2)	(1)	1	0	
5 d	0	1	1	1	1	(3)	2	
6 e	0	0	0	0	0	2	(5)	
7 x		2	2	2	1	1	4	
S								

Fig. 10.17: Smith-Waterman algorithm scoring method

The characteristic features of scoring system in Smith-Waterman Algorithm are depicted in Fig. 10.17.

1. Assigns a score to each pair of bases. It only uses similarity scores positive score for related residues and negative scores for substitutions and gaps.
2. Initializes edges of the matrix with zeros.
3. As the scores are summed in the matrix, any score below 0 is recorded as 0.
4. Begins the trace back at the maximum value found anywhere in the matrix.
5. Continues until the score falls to 0.

The changes in Smith-Waterman algorithm over Needleman-Wunsch are as follows:

1. The value at top row and left column are set as 0. As a result of this either sequence can slide along the other before the alignment starts, without incurring any gap penalty against the residue it leaves behind.
2. Score of the Needleman and Wunsch alignment is the number at the right most col of the lowest ror. In the Smith-Waterman alogirthm it is the optimal value encountered wherever in the matrix it appears.
3. In global alignment the traceback to determine the actual alignment starts at the right most column of the lowest row cell. In Smith-Waterman method it starts at the cell containing the optimal value and continues back only as far as the region of local similarity.

HISTORICAL PROFILE 11

Smith and Waterman for Mixing Mathematics, Biology and Computer Science

Temple Smith is Professor of Biomedical Engineering, Research Professor of Pharmacology, and Director of the Biomolecular Engineering Research Center at Boston University.

Professor Michael S. Waterman is the USC Associates Endowed Professor of Mathematics and Natural Sciences.

The two together are key contributors to bioinformatics. Their research bridges between the information in genome and protein sequence and the structure and function of proteins and other biomolecules. When it comes to mixing mathematics and biology, Professor Michael S. Waterman through his early work that defined the multidisciplinary field of computational biology, and literally, when he published the first textbook on the subject in 1995. Prof. Smith is also involved in the development of database structures that can facilitate the analysis and sequencing of genomes. In 1981, the two published the Smith-Waterman (S-W) algorithm and a related computer software program. The S-W algorithm allowed scientists to search for clues to the function of a newly identified gene (or protein) by comparing its sequence to that of a previously studied gene. They provided the mathematical proof that dynamic programming guaranteed to provide an optimal alignment between two sequences. S-W algorithm and other statistical/computational approaches developed by Prof. Smith are used to identify structure, function, and regulation in biomolecules. This identification has led to the formulation and testing of major hypotheses in the areas of molecular evolution, gene regulation, developmental genetics, and protein structure-function relationships. Among his specific accomplishments is the identification of an anti-oncogen Rb binding site.

Prof. Waterman developed the statistical theory and tools to test whether the match of two sequences was statistically significant, providing a reality check for biologists much like a student's t-test or the relative risk calculations of medical science. As biologists' interest turned to the study of genomes, Waterman began collaborating with Eric Lander, a mathematician-turned-biologist at MIT and later a key leader of the federal Human Genome Project. In 1988, they published the Lander-Waterman formulas, which provided a road map for genome sequencing efforts using the so-called shotgun strategy. The shotgun approach involved cutting up the genome into short fragments of DNA, which were faster and easier to sequence, and then trying to fit them back together in the correct order and orientation, like pieces in a giant puzzle.

EXPERIMENT 10B

To Perform the Global and Localized Alignment in the Given Sequence from a Human Alpha Actinin2 mRNA with Mouse Alpha Actinin2 mRNA using EMBOSS-Align Software

About EMBOSS-Align: The EMBOSS-Align tool contains two programs each using a different algorithm: needle program for global alignment and water program for localized alignment.

Needle program: This program uses the Needleman-Wunsch global alignment algorithm to find the optimum alignment (including gaps) of two sequences when considering their entire length.

Water program. Water uses the Smith-Waterman algorithm (modified for speed enhancements) to calculate the local alignment.

About the Sequences: Two sequences from Human and mouse alpha actin 2 mRNA are selected for the present experiment

Hypothesis: It would be possible to get global and local alignment of these 2 sequences and study the difference in 2 types of alignment.

TESTING THE HYPOTHESIS

Computer set up: A normal PC with internet connection in windows environment:

1. Log on to Web site of EMBOSS-Align. This takes us to web site shown in Fig. 10.18. Load the Sequences Human alpha actinin2 mRNA with Mouse alpha actinin2 mRNA.
2. The 2 sequences are shown in Data Boxes 10.4 and 10.5 in fasta format, which consists of a one-line header starting with a ">" symbol, followed by the sequence name/ description. The sequence is then entered on new line(s) and terminated with a line consisting of "//".
3. Jobs were run, one with the EMBOSS global alignment program (Needle), and other with the local alignment program (Water).
4. As we are comparing 2 nucleotide/RNA sequences, the molecule type was changed to DNA.
5. The default DNA full matrix was used, and the default gap open of "10" and gap extend of "0.5" were also used.

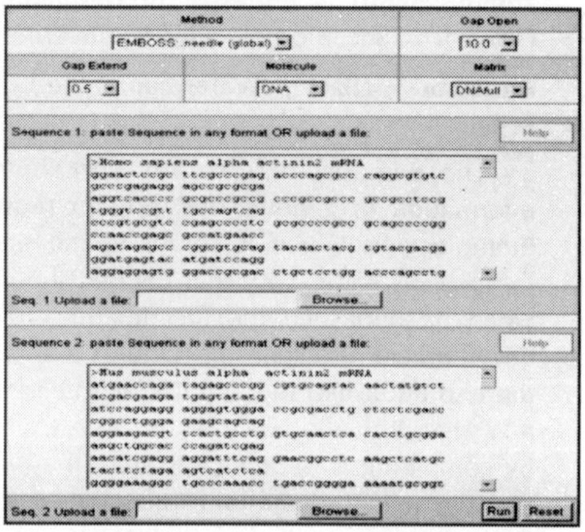

Fig. 10.18: Web view of Emboss

DATA

Data Box 10.4: Human alpha actinin2 mRNA

>Homo sapiens alpha actinin2 mRNA
ggaactccgc ttcgcccgag acccagcgcc caggcgtgtc gcccgagagg agccgcgcga aggtcacccc gcgcccgccg cccgccgccc gccgcctccg tgggtccgtt tgccagtcag
cccgtgcgtc cgagcccctc gcgccccgcc gcagccccgg ccaaccgagc gccatgaacc agatagagcc cggcgtgcag tacaactacg tgtacgacga ggatgagtac atgatccagg aggaggagtg
ggaccgcgac ctgctcctgg acccagcctg ggagaagcag cagaggaaga ccttcactgc ctggtgtaac tcccacctaa ggaaagccgg cacccagatt gagaacatcg
aggaagactt caggaatggc cttaagctca tgctgctttt ggaagtcatc tcaggggaaa ggctgcccaa acctgaccgg ggaaaaatgc ggttccacaa aattgctaat gtcaacaaag
ctttggatta catagccagc aaagggtga aactggtgtc catcggcgct gaagaaattg ttgatggcaa tgtgaaaatg accctgggta tgatctggac catcatcctt cgctttgcta
ttcaggatat ttcggttgaa gaaacatctg ccaaagaagg tctgctgctt tggtgtcaga ggaaaactgc tccttataga aatgtgaaca ttcagaactt ccatactagc tggaaagatg
gccttggact ctgtgccctc atccaccgac accggcctga cctcattgac tactcaaagc ttaacaagga tgaccccata ggaaatatta acctggccat ggaaatcgct gagaagcacc
tggatattcc taaaatgttg gatgctgaag acatcgtgaa cacccctaaa cccgatgaaa gagccatcat gacgtacgtc tcttgcttct accacgcttt tgcgggcgcg gagcaggccg
agacagcggc taacaggata tgtaaggttc ttgctgtgaa tcaagagaat gagaggctga tggaagaata tgagaggcta gcgagtgagc ttttggaatg gattcgtcgc acgatcccct
ggctggagaa ccggactccc gagaagacca tgcaagccat gcagaagaag ctggaggact tccgggatta ccgccggaag cacaagccac ccaaggtgca ggagaaatgc cagctggaga
tcaacttcaa cacgctgcag accaagctgc ggatcagcaa ccgtcctgcc ttcatgccct ccgagggcaa gatggtgtcg gatattgctg gtgcctggca gaggctggag caggctgaga
agggttacga ggagtggttg ctcaatgaga ttcggagact ggagcgcttg gaacacctgg ctgagaagtt caggcagaag gcctcaacgc acgagacttg ggcttatggc aaagagcaga
tcttgctgca gaaggattac gagtcggcgt cgctgacaga ggtgcgggct ctgctgcgga agcacgaggc gttcgagagc gacctggcag cgcaccagga ccgcgtggag cagatcgcag
ccatcgcgca ggagctcaat gaactggact atcacgacgc tgtgaatgtc aatgatcggt gccagaaaat ttgtgaccag tgggaccgac tgggaacgct tactcagaag aggagagaag
ccctagagag aatggagaaa ttgctagaaa ccattgatca gcttcacctg gagtttgcca agagggctgc tcctttcaac aattggatgg agggcgctat ggaggatctg caagatatgt
tcattgtcca cagcattgag gagatccaga gtctgatcac tgcgcatgag cagttcaagg ccacgctgcc cgaggcggac ggagagcggc agtccatcat ggccatccag aacgaggtgg
agaaggtgat tcagagctac aacatcagaa tcagctcaag caacccgtac agcactgtca ccatggatga gctccggacc aagtgggaca aggtgaagca actcgtgccc atccgcgatc aatccctgca ggaggagctg
gctcgccagc atgctaacga gcgtctgagg cgccagtttg ctgcccaagc caatgccatt gggccctgga tccagaacaa gatggaggag attgcccgga
gctccatcca gatcacagga gccctggaag accagatgaa ccagctgaag cagtatgagc acaacatcat caactataag aacaacatcg acaagctgga gggagaccat cagctcatcc
//

Data Box 10.5: Mouse alpha actinin2 mRNA

>Mus musculus alpha actinin2 mRNA
atgaaccaga tagagcccgg cgtgcagtac aactatgtct acgacgaaga tgagtatatg atccaggagg aggagtggga ccgcgacctg ctcctcgacc cggcctggga gaagcagcag
aggaagacgt tcactgcctg gtgcaactca cacctgcgga aagctggcac ccagatcgag aacatcgagg aggatttcag gaacggcctc aagctcatgc tacttctaga agtcatctca
ggggaaaggc tgcccaaacc tgaccgggga aaaatgcggt tccacaagat tgcgaatgtc aacaaggctc tggattacat agccagcaag ggagtgaagc tggtatctat tggagctgaa
gaaatagtcg atggcaatgt gaagatgacc ctgggtatga tctggaccat catcctccgc ttcgccattc aggacatctc agtggaagaa acatctgcca aagaagggct gctgctatgg
tgtcagagga agacagctcc atacagaaat gtgaatatcc agaacttcca caccagctgg aaagatggcc ttggactctg tgcccttatc cacagacacc ggcccgacct cattgactac
tcaaagctta acaaggatga ccccatagga aacattaacc tggccatgga aatagcagag aagcacttgg atatccctaa gatgttggat gctgaagaca ttgtcaacac tcccaaaccc gatgaaagag
ccatcatgac ttacgtttcc tgcttctacc atgctttcgc gggagcggag caggccgaga cagcagctaa cagaatatgt aagggtcttg ctgtgaatca agagaatgag
aggctgatgg aagaatatga gaggctagcg agtgagctgt tggagtggat tcggcgcaca atcccctggc tggagaacag gacgcctgag aagaccatgc aggccatgca gaagaagctg
gaggacttcc gggattaccg tcgtaagcat aagcccccca aggtgcagga gaagtgccag ctggagatca acttcaacac gctgcagacc aagcttcgga tcagcaaccg ggccgccttc

```
atgccctccg agggcaagat ggtgtcggac attgctggtg cctgccagag gttggagcag gctgagaagg gctatgagga gtggctattg aatgagattc gaagactgga gcgcttggag
cacttggctg agaagttcag gcagaaggcc tccacgcacg agacatgggc ttatggcaaa gaacagatct tgctgcagaa ggactacgag tctgcgtccc tgacggaagt tcgagcccta
cttcggaagc acgaggcctt tgagagcgac ctggctgcac accaggaccg cgtggagcag atcgctgcca ttgcgcagga gctcaatgaa ctggactatc acgatgctgt gaacgtcaat
gatcggtgcc agaaaatttg cgaccagtgg gatagattgg ggacgctcac ccagaagagg agggaggcct tggagagaac tgagaagttg ctggaaacaa tcgaccaact gcacctggag
tttgccaaga gagccgcgcc tttcaacaac tggatggagg gagccatgga ggacctgcag gacatgttca ttgtccacag catcgaggag atccagagtc tgatcactgc acatgagcag
ttcaaggcca cactgcccga ggccgacggg gagcggcagt ccatcctggc catccagaac gaggtggaga aggtgatcca gagctacagc atccggatca gctccagcaa cccctacagc
actgtcacca tggatgagct gcgcaacaag tgggacaagg tgaagcagct ggtgccggtc cgggatcagt ccctgcagga ggagctggct cgccaacatg ctaatgagcg tctgcggcgc
cagtttgctg ctcaggccaa tgccattggg ccctggatcc agaacaagat ggaggagatc gctcggagct ccatccagat caccggggcc ctggaggacc agatgaacca gctgaagcag
//
```

```
########################################
# Program:  needle
# Rundate:  Wed Oct 06 07:18:56 2004
# Align_format: srspair
# Report_file: /ebi/extserv/old-work/needle-20041006-07185370444033.output
########################################
#=======================================
#
# Aligned_sequences: 2
# 1: Homo
# 2: Mus
# Matrix: EBLOSUM62
# Gap_penalty: 10.0
# Extend_penalty: 0.5
#
# Length: 4182
# Identity:    2489/4182 (59.5%)
# Similarity:  2489/4182 (59.5%)
# Gaps:        1434/4182 (34.3%)
# Score: 14688.0
#
#
#=======================================
```

(a)

```
########################################
# Program:  water
# Rundate:  Wed Oct 06 07:21:23 2004
# Align_format: srspair
# Report_file: /ebi/extserv/old-work/water-20041006-07211952164084.output
########################################
#=======================================
#
# Aligned_sequences: 2
# 1: Homo
# 2: Mus
# Matrix: EBLOSUM62
# Gap_penalty: 10.0
# Extend_penalty: 0.5
#
# Length: 2752
# Identity:    2488/2752 (90.4%)
# Similarity:  2488/2752 (90.4%)
# Gaps:           6/2752 ( 0.2%)
# Score: 14691.0
#
#
#=======================================
```

(b)

Fig. 10.19: Summary of result obtained using program (a) Needle and (b) water

(a) (b)

Fig. 10.20: Graphical display of result through needle (a); water (b)

INTERPRETATION OF DATA

Result of alignment of given 2 sequences shown in Fig. 10.20. Needle finds an alignment with the maximum possible score where the score (here it is 14688.0) of an alignment is equal to the sum of the matches taken from the scoring matrix. An important problem is the treatment of gaps, i.e., spaces inserted to optimize the alignment score. A penalty is subtracted from the score for each gap opened (the 'gap open' penalty) and a penalty is subtracted from the score for the total number of gap spaces multiplied by a cost (the 'gap extension' penalty). Typically, the cost of extending a gap is set to be 5-10 times lower than the cost for opening a gap. This is a true implementation of the Needleman-Wunsch algorithm and so produces a full path matrix. It, therefore, cannot be used with genome-sized sequences unless you've a lot of memory and a lot of time.

Needle is for aligning two sequences over their entire length. This works best with closely related sequences. If you use needle to align very distantly related sequences, it will produce a result but much of the alignment may have little or no biological significance. A true Needleman Wunsch implementation like needle needs memory proportional to the product of the sequence lengths. For two sequences of length 10,000,000 and 1,000 it, therefore, needs memory proportional to 10,000,000,000 characters. Two arrays of this size are produced, one of integers and one of floats so multiply that figure by 8 to get the memory usage in bytes. That doesn't include other overheads. Therefore only use water and needle for accurate alignment of reasonably short sequences.

Water finds an alignment with the maximum possible score (14691.0) where the score of an alignment is equal to the sum of the matches taken from the scoring matrix. The identities and similarity using water program is much higher because it considers a shorter segments of sequences of DNA (2752 in Fig. 10.18) compared to sequence length of 4182 in needle program. Water is a true implementation of the Smith-Waterman algorithm and so produces a full path matrix. It, therefore, cannot be used with genome-sized sequences unless you have a lot of memory and a lot of time.

10.8 WORD OR K-TUPLE OR APPROXIMATE METHODS

As mentioned earlier that dot matrix and dynamic, programming methods have their own limitations and are of limited application in routine database screening of genes from a new genome against the database for similarity to other sequences. Word or k-tuple or approximate methods can perform these similarity searches fairly well and quickly. This approach is called k-tuple because it takes a small integer and determines all instances of each k-tuple (i.e. word of length k) residue in the probe sequence in the database.

The easiest way to think about a word- or k-tuple-based search is that one makes an index of simple words occurring in your sequence to compare to your database. For example, for proteins one uses residue 1, 2, 3 to make one word, then 2, 3, 4 to make a second word. This is a little like humans actually search for words in those "wordsearch" puzzles. If you need to find WASHINGTON in a word search puzzle, it's very hard to look precisely for that word, but you don't want to look for every site that begins W. So, you look for may be 3 letters at a time...say WAS. If you find WAS, you try to extend the "hit" by seeing if H is the next letter.

BLAST and FASTA algorithms are based on these methods. These algorithms are very fast when searching large databases, so they are the most common of the database search algorithms. Other programs like the Waterman-Smith algorithm take as much as 50 times the computer time to run, which can be limiting, especially if doing many searches. The subject is taken in details in chapter 12.

SUMMARY

Sequence alignment is identification of residue residue correspondence in two sequences. Based on presence of mismatches or gaps every alignment is given a score. The score indicates a qualitative measure of significance of the alignment. Higher the score better is the sequence alignment. Sequence alignment can be local, that matches segments of the two sequences, or global that matches the two sequences over their entire length. Sequence alignment can be between two or many sequences at a time. The latter is called as multiple sequence alignment. A table that gives a score for substitution of a nucleotide or amino acid in a DNA-to-DNA or protein-protein comparison respectively is called scoring matrix. There are several popularly used scoring matrices are PAM, BLOSUM or specialised matrices.

There are three methods of sequence alignment studies. First the Dot matrix analysis provides easy comparison. Dot matrix programs are very useful for finding repeated domains, conserved domains, exons and introns in a sequence. The second method for studying sequence alignment is dynamic programming method. Needleman and Wunsch in 1970 and Smith and Waterman 1981 gave popular algorithms for global and local alignment respectively. The last, word method or approximate method is popularly used in similarity searches. BLAST and FASTA algorithms are based on this method.

EXERCISE

1. Using the PAM scoring matrix from above find score the following DNA sequence alignment:
   GGACTCATCGTAA
   GAACCTCGTAA.
   Also compute the new, optimal score.
2. We have used DNA sequence egfr.doc in Experiment 10A using DOTLET program to generate a matrix comparison of the sequence to itself. Please explain in a simple phrase or sentence the following observations:
   (a) Why is there a perfect diagonal?
   (b) What is a biological meaning or interpretation of these sequences?
3. Why it is different to study sequence alignment in proteins than amino acids?
4. Differentiate between local and global sequence alignment.
5. Explain Smith-Waterman algorithms for dynamic programming. Discuss why variants to smith-waterman algorithms are needed?
6. Explain alignments methods working for BLAST and FASTA similarity searches.

11

MULTIPLE SEQUENCE ALIGNMENT

In chapter 10, the theory and methods of sequence alignments were explained. The purpose of this chapter is to familiarize the readers with concepts in practically aligning the sequences in the two types of alignments: one-to-one referred to as pairwise sequence alignment; one-to-many as in database, also called multiple sequence alignment. This familiarity should encompass several aspects:

1. The basis of alignment methods.
2. Commonly used, tools for alignment.
3. Interpretation of alignments.
4. Relating alignments to biological structure and function.

Alignment is a judgment of similarity. The result of such an assessment depends upon the criteria and upon the algorithm. An alignment is also hypothesis of positional homology between nucleotides or amino acids. Homology refers to the similarity that is derived from a common ancestor. Identification of homology is central to phylogenetic analysis. Biologists often assume that there is a single correct alignment, when in fact there are many possible alignments and several alternatives may have particular statistical and biological significance. The statistical significance of one alignment compared to another alignment is one important way to assess the value of the alignment. And, finally one must realize that biological significance of a particular alignment is usually the reason for which the alignment was performed.

Sequence alignment of proteins for judgment of similarity is more popular in comparison to nucleic acid sequences. In case of DNA sequence there are only 4 bases namely A, T, G and C. Assuming that in all, its letters in the sequence occur in equal proportion. The probability of one base occupying any position in random sequence is 1/4. In comparison there are 20 different amino acids available and the chance of one amino acid occupying a position in one sequence is 1/20. Because of this reason sequence alignments are more popularly done with amino acid sequences.

In the previous chapter we looked at several examples of alignment scoring and discussed it in detail. Let's study alignment of two amino acid sequences given in Fig. 11.1 using default scoring scheme and examine the relationship between two amino acid sequences. Three possible cases of gap insertion and gap extension are considered for score calculation using the scoring scheme given in Table 11.1. The gapped alignment (case 2) is not the optimal alignment, despite the fact that it has maximum matches. How do we decide that the alignment is optimal? This is discussed in the present chapter with reference to pairwise and multiple sequence alignment.

Align the Following two Sequences:

```
Sequence 1:  5'-GGGATATCC-3'          10–9–9–9–9=–26
Sequence 2:  5'-GATTC-3'
```

Case 1–After Inserting a Gap at Position 1

```
Sequence 1:  5'- GGGATATCC-3'   10–9–9+10–9–50=–57
Sequence 2:  5'- - GAT TC-3'
```

Case 2–After Inserting More Gaps

```
Sequence 1:  5'-GGG AT AT CC- 3'
                                  10(5)–50–3=-3
Sequence 2:  5'- - - GA- TTC - - 3'
```

Fig. 11.1: An example of alignment

Table 11.1: The scoring scheme

Criterion	*Score*
Identity	+10
Mismatch	-9
Gap creation	-50
Gap extension	-3
Terminal Mismatch	0

11.1 SOFTWARE FOR PAIRWISE SEQUENCE COMPARISON

Almost all the examples given in chapter 10 including dot matrix and dynamic programming methods and the example discussed above are for pairwise alignment where we consider alignment between a pair of sequences. In Fig. 11.1, the two small sequences are aligned. As seen in the Fig. 11.1 that matching of identical residue of two sequences can be improved a lot by placing a gap corresponding to insertions and deletions in correct positions. Putting gaps in the best places, in order to line up as many equivalent positions as possible, is an extremely difficult task to perform by eye. Furthermore in very distant homologs these residue insertions and deletions can be very extensive. Several softwares and web sites now offer tools for easy and effective pairwise alignment, some of which are given in Table 11.2.

Table 11.2: Web sites offering tools for pairwise alignment

LFASTA	LFASTA is a part of FAST2 package (chapter 12). LFAST shows the actual local alignments between the two sequences and their scores.
SIM4	align cDNA and genomic DNA. It is based on blast searches and is entirely different from SIM.
SIM	alignment tool at ExPASy. SIM is a program which finds a user defined number of best non-intersecting alignments between two protein sequences or with a sequence.
BLAST	align two sequences at NCBI. It is discussed in details in chapter 12.
LALIGN	finds multiple watching subsegments in two sequences. This program is developed by William Pearson. This program is a part of FASTA Package.
JALIGNER	open source Java.

11.2 MULTIPLE SEQUENCE ALIGNMENT

In multiple sequence alignment we consider alignment of more than two homologous sequences at one time. In multiple sequence alignment the sequences are aligned optimally by bringing the greatest number of similar characters into register in the same column of alignment just as described for pairwise alignment. Biological goals of multiple sequence alignments include: to generate concise and information such summary of sequence data, to illustrate similarity/dissimilarity among a group of sequences and to develop a model useful for testing evolutionary hypothesis. However, multiple sequences alignment is much more complicated and presents several difficult challenges:

1. The dynamic programming methods when used for optimal alignment of more than three sequences pose the limitation of computer time. Finding matches, mismatches, and gaps and taking into account all the variations in all the sequences at the same time poses a very difficult challenge.

2. Identifying a reasonable method of obtaining a cumulative score for substitution in the columns in multiple sequence alignment is another computational challenge.
3. The biggest problem in multiple alignments is probably that it is very difficult to use justifiable substitution matrices and gap penalties. Basically, when sequences are close related almost any scoring system works. When polypeptides are divergent, alignment results become very dependent on substitution matrix and gap penalty. Since there is not a good general theory or test of these parameters, the method has a fundamental weakness.

Present day softwares for multiple sequence alignment solve some of these problems. Most DNA or protein sequence analysis suites have multiple sequence alignment tools. The commonly used programs for multiple sequence alignment are PILUP and CLUSTAL. Probably the most widely used multiple alignment software is CLUSTAL. This program quickly grew in popularity and has evolved through a long series of jumps and steps to become the world-wide standard software for this analysis. It is available for IBM PC, Mac, Unix, and VAX/VMS computers. CLUSTALW can be used via the WWW or the program can be downloaded to your PC. A web view of clustal W is shown in Fig. 11.2.

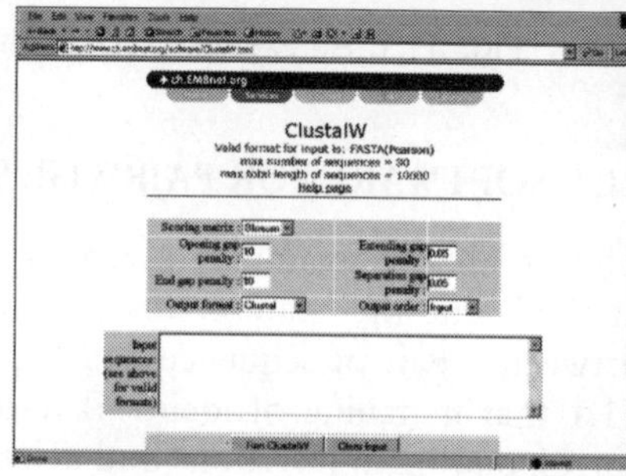

Fig. 11.2: Web view of clustalW

Table 11.3: Software for multiple alignments

Sl. No	*Name of software*	*Platform*
1.	ClustalW	Unix, Mac, PC, VMS
2.	ClustalX	Unix, Mac, PC, VMS
3.	MSA	Unix
4.	DIALIGN	Unix
5.	DCA	Unix
6.	MACAW	Mac, PC
7.	T-Coffee	Unix
8.	MAFFT	Linux, Unix, Windows XP, Mac OSX

Table 11.4: Multiple alignment editors

Sl.No.	*Software*	*Comments*
1.	Jalview	Java multiple alignment editor. Colour by amino acid physico-chemical property, similarity to consensus sequence, hydrophobicity or secondary structure. Gaps can be inserted/ deleted using the mouse. UPGMA and NJ trees calculated and drawn based on percent identity distances.
2.	CINEMA 2.1	Colour interactive Editor for Multiple Alignments. Alignments may be loaded from the PRINTS fingerprint database. And much more....
3.	JavaShade	Java multiple alignment editor.
4.	SEAVIEW	A colour graphical interface to manually edit the alignment, and also to run dot-plot or ClustalW programs to locally improve the alignment. And much more...can be used in combination with PHYLO_WIN for phylogenetic analyses.
5.	MPSA	Colour multiple alignment editor for secondary structure prediction.
6.	Se-Al	Particularly useful for manipulating protein coding DNA/RNA sequences.
7.	STRAP	Interactive extendable and scriptable editor for large protein alignments.

It is also possible to download the software on stand-alone machine and work with it. In Table 11.3 few commonly used softwares are listed. Graphical interfaces have been developed to manipulate and edit multiple alignments (Table 11.4). Generally these interfaces allow user to colour or shading residues (amino acids or bases) according to visual retrain such as physico-chemical properties, degree of conservation hydrophobicity or secondary structure. Uses of colours are very useful to interpret a multiple alignment. It gives a multiple alignment a much more comprehensive information embedded look than a rather simple textual representation.

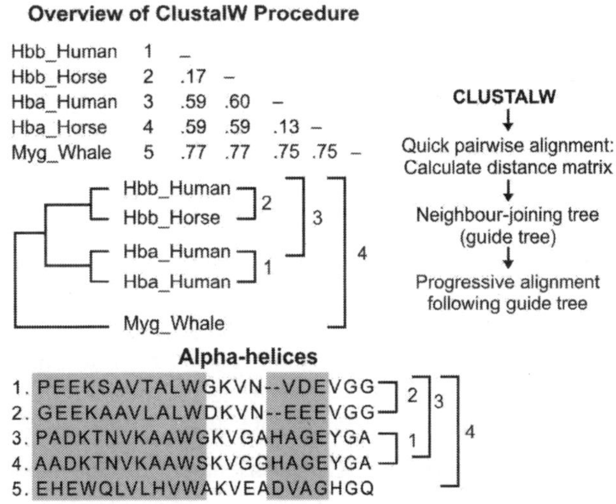

Fig. 11.3: Overview of clustalW procedure

CLUSTAL is an example of a heuristic algorithm for progressive multiple alignment. Heuristic refers to a procedure that progresses along a empirical lines by using a thumb rule to reach to a solution. The solution is not guaranteed to be optimal. Progressive alignment consists of building a multiple alignment using pairwise alignment in three steps: First, compute the alignment scores (or distances) between all pairs of sequences; Second, build a tree that reflects the similarities between sequences using the pairwise alignment distances, Thirdly, align the sequence following the guide the progressive alignment is the most commonly uses method to align biological sequence. This approach is very rapid and requires low memory space, and offers good performance on relatively well conserved homologous sequences. Block based global alignments depend on conserved blocks shared by the sequences. Motif based local multiple alignment is tried for the regions of sequences which are similar, without necessarily being globally related.

CLUSTAL program was developed by Higgins and Sharp in 1987 (Historical Profile 12). In 1996 Julie Thompson while working with Higgins and Gibson developed ClustalW version of software. Basically, one aligns each sequence pairwise to every other sequence. Then the multiple alignment is built by first aligning the most closely related pair. The next sequence is added to the alignment, and new gaps may be introduced. Overview of clustal procedure is depicted and explained in Fig. 11.3. Some refinements of the CLUSTALW/X algorithm are that nearly identical sequences are actually down weighted as they presumably donot represent the entire evolutionary spectrum. Gap penalties are also adjusted dynamically, with it easier to add to a gap than to open a new one.

Disadvantage with Clustal usage is that there is no objective function defined. Thus it is difficult is qualify an alignment whether it is good or bad. Also there is no way of knowing if the alignment is correct. Further finding local minimum with Clustal W is a very peculiar problem. If an error is introduced early in the alignment process it is impossible to correct this later in the procedure. This is explained with the help of a data set in Fig. 11.4. The clustal alignment in the given case opted not to align BAT below BAT. Despite the fact that with later alignments it is known that BAT refers to the conserved region of the sequence, still it cannot be corrected at this stage. These sorts of situations sometimes makes the alignment an arbitrary alignment. Clustal still is most widely used alignment software because of its good performance, fast speed and low memory space.

Efficacy of multiple alignment depends a lot on the per cent sequence identity. If a sequence is similar in the range of 75 to 100 per cent any program including automatic methods will work good. This fact is explained in Fig. 11.5. If sequence similarity is between 40 to 60 then the consensus

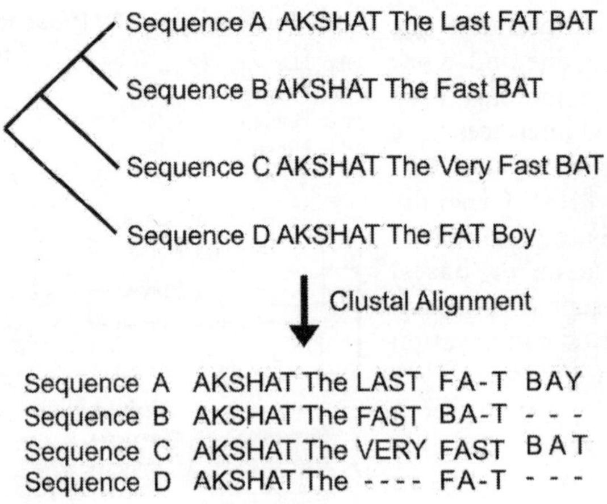

Fig. 11.4: Clustal alignment of four hypothetical sequences A, B, C and D

methods will work good. Consensus methods are those methods which are generally agreed to give good results for multiple sequence alignment. For low similarity range 20 to 40 per cent profile methods are useful in comparing and aligning the sequence. In such case finding local alignment is more relevant than a global alignment. If the similarity is lower than this range that is called is a mid night zone. It is difficult to align these sequences. Only structural information about these sequences such as threading or structure-structure alignment is possible.

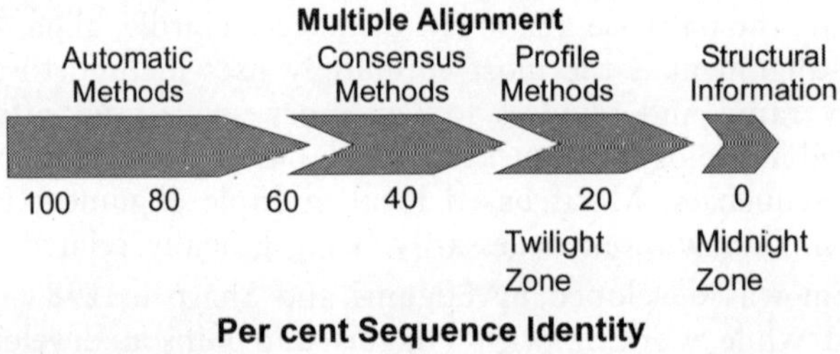

Fig. 11.5: Classification of alignment methods based on per cent sequence identity

11.3 INTERPRETATION OF A SEQUENCE ALIGNMENT

Let us use the multiple alignments of 17 sequences for cellulose-binding domain of cellobiohydrolase I (CBD-CBH1) as an example to understand interpretation of a multiple sequence alignment (Fig. 11.6). This is a small (about 30-35 residues) disulfide-bonded domain of known 3-D structure (PDB code 1CBH). It is believed that the function of the domain is either to bind with high affinity to the cellulose fiber to allow the adjacent enzymatic domain to hydrolyse the cellulose. Another possibility is that the CBD domain wedges itself in between cellulose chains, making it easier for the enzymatic domain to attack the fiber.

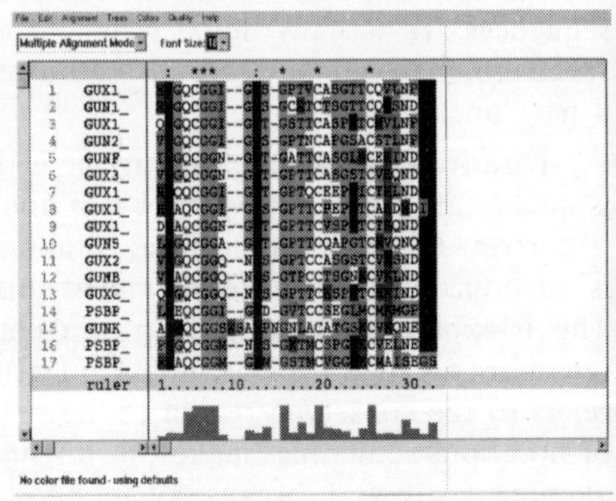

Fig. 11.6: Alignment study of 17 sequences for cellulose-finding domain of cellobiohydrolaseI

The positions that are nearly completely conserved are easy to spot. These are depicted by at help in amino acid position 2 and 12, but with more varied patterns, it

HISTORICAL PROFILE 12

Des Higgins and Dr. Julie Dawn Thompson: For Significant Contribution in Multiple Sequence Alignment Software Development

Professor Des Higgins

Dr. Julie Dawn Thompson

Dr. Des Higgins has been working in the areas of bioinformatics and molecular evolution since 1985. His main work has been on methods and software of DNA and protein sequence alignment. His first work in this area was writing a series of programs, Clustal, that could carry out reasonably accurate multiple alignments but could do this quickly and simply. In 1994, with Julie Thompson and Toby Gibson he produced ClustalW which is now the main software for carrying out multiple alignments along with ClustalX, which is essentially ClustalW with a nice graphical interface and many extra visual features. ClustalX was also made in collaboration with Francois Jeanmougin and Frederic Plewniak. More recently, he has been working on a new method, called T-Coffee which was invented by a Ph.D. student in his group at the EMBL, Cedric Notredame. T-Coffee is slower than ClustalW, but it produces more accurate alignments and allows us to mix different types of data together such as protein structures, ESTs, etc. He is now teaching biochemistry in University College, Cork, Ireland, and his group works on not only multiple alignments but also on bacterial genomics and the application of multivariate analysis microarray data.

Dr. Julie Dawn Thompson did her BSc. from University of Sheffield, England in 1994. She did her Ph.D. from Louis Pasteur University in Strasbourg, Franch. She was working with Higgins and Gibson in EMBL, Germany in 1996 when she gave ClustalW for multiple sequence alignments. Since 1995 onwards she has been working with Institut de Génétique et de Biologie Moléculaire et Cellulaire, France. In 2003 she gave a number of software for multiple sequence construction and analysis including DBclustal, LEON and RASCAL for rapid scanning and correction of multiple sequence alignments. Recently her group has developed a multiple sequence alignment ontology MAO and information managment system MACSIMS for efficient exploitation of multiple sequence alignment.

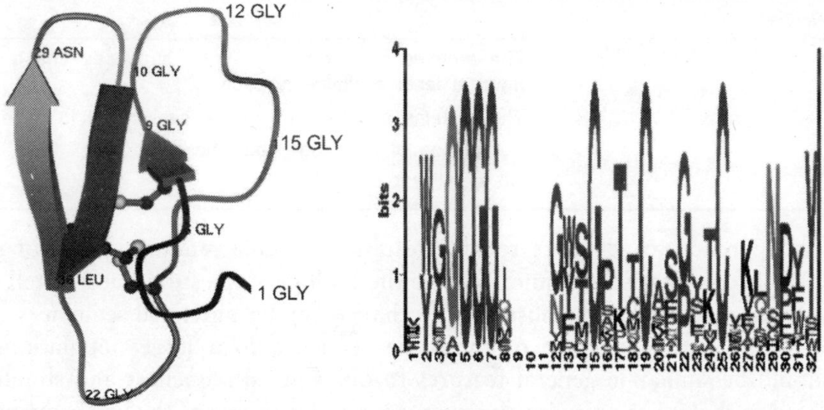

Fig. 11.7: (a) 3-D struture cellulose binding domain of cellobiohydrolase

(b) The plate shows a summary of alignment as a sequenced logo, in which letters of different sizes indicate different proportions of amino acids becomes more difficult to predict alignment. For example, in Fig. 11.6 at position 24 there is a threonine in nine sequences, and lysine in five. There are several other observations one might make about multiple alignments. For example, the fact that residues are aligned in a column does not necessarily mean that they are actually aligned structurally or in any other way. The correlation between residues far apart in the sequence in a protein family is usually difficult to spot in a multiple alignment. Other methods must be used to visualize this. The combination of a known 3-D structure and a multiple alignment can be very powerful for understanding a protein domain. Of course, knowledge of the biology and chemistry of the proteins increases the understanding. On closer examination of this multiple alignment (together with the known 3-D structure), there are a number of features in the multiple sequence alignment CBD_1 that stand out. For example, one of the sequences (GUNK_FUSOX) contains an insertion. The location of this insertion corresponds to a turn in the 3-D (Fig. 11.7a) structure. There are three strongly conserved Gly residues: numbers 6, 7, 15: There could be several reasons for this; the internal angles may have values forbidden or highly unfavourable for other residues, or the positions in the structure may not allow for any sidechain without destroying the function. Inspection of the conserved residues shows that several of them (2 Trp/Phe, 4 Gln, 31 Tyr/Phe/Trp, 32 Tyr) are located on the same side of the 3-D structure. This example highlights applications the role of multiple sequences alignment in understanding the protein structure and function better. Similarly these sequences when used to draw a phylogenetic tree solves several problems evolution and ancestry.

It is an important question to determine the significance of alignment. Many alignments give optimal alignment but there is no structural similarity. How do we determine whether the similarity is significant or not or it could have arisen by chance, there are many rules of thumb for expression of similarity in terms of optimal alignment.

Below are some common characteristics of good alignment.

1. Use alignment methods/softwares that are known to produce quality alignments.
2. Globally aligned sequences should be the same length or nearly so.
3. Scoring matrices should be appropriate. Defaults are usually best if you don't know what else to use.

Table 11.7: Interpretation of different levels of identical residues in sequence alignment

% Identical Residues	*Inference*
> 45%	The proteins will have very similar structures and are very likely to have at least a similar function
> 25%	The proteins are likely to have a similar folding pattern
18-25%	Suggestion of homology could be risky
< 18%	No significant similarity

A practical approach to get more reliable results is to take random strings of nucleotides or amino acids as controls. Statistics cannot answer the biological question but can tell the likelihood that a similarity as good as the one observed by chance among unrelated sequences. To do this we compare our results with alignment of the same sequence to a large population. The control population should be similar in general features to our aligned sequences and should contain few sequences related to them. For pairwise alignments we can pick one of the two sequence make many scrambled copies of it using a random number generator and align each permitted copy to second sequence.

EXPERIMENT 11A

Perform a Meaningful Multiple Sequence Alignment of given ten DNA sequences for the Enzyme RUBPcase from ten Different Organism using ClustalX Software

About ClustalX: ClustalX is a new windows interface for the ClustalW multiple sequence alignment program. It provides an integrated environment for performing multiple sequence and profile alignments and analysing the results. The sequence alignment is displayed in a window on the screen. A versatile colouring scheme has been incorporated allowing you to highlight conserved features in the alignment. The pull-down menus at the top of the window allow you to select all the options required for traditional multiple sequence and profile alignment. You can cut-and-paste sequences to change the order of the alignment; you can select a subset of sequences to be aligned; you can select a sub-range of the alignment to be realigned and inserted back into the original alignment. Alignment quality analysis can be performed and low-scoring segments or exceptional residues can be highlighted.

ClustalX is available for a number of different platforms including: SUN Solaris, IRIX 5.3 on Silicon Graphics, Digital UNIX on DECStations, Microsoft Windows (32 bit) for PC's, Linux ELF for x86 PC's and Macintosh PowerMac. The alignment is displayed on the screen with the sequence names on the left-hand side. The sequence alignment is for display only; it cannot be edited here (except for changing the sequence order by cutting-and-pasting on the sequence names).

The Hypothesis: It would be possible to align the given ten sequences using the software clustalX. For sake of convenience the nucleotide sequence is chosen her, instead of amino acid sequence.

TESTING THE HYPOTHESIS

Computer Set up: ClustalX 1.81 installed at Windows interface.

Starting material: Nucleotide sequence of gene encoding for the larger sub-unit of enzyme Rubpcase may be collected from web site of EMBL and contained in a notepad called as sequence Experiment 11A 1.doc. For the sake of convenience only 120 nucleotides were selected from each of the ten sources.

1. Start to clustalX 1.81 and load sequence of file Experiment 11A.
2. In the alignment menu select complete alignment and save the output alignment file as Experiment 11A (Fig. 14.10).
3. Study the various parameters of alignment.

DATA

Data Box 11.1: Structure of input file

```
>Halimeda_opuntia_#10
cggagctggatttaaagctggtgttaaagattatcgtttaacgtattatacaccagactatcaagtattagatacagacattcttgcagcctttcgaatgacacctcaaccaggtgtgcc
>Dichotomosiphon_tuberosus_#9
aggagcaggatttaaagcaggtgtaaaagattatcgtttaacctattacactccagattatcaagtaaaagatactgatgttctagctgctttcagaatgactcctcaacctggtgtacc
>Halimeda_discoidea_#8
tggagctggatttaaagctggtgttaaagattatcgtttaacatactatacaccagaataccaagtcttagatactgacatccttgcagcctttcgaatgacacctcaaccaggtgttcc
>Caulerpa_racemosa_#7
tggtgctggatttaaagctggtgtaaaagattatcgtttaacatattatactcctgattataaagttttagaaacaggcattcttgcagcatttcgtatgacacctcaacctggtgtacc
>Caulerpa_racemosa_#6
aggtgctggatttaaagctggtgtaaaagattatcgtttaacatattatactcctgattataaagttttagaaacagacattcttgcagcatttcgtatgacacctcaacctggtgtacc
>Caulerpa_okamurae_#5
tggtgctggatttaaagccggtgtaaaagattatcgtttaacatattatactcctgaatataaagttttagagacagatattcttgcagcatttcgtatgacacctcaacccggcgttcc
>Caulerpa_brachypus_#4
aggtgctggatttaaagctggtgtaaaagattatcgtttaacatattatactcctgattataaagttttagaaacagatatcttgcagcctttcgtatgacacctcaacctggtgtacc
>Pedobesia_ryukyuensis_#3
tgccggctttaaagcaggggtaaaagattatcgtttaacttattatactcctgattatcaagttaaagaaactgatatttagctgcttttcgtatgacacctcaaccaggtgttccacc
>Codium_lucasii_#2
tgcaggatttaaagcaggagttaaagattatcgactaacttattatacaccagattatcaagttaaagaaactgatattcttgccgcttttcgtatgactccacaacctggtgttcctcc
>Bryopsis_plumosa_#1
tgcaggatttaaagcgggggtaaaagactatcgtctaacttattacacacctgattaccaagtaaaagatacagatattcttgccgcctttcgtatgactccacaacctggtgttccacc
```

Data Box 11.2: Classification of the species studied

Eukaryota
↓
Viridiplantae
↓
Chlorophyta
↓
Ulvophyceae
↓
Caulerpales

Bryopsidaceae Bryopsidaceae Caulerpaceae Udoteaceae
Dichotomosiphonaceae

#1 *Bryopsis plumose* Bryopsidaceae
#2 *Codium lucasii* Bryopsidaceae
#3 *Pedobesia* Bryopsidaceae
#4 *Caulerpa* brachypus Caulerpaceae
#5 Caulerpa *okamurae* Caulerpaceae
#6 Caulerpa racemosa Caulerpaceae
#7 *Halimeda discoidea* Udoteaceae
#8 *Dichotomosiphon tuberous* Dichotomosiphonaceae
#9 *Halimeda* opuntia Udoteaceae

Data Box 11.3: Before complete alignment

```
Halimeda_opuntia_#10            CGGAGCTGGATTTAAAGCTGGTGTTAAAGATTATCGTTTAACGTATTATA 50
Dichotomosiphon_tuberosus_#9    AGGAGCAGGATTTAAAGCAGGTGTAAAAGATTATCGTTTAACCTATTACA 50
Halimeda_discoidea_#8           TGGAGCTGGATTTAAAGCTGGTGTTAAAGATTATCGTTTAACATACTATA 50
Caulerpa_racemosa_#7            TGGTGCTGGATTTAAAGCTGGTGTAAAAGATTATCGTTTAACATATTATA 50
Caulerpa_racemosa_#6            AGGTGCTGGATTTAAAGCTGGTGTAAAAGATTATCGTTTAACATATTATA 50
Caulerpa_okamurae_#5            TGGTGCTGGATTTAAAGCCGGTGTAAAAGATTATCGTTTAACATATTATA 50
Caulerpa_brachypus_#4           AGGTGCTGGATTTAAAGCTGGTGTAAAAGATTATCGTTTAACATATTATA 50
Pedobesia_ryukyuensis_#3        TGCCGGCTTTAAAGCAGGGGTAAAAGATTATCGTTTAACTTATTATACTC 50
Codium_lucasii_#2               TGCAGGATTTAAAGCAGGAGTTAAAGATTATCGACTAACTTATTATACAC 50
Bryopsis_plumosa_#1             TGCAGGATTTAAAGCGGGGGTAAAAGACTATCGTCTAACTTATTACACAC 50
                                 *  *           *  *      *  **             **

Halimeda_opuntia_#10            CACCAGACTATCAAGTATTAGATACAGACATTCTTGCAGCCTTTCGAATG 100
Dichotomosiphon_tuberosus_#9    CTCCAGATTATCAAGTAAAAGATACTGATGTTCTAGCTGCTTTCAGAATG 100
Halimeda_discoidea_#8           CACCAGAATACCAAGTCTTAGATACTGACATCCTTGCAGCCTTTCGAATG 100
Caulerpa_racemosa_#7            CTCCTGATTATAAAGTTTTAGAAACAGGCATTCTTGCAGCATTTCGTATG 100
Caulerpa_racemosa_#6            CTCCTGATTATAAAGTTTTAGAAACAGACATTCTTGCAGCATTTCGTATG 100
Caulerpa_okamurae_#5            CTCCTGAATATAAAGTTTTAGAGACAGATATTCTTGCAGCATTTCGTATG 100
Caulerpa_brachypus_#4           CTCCTGATTATAAAGTTTTAGAAACAGATATCTTGCAGCCTTTCGTATGA 100
Pedobesia_ryukyuensis_#3        CTGATTATCAAGTTAAAGAAACTGATATTTTAGCTGCTTTTCGTATGACA 100
Codium_lucasii_#2               CAGATTATCAAGTTAAAGAAACTGATATTCTTGCCGCTTTTCGTATGACT 100
Bryopsis_plumosa_#1             CTGATTACCAAGTAAAAGATACAGATATTCTTGCCGCCTTTCGTATGACT 100
                                *     *  *                    *

Halimeda_opuntia_#10            ACACCTCAACCAGGTGTGCC 120
Dichotomosiphon_tuberosus_#9    ACTCCTCAACCTGGTGTACC 120
Halimeda_discoidea_#8           ACACCTCAACCAGGTGTTCC 120
Caulerpa_racemosa_#7            ACACCTCAACCTGGTGTACC 120
Caulerpa_racemosa_#6            ACACCTCAACCTGGTGTACC 120
Caulerpa_okamurae_#5            ACACCTCAACCCGGCGTTCC 120
Caulerpa_brachypus_#4           CACCTCAACCTGGTGTACC- 119
Pedobesia_ryukyuensis_#3        CCTCAACCAGGTGTTCCACC 120
Codium_lucasii_#2               CCACAACCTGGTGTTCCTCC 120
Bryopsis_plumosa_#1             CCACAACCTGGTGTTCCACC 120
                                   *        *     *
```

Data Box 11.4: After Complete Alignment

```
limeda_discoidea_#8           ACACCTCAACCAGGTGTTCC 120
ulerpa_racemosa_#7            ACACCTCAACCTGGTGTACC 120
ulerpa_racemosa_#6            ACACCTCAACCTGGTGTACC 120
ulerpa_brachypus_#4           ACACCTCAACCTGGTGTACC 119
ulerpa_okamurae_#5            ACACCTCAACCCGGCGTTCC 120
dium_lucasii_#2               ACTCCACAACCTGGTGTTCC 117
yopsis_plumosa_#1             ACTCCACAACCTGGTGTTCC 117
dobesia_ryukyuensis_#3        ACACCTCAACCAGGTGTTCC 117
chotomosiphon_tuberosus_#9    ACTCCTCAACCTGGTGTACC 120
                              ** ** ***** ** ** **

USTAL X (1.81) multiple sequence alignment

limeda_opuntia_#10            CGGAGCTGGATTTAAAGCTGGTGTTAAAGATTATCGTTTAACGTATTATA 50
limeda_discoidea_#8           TGGAGCTGGATTTAAAGCTGGTGTTAAAGATTATCGTTTAACATACTATA 50
ulerpa_racemosa_#7            TGGTGCTGGATTTAAAGCTGGTGTAAAAGATTATCGTTTAACATATTATA 50
ulerpa_racemosa_#6            AGGTGCTGGATTTAAAGCTGGTGTAAAAGATTATCGTTTAACATATTATA 50
ulerpa_brachypus_#4           AGGTGCTGGATTTAAAGCTGGTGTAAAAGATTATCGTTTAACATATTATA 50
ulerpa_okamurae_#5            TGGTGCTGGATTTAAAGCCGGTGTAAAAGATTATCGTTTAACATATTATA 50
dium_lucasii_#2               ---TGCAGGATTTAAAGCAGGAGTTAAAGATTATCGACTAACTTATTATA 47
yopsis_plumosa_#1             ---TGCAGGATTTAAAGCGGGGGTAAAAGACTATCGTCTAACTTATTACA 47
dobesia_ryukyuensis_#3        ---TGCCGGCTTTAAAGCAGGGGTAAAAGATTATCGTTTAACTTATTATA 47
chotomosiphon_tuberosus_#9    AGGAGCAGGATTTAAAGCAGGTGTAAAAGATTATCGTTTAACCTATTACA 50
                                  ** ** ******** ** ** ***** *****  **** ** ** *

limeda_opuntia_#10            CACCAGACTATCAAGTATTAGATACAGACATTCTTGCAGCCTTTCGAATG 100
limeda_discoidea_#8           CACCAGAATACCAAGTCTTAGATACTGACATCCTTGCAGCCTTTCGAATG 100
ilerpa_racemosa_#7            CTCCTGATTATAAAGTTTTAGAAACAGGCATTCTTGCAGCATTTCGTATG 100
ilerpa_racemosa_#6            CTCCTGATTATAAAGTTTTAGAAACAGACATTCTTGCAGCATTTCGTATG 100
ilerpa_brachypus_#4           CTCCTGATTATAAAGTTTTAGAAACAGATAT-CTTGCAGCCTTTCGTATG 99
ilerpa_okamurae_#5            CTCCTGAATATAAAGTTTTAGAGACAGATATTCTTGCAGCATTTCGTATG 100
dium_lucasii_#2               CACCAGATTATCAAGTTAAAGAAACTGATATTCTTGCCGCTTTTCGTATG 97
yopsis_plumosa_#1             CACCTGATTACCAAGTAAAAGATACAGATATTCTTGCCGCCTTTCGTATG 97
dobesia_ryukyuensis_#3        CTCCTGATTATCAAGTTAAAGAAACTGATATTTTAGCTGCTTTTCGTATG 97
chotomosiphon_tuberosus_#9    CTCCAGATTATCAAGTAAAAGATACTGATGTTCTAGCTGCTTTCAGAATG 100
                              * ** ** **  ****   *** ** *   *  * ** ** **  * ***
```

INTERPRETATION OF DATA

Output of clustal before and after complete alignment is given in Databoxes 11.3 and 11.4. A ruler is displayed below the sequences, starting at 1 for the first residue position (residue numbers in the sequence input file are ignored). The line above the ruler is used to mark strongly conserved positions. Three characters ('*', ':' and '.') are used: '*' indicates positions which have a single, fully conserved residue ':' indicates that one of the following 'strong' groups is fully conserved:–

STA
NEQK
NHQK
NDEQ
QHRK
MILV
MILF
HY
FYW

'.' indicates that one of the following 'weaker' groups is fully conserved:

CSA
ATV
SAG
STNK
STPA
SGND

SNDEQK
NDEQHK
NEQHRK
FVLIM
HFY

These are all the positively scoring groups that occur in the Pam250 matrix. The strong and weak groups are defined as strong score >0.5 and weak score = < 0.5 respectively. As seen in Data Boxes 11.3 and 11.4 gaps are introduced after complete alignment. This leads to an increase in number of '*', thus more number of conserved residues are detected. The phylogenetic tree drawn on the basis of this multiple sequence alignment is shown in chapter 14.

An alignment with display by default the following symbols denoting the degree of conservation observed in each column:

"*" means that the residues or nucleotides in that column are identical in all sequences in the alignment.

":" means that conserved substitutions have been observed, according to the colour table.

"." means that semi-conserved substitutions are observed.

SUMMARY

Pairwise and multiple sequence alignment refer to alignment of one-to-one and one-to-many sequence alignments respectively. Alignment is a judgment of similarity. Study of pairwise alignment between two sequence yields information about similarity, presence of indels and evolutionary relationship in the sequences. Important web sites for pair wise alignment are LFASTA, SIM4, WISE, SIM, BLAST. Multiple sequence alignment is much more complicated and presents several difficult challenge. PILUP and CLUSTAL are most widely used programs for multiple sequence alignment. CLUSTAL is an example of heuristic algorithm for progressive multiple alignment. There are 4 approaches of obtaining a multiple sequence alignment: optimal alignment; progressive pair wise alignment, block based optimal alignments and motif-based local multiple alignment. There are many rules of thumb for expression of similarity in terms of alignment.

EXERCISE

1. Use the following set of accession numbers to obtain some protein sequences of glutamate synthetase to use.

   These were chosen because they have similar enzymatic function and are therefore conserved, while being drawn from five kingdoms. These numbers can be used to obtain the protein sequences from NCBI. FASTA format is recommended for easy transfer into your log and uploading

A38596	Maize	Plant
CAC05496	*Arabidopsis*	Plant
AAC08261	*Porphyra*	Algae
CAA76602	*Plasmodium*	Protozoa
AAF49409	*Drosophila*	Animal
CAB92626	*Neurospora*	Fungi
CAA61505	*Saccharomyces*	Fungi
BAB05447	*Bacillus*	Bacteria
AAA58014	*E. coli*	Bacteria

AAK94787	*Klebsiella*	Bacteria
CAB64595	*Nostoc*	Bacteria
AAG44102	*Staphylococcus*	Bacteria

Do the multiple sequence alignment using clustalX.

2. Use a server-based version at EBI for ClustalW for sequences given in exercise 1. Make note of conserved regions.
3. Compare some select pairwise alignments to see how they compare to your Multiple Sequence alignments. Use either BLAST2 or ALIGN to compare pairs of sequences. Try both pairs which appear to be closely related and pairs which are distantly related. Do your alignments agree?
4. Extract information about the structure of RUBPcase subunit given in Experiment 11A. Link the multiple sequence alignment with structure of the protein.

12

SIMILARITY SEARCHES OF BIOLOGICAL DATABASES

Similarity searches of biological databases is mainstay of bioinformatics. The sequence of gene of interest is compared to every sequence in the sequence database and similar ones are identified if a query sequence can be readily aligned to a database sequence of known function, structure or biochemical activity, the query sequence is predicted to have same function, structure or biochemical activity. Finding similar sequences from different species where complete genome is already sequenced, is very useful is predicting evolutionary relationship. Protein sequence is also searched to find to which family of protein it belongs. Similarly genomic DNA sequences may also be searched to locate to which category these sequences belongs viz. transcription factor binding site, promoter recognition signals identification.

The situation is a little more common in molecular biology: proteins and DNA can be similar with respect to their function, structure, or primary sequence. The general rule is that sequence determines shape, and shape determines function. So when we study sequence similarity, we eventually hope to discover or validate similarity in shape and function. This approach is often successful. However, there are many examples where two sequences have little or no similarity, but still the molecules fold into the same shape and share the same function.

Let us consider the analogy displayed in Fig. 12.1, in which a sequence collected from blood sample available at the site of a murder. From the circumstantial evidences of the crime, it is clear that any one of the four persons A, B, C and D are involved in the murder. Our job is to match this sequence collected from the crime site to the sequence collected from four suspects. The sequence in the blood sample at site was ETESEDEDEED and its needs be matched to one of the four candidate sequences:

```
1. WEATHER IS BEAUTIFUL, WISH YOU WERE HERE
2. WEATHER IS HERE, WISH YOU WERE BEAUTIFUL
3. WISH THERE WAS BEAUTIFUL WEATHER HERE
4. ENTERING SAFEMODE NEED HELP
```

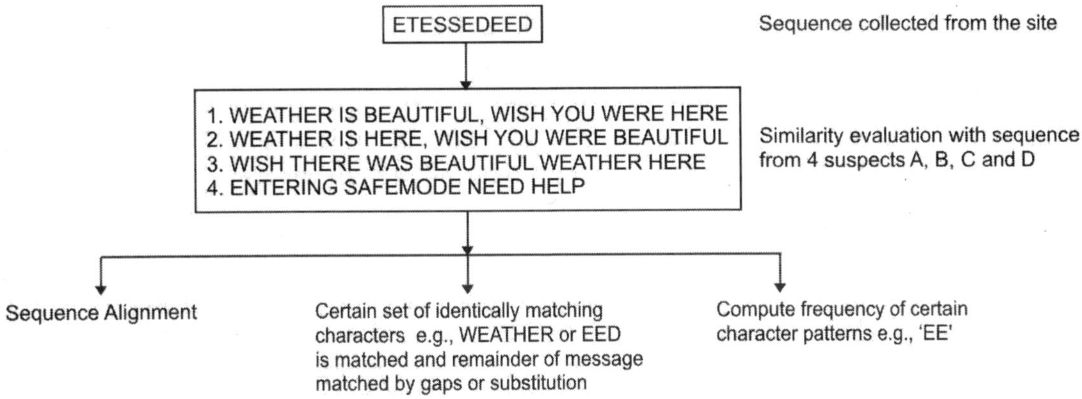

Fig. 12.1: Process of similarity evaluation in the four sequences

There can be several ways to evaluate the similarity between the sequence from sample at crime spot and the sequences of the prime suspects:

1. Consider all possible alignments of the crime spot sequence and suspects sequences. By using a system that rewards matches and penalizes mismatches and gaps, we can assign a score to each alignment. Using the highest of those alignment scores, we find the most likely person with related sequence. This method has a very good chance of finding the criminal sequence but is computationally expensive because all possible alignments need to be evaluated.
2. Another way to evaluate is to assume that the sequence A, B, C and D must have some set of characters that identically match a set of characters in the query sequence. The remainder of the sequence should match by applying substitutions and gaps as needed. This method is much faster and computationally cheaper than the first but less sensitive.

   ```
   E N T E R I N G S A F E M O D E N E E D H E L P
   E - T E - - - - S - - E - - D E dE E D - - - -
   ```

   In our example notice that query sequence ends in "EED". There is one place in fourth message that EED matches exactly. There are no double or triple character strings in the query sequence that exactly match any position in the first three candidate messages. Ignoring spaces, if we align "EED" in the query sequence to the sequences we get an alignment that is easily explained by applying the error set: where the lower case "*d*" in the received message indicates a substitution error and the "-" indicates a character loss.
3. Third method for evaluating similarity is to compute the frequency of certain character patterns in each suspect sequence and then compare those frequencies to the patterns in the query sequence. For example, the pattern "EE" occurs once in the sequenced but not at all in the first three. "EE" also occurs once in the query sequence. This method is computationally inexpensive, but lacks sensitivity to the context of the characters.

This simple example discussed above helps to highlight following features about similarity searches:

1. There are often multiple ways to perform a similarity evaluation. Each method has advantages and disadvantages in terms of accuracy and computational cost.
2. It is not possible to know with 100% certainty whether any single method will find the right match. In some cases, a combination of methods will give us higher confidence in an answer than any individual method.

3. Similarity comparisons provide evidence of sequence closeness only. Ultimately, investigators need to review the results and make final decisions based on their domain knowledge.
4. Similarity comparison would indicate the changes during evolutionary process.

Biological sequences undergo transformations that alter their structure and meaning. Transformations may be the result of evolutionary processes such as species-specific variations of a protein or may be the result of errors introduced in preparing or reading the composition of sequences. Most biological similarity searches assume a simple error model i.e. independent processes may result in the insertion, deletion, or substitution of characters in the text string that represents the residues of a peptide or nucleotide molecule. Evolutionary mutation processes are strongly regulated by natural selection since unfavourable alterations eventually disappear.

Many biological questions require the comparison of one or more sequences to each other. The nature of these comparisons depends on the question being asked, the time allowed to answer the question, the manner in which the answers will be used in subsequent analyses, the required accuracy of the answer, and so on. Fundamentally, the purpose of all similarity searches is to measure the edit distance between sequences. However, the meaning of edit distance changes depending on the purpose of investigation. For example, a question in which protein hydrophobicity is the basis for comparison, will use different matrices and a different algorithm than one, in which the presence or absence of a specific binding domain is in question.

Genome projects of several organisms are already complete and all genomic sequences of these organisms is already available (Section 3). In addition to genomic DNA sequences, complete cDNA copies of mRNAs that carry all the sequence information for the protein products have also been obtained for some of the expressed genes of various organisms. Similarity searches through the databases from several organisms are used to locate similar sequences that are predicted to have similar biological function through close evolutionary relationship. Sequence similarity search can also be used to find function of genes whose sequence has been determined in the laboratory. The sequence of the gene of interest is compared to every sequence in a sequence database and similar ones are identified.

12.1 SIMILARITY AND HOMOLOGY

It is important not to confuse the terms homology and similarity. Homology implies a common ancestry. In many cases similarity is due to homology, but it is not the case always. Short genetic sequences may be similar to each other but the two genes are not evolutionary related. For example, many non homologus bacterial genes contain similar promoter sequences at -35 and -10 regions. Homologous sequences can be compared by multiple sequences alignment. However unlike similarity, homology can never be expressed in terms of percentage.

Corresponding to the modes of biology sequence variation, there are two types of sequence comparison in terms of alphabets assessment: homology evaluation and contextual analysis. Although these evaluations have much in common, they are fundamentally asking different questions. In contextual analyses we look for the common features among sequences without concern for whether the sequences have a common ancestor. These comparisons determine whether sequences overlap or are contained within another sequences. Contextual analyses are commonly part of the processing needed to join together many, small sequences into fewer, long, sequences. Contextual comparison is used to find vector contamination (chapter 13) in sequences, evaluate primer candidates, do biochip design (chapter 17), and so forth. The example discussed in Fig. 12.1 is an example of contextual analysis.

Homology evaluation looks for evidence that biological sequences are evolutionary related but have been altered relative to each other by evolution. Homologue sequences can be classified as orthologs or paralogs. Orthologs are related molecules that have been changed due to speciation while paralogs are replicated molecules in the same organism that have been altered through generations of independent mutation. Orthologs sequences in different organisms generally have same function. While paralogs are similar sequences in same organism having same or different function paralog sequences are due to gene duplications.

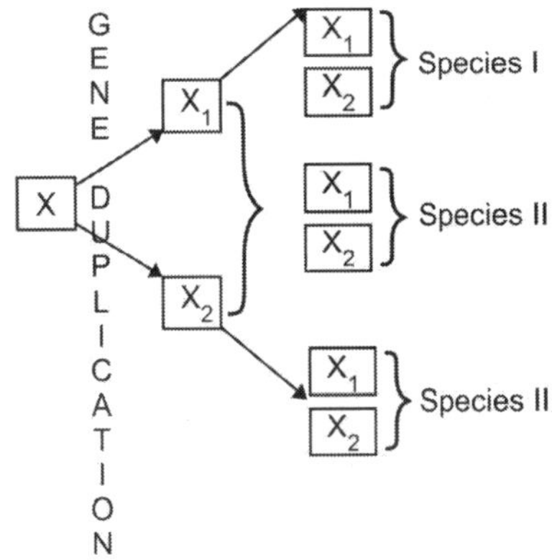

Fig. 12.2: Concept of orthologous and paralogous sequences

Concept of orthologs and paralogs is explained in Fig. 12.2 in any hypothetical species the duplication of a given gene X produces 2 copies called X_1 and X_2 here. The process of speciation leads to occurrence of gene X_1 and X_2 and different species I, II and III. Sequence X_1 and X_2 are orthologous to each other in species I, II and III. Thus even if there are evidences for gene duplication similar sequence in different species performing same function is called orthologous sequences.

Genes X_1 and X_2 now will evolve in different direction in different direction in different species. This might also result in difference in function of these genes. Gene sequences X_1 and X_2 are example of paralogous sequences in species I, II and III. These give size to two gene families ultimately. In another example of orthologous proteins sequences discussed in details below Fig. 12.3. The seven similar sequences from seven different organisms refer to 7TM (bovine seven transmembrane) protein. The seven species sequences have diverged from a common ancestor.

Evidence of homology is sought by using a similarity objective function and set of rewards and penalties that model evolution. A comparison algorithm scores sequence similarity using the evolutionary model from randomly related sequences. Evolutionary models used in sequence analyses are fairly simple and based on collections of sequences that are related to each other in function and years of separation. A detail of evolutionary models is beyond the scope of this book. Let us consider creating an evolutionary model for a given transmembrane protein that could be used to evaluate newly found proteins. A common feature in these protein is at least one transmembrane region comprising a α-helix stretch of between 21 and 26 hydrophobic amino acid residues. Human, mouse, rat, chicken, and bovine seven transmembrane (7TM) proteins have been found and sequenced. These orthologous sequences are separated in time species diverged from a common ancestor.

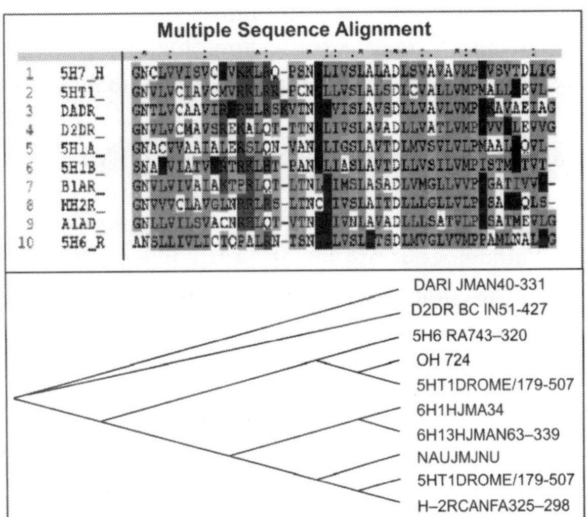

Fig. 12.3: (a) Multiple sequence alignment of seven transmembrane proteins
(b) Evotiutonary tree of these sequences based on multiple sequence alignment

To create a suitable evolutionary model for 7TM proteins, it first requires to develop a multiple sequence alignment such that it is possible to determine where substitutions and gaps have occurred. A portion of one such alignment is shown in Fig. 12.3a. It is to be noticed that the first residue is almost always a G but can be either an S or an A. The second character is always an N, and the next character can be A, C, V, or L. The '-' in the alignment indicates positions in which gaps are needed to properly align the sequences. Since mutations occur at more or less a fixed rate, then any protein compared to a 7TM protein should have roughly the same per cent identity as the sequences in Fig. 12.3 have to each other. Interestingly, per cent identity might be higher in orthologous sequences than in paralogous sequences. For example, human and mouse diverged around 35 million years ago meaning that orthologous human and mouse proteins may differ only by the number of successful mutations that could occur in that time period. Human and mouse orthologs therefore have approximately 90% identity to each other.

12.2 ASSESSING SIGNIFICANCE OF SEQUENCE SIMILARITY

General rules to be considered before assessing sequence similarity are given below:

1. Translate DNA sequences that encode proteins into proteins sequences before performing a database search. DNA sequences comprise only four nucleotides whereas protein sequences comprise 20 amino acids. Due to five fold larger variety of sequence characters in proteins it is much easier to detect patterns of sequence similarity between protein sequences than between DNA sequences.
2. Protein sequence comparisons typically double the evolutionary look-back time over DNA sequence comparisons.
3. If homology has been inferred due to significant similarity scores between two proteins, A and B, that align over their entire lengths and between protein B and a third protein, C, then proteins A and C must also be homologous, even if they share no significant similarity.
4. Matches that are more than 50% identical in a 20-40 amino acid region occur frequently by chance.
5. The requirement for a common folded structure in homologous proteins usually causes these proteins to be similar over the entire length of the gene product (or domain). Therefore, most sequences that share statistically significant similarity throughout their entire lengths are homologous are might have common folded structure.
6. Low complexity regions e.g. transmembrane regions and coiled-coil regions frequently display significant similarity in the absence of homology. Transmembrane and coiled-coil regions should be identified and masked (by eliminating these regions from the query) by the user.

Statistical analysis compares the score by aligning a protein sequence with a putative relative to distribution of scores obtained by aligning that protein with unrelated sequences. There are several parameters that can be derived from the alignment. One parameter is per cent identity already discussed in earlier chapters, which is the number of residues that identically match in an alignment divided by total number of residues in the alignment. In Table 11.7 interpretation of different levels of identical residues in sequence alignment is given. The proteins and DNA sequences that align should certainly show more than 25 per cent identity. Also a practical approach to get more reliable result is to take a random string of nucleotides or amino acid as controls. The larger the database is more relatable is the statistical analysis. Several scores that are used studied in similarity and homology analysis are given in Table 12.1.

Table 12.1: Range and significance of Z-score

Z-score value	*Significance*
15	considered to be very significant and the sequences can be confidently described as homologues.
5-15	are probably homologous though more distant relatives and there may be a significant variation in structures although the structural core would probably be conserved between them
<5	there is less certainty in the relationship of two sequences.

12.2.1 Z-score Calculation

The Z-score is the simplest way for statistically assessing the significance of the similarity between two sequences. The Z-score is a measure of how unusual our original match is, in terms of the mean and standard deviation of the population score.

$$\text{Z-Score} = \frac{\text{Score–mean}}{\text{Standard deviation}}$$

The mean score and deviation are determined from distribution of scores of unrelated sequences. Range of Z-scores and their significance is given in Table 12.1.

12.2.2 P-value

Many programs report the probability or the p-value. A rough guide to interpreting p-value is given in Table 12.2.

Table 12.2: Range and significance of P-value

P-value	*Significance*
10–100 to 10–50	sequences are very identical for example alleles.
10–5 to 10–10	closely related sequences, can be homologous.
10–5 to 10–1	usually distant relatives.
P > 10–1	unsignificant.

12.2.3 E-value

The E-value is called as expectation value. It is a very popular method of reporting similarity in several similarity search programs including PSI-BLAST. The E-value of an alignment score between sequences gives the statistical chance that an unrelated sequence in a database or a random sequence could have achieved such a score with a query sequence given as many sequences as there are in the database. Smaller the E-value, more significant is the alignment. A cut off value in the range of 0.01–0.05 is used is most similarity searches database. These are obtained by simply multiplying the P-value (probability) by the size of the database. Lower the E-value, greater is the probability that the query sequences are matched. Significance of different E-values ranges is given in Table 12.3.

Table 12.3: E-values range and significance

E-value	*Significance*
≤ 0.02	typically obtained for very close relatives.
< 0.02 to 1	suggests an evolutionary relationship
1 >	suggests that no relatives could be detected.

12.3 SIMILARITY SEARCH PROGRAMS

Similarity searches programs have traditionally been classified on the basis of the type of inputs used, the manner in which rewards and penalties are accessed, algorithm rigour, complexity, and so forth. The present day similarity search tools are classified in three broad types:

1. Fast searching methods e.g. BLAST and FASTA
2. Dynamic programming searching methods
3. Profile based methods

In last few years biotechnology companies such as Compugen and Paracel have developed special purpose hardware that accelerates the dynamic programming algorithms. This special purpose hardware has made dynamic programming algorithms competitive with FASTA and BLAST both in speed and in simplicity of use. Fig. 12.4 is a conceptual representation of the similarity search method based on the expected per cent identity of the sequences being compared and the type of search being performed. The y-axis is a search type starting with contextual analysis at the bottom and moving upward to closely related homology searching and distantly related homology searching. Equivalently, the x-axis is a starting with highly similar sequences on the left and more dissimilar sequences on the right. Algorithms shown toward the bottom, left in Fig. 12.4 tend to be simpler and faster than algorithms shown toward the upper right. The BLAST and Smith Waterman variants overlap each other and span a wide range of applicability. Profile variants tend to be best used in low identity, homology applications while regular expression searches are best used in high identity, contextual evaluations.

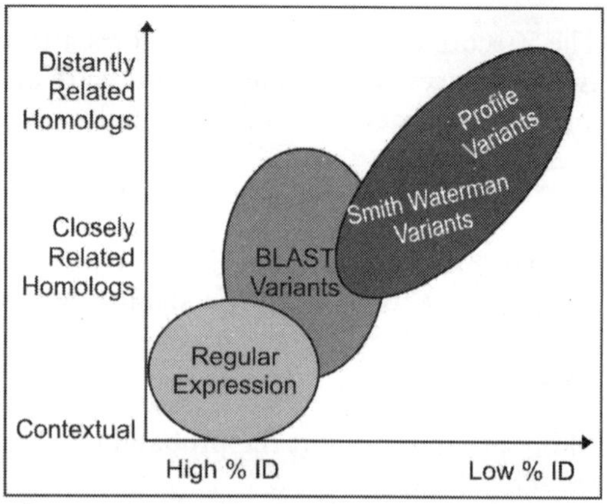

Fig. 12.4: Search domain classification for various algorithms

However FASTA and BLAST have become standard in the field of similarity searches and are being extensively used by biologists over entire world. It is almost established that BLAST works better for proteins and FASTA for DNA alignments.

12.3.1 FAST Searching Methods

FAST searching methods are also called as WORD or KTUPLE or approximate methods. They are already briefly introduced in chapter 10. Here an index of simple words occurring in given querry sequence is made and these words are used for comparison to other sequences. Both BLAST and FASTA methods belong to this category.

12.3.1.1 *Basic Local Alignment Search Tool (BLAST)*

BLAST is used to find multiple high scoring local similarities among sequences having moderate to high similarity. The BLAST programs are a set of sequence comparison algorithms introduced in 1990 by David J. Lipman and Stephen Altschul and coworkers at NCBI (Historical Profile 13) that are used to search sequence databases for optimal local alignments to a query. BLAST finds region of high similarity in alignments without gaps. The high similarity regions are evaluated by substitution matrix.

Several terms in BLAST terminology are given in Box. 12.1. The fundamental unit of BLAST algorithm output is the High-scoring Segment Pair (HSP). An HSP consists of two sequence fragments of arbitrary but equal length whose alignment is locally maximal and for which the alignment score meets or exceeds a threshold or cutoff score.

BOX 12.1: BLAST Terminology

Segment Pair: A pair of equal length substrings within strings S_1 and S_2 which are aligned without gaps.

Locally maximal Segment 1: A segment pair whose alignment score (without gaps) cannot be increased by either extending or shortening it.

Maximal Segment Pair (MSP): That segment pair within strings S_1 and S_2 which has the highest score among all locally maximal segments.

High Scoring Pair (HSP): When a database is being searched BLAST attempts to find an MSP with an alignment score higher than a threshold value S. These MSPs are called HSPs. The value of S is chosen such that random hits are minimized.

Table 12.4: The BLAST family of program given at BLAST website

Sl.No.	*Name of program*	*Description*
1.	**BLASTN**	Compares nucleic acid sequences to each other. BLASTN uses one match score and one mismatch score for all characters as well as an affine gap model. The default seed word size is 11 characters so sequences under investigation must be fairly similar in order for BLASTN to report any hits.
2.	**BLASTP**	BLASTP is the protein equivalent of BLASTN. BLASTP permits the use of a limited number of evolution-based scoring matrices such as PAM and BLOSUM along with a limited set affine gap parameters. The default seed word size is 3 for BLASTP which means that BLASTP searches will run slower than BLASTN but have the advantage of allowing more diversity in the sequences under study.
3.	**BLASTX**	BLASTX compares nucleic acid coding sequences to protein databases. The algorithm begins by translating the queries into six proteins corresponding to a three-frame forward translation and a three-frame reverse-compliment translation. The six proteins are then searched against the protein data using the BLASTP algorithm. Results are linked to the frame but there is no frame-to-frame linkage.
4.	**TBLASTN**	TBLASTN compares protein queries to nucleic acid genomic or coding data. Nucleic acid data are converted into six proteins as described for BLASTX and then BLASTP is executed. As with BLASTX comparisons, TBLASTN reports results linked to the protein frame.
5.	**BLASTX**	TBLASTX compares nucleic acid coding sequences to other nucleic acid coding sequences through translated proteins. The query and data sequences are each converted into six protein frames, which are then searched using BLASTP, and the results are linked to the query and database frames. This search is particularly useful when nucleic acid sequences are too divergent for BLASTN to report hits even though those sequences code for similar proteins.
6.	**PSI-BLAST**	Position specific iterative BLAST (PSI-BLAST) initially creates a profile from the results of a BLASTP search. Results above a threshold are aligned to each other and the residue substitution frequencies are calculated at each character position. Highly conserved positions contribute to higher match rewards and mismatch penalties than weakly conserved positions. The profile is used as the input to the next iteration and the returned results above a threshold are again multiple aligned and another profile computed. PSI-BLAST increases search sensitivity but care must be taken when selecting search results at each iteration to prevent specificity degradation. PSI-BLAST is useful in low to moderate homology evaluations.
7.	**PHI-BLAST**	PHI-BLAST (Pattern-Hit Initiated BLAST) finds sequences that contain an occurrence of a regular expression that are also homologous to another sequence in the region(s) near the regular expression. In many cases, protein domains, for example, can be represented by regular expressions. The advantage of doing regular expression searches is that they are computationally inexpensive and easy to implement. The disadvantage of regular expression searches is that by themselves are not usually sufficiently specific to remove random matches. PHI-BLAST improves the specificity of regular expression searches.
8.	**MEGABLAST**	MEGABLAST is used for evaluations of highly similar, large nucleotide sequence sets. MEGABLAST will combine smaller queries internally to reduce execution time, however, the results obtained by combining queries is different than if the queries were not combined.

BLAST is a fast, general-purpose similarity search tool that may be used in both contextual and homology analyses. BLAST characteristically has good sensitivity and very good specificity meaning that it almost always finds true matches and virtually always does not report false positives. An important feature of BLAST is its ability to report multiple local alignments between sequences. This feature allows an investigator to look for repetitive elements, domain ordering, evidence of gene transposition, and so forth.

There are 2 versions of blast available at the EBI, namely WU-Blast2 and NCBI-Blast2. WU Blast 2 is used in Experiment 12.1A. In Table 12.3 other BLAST programs are described. These are distinctly different software packages, although they have a common lineage for some portions of their code, so the two packages do their work differently and obtain different results and offer different features.

The BLAST program improved the overall speed of searches while retaining good sensitivity by breaking the query and database sequences into fragments ("words"), and initially seeking matches between fragments (chapter 10). BLAST are the most popular similarity search program today. The initial search is done for a word of length "W" that scores at least "T" when compared to the query using a given substitution matrix. Word hits are then extended in either direction in an attempt to generate an alignment with a score exceeding the threshold of "S". The "T" parameter dictates the speed and sensitivity of the search. The BLAST algorithm is briefly explained in box 12.2 below.

BOX 12.2: BLAST Alogrithm

Given a length parameter w and a threshold score t, find all words in the database sequence that aligns with words in the query sequence with a gapless alignment score greater than t. Such words are called hits.

↓

Recommended word size w is 3–5 for amino acids and ~12 for nucleotides.

↓

If t is increased number of hits are reduced and the program runs faster. Reducing t allows more distant relationships to be found.

↓

Extend each hit to a locally maximal segment and check if its score is above S i.e., if this sequence is an HSP.

↓

Since pair score matrices typically include negative values, extension can either increase or decrease the score. The extension process is terminated when the reduction in score exceeds certain preset threshold.

↓

After BLAST has looked up all possible words and extended them maximally, it assembles the best scoring query sequence pair and stores them in a data structure called SeqAlign.

↓

A special formatter uses this data structure to format the output in one several possible ways (user specified) and presented to the user.

↓

Extend each hit to a locally maximal segment and check if its score is above S. i.e., if this sequence is an HSP. Since pair score matrices typically include negative values, extension can either increase or decrease the score. The extension process is terminated when the reduction in score exceeds certain preset threshold.

EXPERIMENT 12A

To find the nucleotide sequences similar to given sequence using Wu-blast similarity research program

About Wu-BLASTN: A detailed description about Wu-BLASTN is given in Table 12.3.

About the Sequence: If the given sequence is a random nucleotide sequence whose name and function is not known.

Hypothesis: It is assumed that the given sequence is carrying out an important biological function. It would be possible to find the sequences similar to this using Wu-BLASTN program.

TESTING THE HYPOTHESIS

1. The Wu-blastn program is chosen, which is designed to search a nucleotide query sequence against a DNA databank, in this case we have selected the EMBL database (Fig. 12.5).
2. The given sequence, sequence 3 (given in data box 12.1) is entered into the textbox in fasta format, which consists of a one-line header starting with a ">" symbol, followed by the sequence name. The sequence is then entered on new line(s). "interactive" is chosen so that we will have the results delivered to the browser as soon as they are available. Alternatively you can chose "email" and fill in your email and have the results delivered via email.
3. The title of the search is left as "_Sequence" although you can give your search title any name you wish to help you identify the results.
4. The number of scores (hits to the database) is limited to 10 and the number alignment of these against the query sequence is limited to 5. This is done to limit the size of the output results.

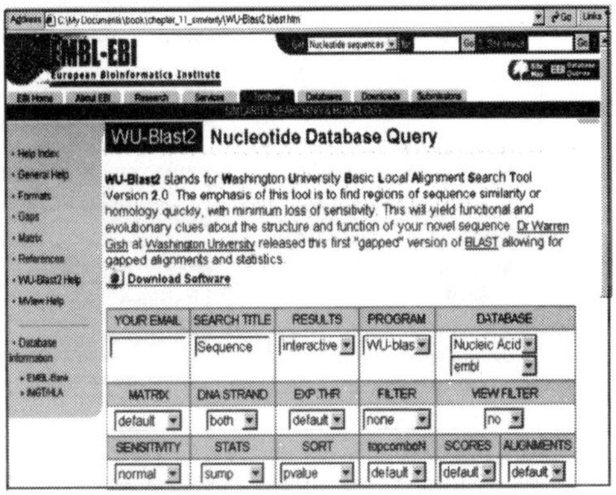

Fig. 12.5: Webview of EMBL-EBI

DATA

Data Box 12.1: The given sequence

```
>sequence 3
cctcactcac ttgcccctta caggactcag ctcttgaagg caatagcttt atagaaaaaa
cgaataggaa gacttgaagt gctatttttt tttttttttt tgtcaaggct gctgaagttt
attggcttct catcgtacct aagcctcctg gagcaataaa actgggagaa acttttacca
agatttttat ccctgccttg atatatactt tttcttccaa atgctttggt gggaagaagt
agaggactgt tatgaaagag aagatgttca aaagaaaaca ttcacaaaat ggataaatgc
acaattttct aagtttggaa agcaacacat agacaacctc ttcagtgacc tgcaggatgg
aaaacgcctc ctagacctct tggaaggcct tacagggcaa aaactgccaa aagaaaaggg
atctacaaga gttcatgccc tgaacaatgt caacaaggca ctgcgggtct tacagaaaaa
taatgttgat ttagtgaata taggaagcac tgacatagtg gatggaaatc ataaactcac
tcttggtttg atttggaata taatcctcca ctggcaggtc aaaaatgtga tgaaaactat
catggctgga ttgcagcaaa ccaacagtga aaagattctt ctgagctggg ttcgacagtc
aacacgtaat tatccacagg ttaacgtcat caacttcacc tctagctggt ccgacgggtt
ggctttgaat gctcttatcc atagtcacag gcccgacctg tttgattgga atagtgtggt
ttcacagcac tcagccaccc aaagactgga acatgccttc aacattgcaa aatgccagtt
aggcatagaa aaacttcttg atcctgaaga tgttgctacc acttatccag acaagaagtc
```

catcttaatg tacatcacat cactctttca agttttgcca caacaagtga gcattgaagc
cattcaagaa gtggaaatgt tgcccaggac atcttcaaaa gtaactagag aagaacattt
tcaattacat caccagatgc attactctca acagatcaca gtcagtctag cacagggcta
tgaacaaact tcttcatctc ctaagcctcg attcaagagt tatgccttca cacaggctgc
ttatgttgcc acctctgatt ccacacagag cccctatcct tcacagcatt tggaagctcc
cagagacaag tcacttgaca gttcattgat ggagacggaa gtaaatctgg atagttacca
aactgcttta gaagaagtac tttcatggct tctttctgcc gaggatacat tgcgagcacaaggagagatt tcaaatgatg ttgaagaagt gaaagaacag tttcatgctc atgagggatt
catgatggat ctgacatctc atcaaggact tgttggtaat gttctacagt taggaagtca
//

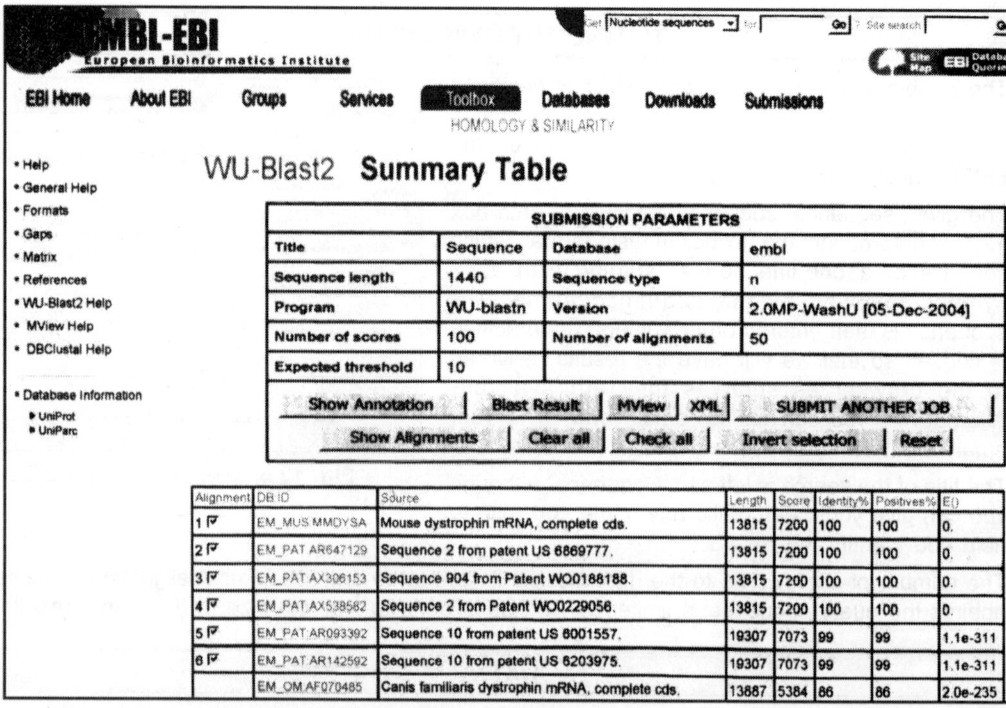

Fig. 12.6: Output of WuBlast 2 search

Data Box 12.2

Sequences producing High-scoring Segment Pairs:	Score	P(N)	N
EM_MUS:MMDYSA M68859 Mouse dystrophin mRNA, complete cds.	69075	0.	1
EM_PAT:AX306153 AX306153 Sequence 904 from Patent WO0188188.	69075	0.	1
EM_PAT:AR093392 AR093392 Sequence 10 from patent US 6001557.	68939	0.	1
EM_PAT:AR142592 AR142592 Sequence 10 from patent US 6203975.	68939	0.	1
EM_OM:AF070485 AF070485 Canis familiaris dystrophin mRNA,...	46390	0.	2
EM_HUM:HSDYS M18533 Homo sapiens dystrophin (DMD) mRNA, c...	28291	0.	8
EM_HUM:HSDMDR X14298 Human mRNA for dystrophin	28228	0.	4
EM_PAT:AX114289 AX114289 Sequence 1 from Patent WO0129243.	17004	0.	2
EM_MUS:MMDYS M18025 Mouse dystrophin mRNA, partial cds.	16348	0.	1
EM_OV:GGDYS X13369 Chicken mRNA for dystrophin (Duchenne ...	16291	0.	6

```
Query:1  CCTCACTCACTTGCCCCTTACAGGACTCAGCTCTTGAAGGCAATAGCTTTATAGAAAAAA60
         ||||||||||||||||||||||||||||||||||||||||||||||||||||||||||||
Sbjct:1  CCTCACTCACTTGCCCCTTACAGGACTCAGCTCTTGAAGGCAATAGCTTTATAGAAAAAA60
.
.
.
etc.
Query:    3TCACTCACTTGCCCCTTACAGGACTCAGCTCTTGAAGGCAATAGCTTTATAGAAAAAAC61
          ||||||||||| |||||||||||||||||| | ||| |||||||| |||||| || ||||
Sbjct:    80 TCACTCACTTTCCCCTTACAGGACTCAGATTTTGGAGGCAATTAGCTTCATC-AGAGAA- 137
.
.
.
etc.
```

Fig. 12.7: Pairwise alignment in BLAST output

INTERPRETATION OF DATA

The one-line sequence descriptions in Data Box 12.2 and summary table (Fig. 12.6) are useful for identifying biologically interesting database matches. Unless otherwise requested, the database sequences are sorted by increasing P-value. We will first consider the score list, the search was limited to 10° scores in this search. In Data Box 12.2 only upper ten sequence details are shown.

Let us consider the top (best) match with our query sequence: EM_MUS: MMDYSA: EM is the EMBL database division that the entry is in and MUS refers to mouse (Musculus). MMDYSA is the entry name in the database and the link to the entry, mouse dystrophin mRNA, complete cds is the description of the database entry. 69075 is the high score of the entry, the greater this value, the better the match between the query sequence and the database entry. P = 0 is the probabilities value. The "P(N)" column contains the lowest P-value ascribed to any set of HSPs for each database sequence. P-values reported for HSPs are dependent on several factors including the scoring system employed, the residue composition of the query sequence, an assumed residue composition for a typical database sequence, the length of the query sequence, and the total length of the database. N = 1 means there is 1 high scoring segment pair for this sequence alignment. This "N" column displays the number of HSPs in the set which was ascribed the lowest P-value. The second entry EM-PAT: AX 306153 also share the same score.

Pressing the alignment button in the summary table, we view the pairwise alignments of sequences from the score list against your query sequence. The first example is shown in Fig. 12.7. Note that matching sequences are connected with a "|" symbol. As this is a perfect match, all of the nucleotides are connected with a "|" symbol, mismatches would be connected with a space. A gap would be represented with a "-" symbol. You will notice from the score list that dystrophin mRNA from different species is quite similar, and are obviously evolutionarily related/conserved. In this list of top ten hits against our query sequence, we have discovered similarities between this mouse dystrophin mRNA and dog, human and chicken dystrophin mRNA:

EM_OM: AF070485 Canis familiaris dystrophin mRNA.
EM_HUM: HSDMDR Human mRNA for dystrophin.
EM_OV: GGDYS Chicken mRNA for dystrophin.

These other dystrophin mRNA's have evolved differently and will contain substitutions, insertions and deletions which produce mismatches with our query sequence.

12.3.1.2 FASTA

FASTA (pronounced FAST-Aye) stands for FAST-All, reflecting the fact that it can be used for a fast protein comparison or a fast nucleotide comparison. This was developed by Pearson and Lipman in 1988 (Historical Profiles 13 and 14). This program achieves a high level of sensitivity for similarity searching at high speed. This is achieved by performing optimized searches for local alignments using a substitution matrix. The high speed of this program is achieved by using the observed pattern of

HISTICAL PROFILE 13

Dr. David J. Lipman for Developing Similarity Search Programs

David J. Lipman is an American biologist who since 1989 has been the Director of NCBI (the National Center for Biotechnology Information) at the National Institutes of Health. Dr. Lipman received his undergraduate degree from Brown University and his M.D. in 1980 from the State University of New York at Buffalo. Lipman is one of the original authors of the BLAST sequence alignment program and a widely respected figure in Bioinformatics. He also leads an intramural research program, including groups led by Stephen Altschul (another BLAST co-author), David Landsman, Eugene Koonin (a prolific author on comparative genomics), and L. Aravind. Lipman has an Erdös number of 3. He received the Association of Biomolecular Resource Facilities Award for outstanding contributions to Biomolecular Technologies in 1996. He was awarded the Senior Scientist Accomplishment award from the International Society for Computational Biology in 2004. Dr. Lipman was one of the originators of the Influenza Genome Sequencing Project, a project to sequence and make available the genomes of thousands of influenza virus isolates. Dr. Lipman was elected to the U.S. National Academy of Sciences in 2004.

HISTORICAL PROFILE 14

Dr. Willam R. Pearson for Developing Similarity Search Programs

William R. Pearson

William R. Pearson is Professor of Biochemistry, California Institute of Technology. He working with David J. Lipman gave the FASTA programs can be used to search protein and DNA sequence databases, and to confirm the statistical significance of a match by comparing the alignment score to a distribution of scores produced by shuffled sequences. He is also known for a long-standing interest in exploiting protein sequence information, both for understanding better how new protein sequences arise and for understanding the relationship between protein sequence and protein structure.

word hits to identify potential matches before attempting the more time consuming optimized search. The trade-off between speed and sensitivity is controlled by the ktup parameter, which specifies the size of the word. Increasing the ktup decreases the number of background hits. Not every word hit is investigated but instead initially looks for segment's containing several nearby hits.

You can cut and paste or type a sequence into the large text window. FASTA program is used for study in Experiment 12B. You may also paste a sequence in GCG, FASTA, EMBL, GenBank, PIR, NBRF, Phylip or UniProt/Swiss-Prot format (chapter 8). Partially formatted sequences are not accepted. Copying and Pasting directly from word processors may yield unpredictable results, as hidden/control characters may be present. Adding a return to the end of the sequence may help certain applications understand the input.

If your sequence is DNA and contains more than 50% ambiguity codes it will be rejected by the system (both e-mail and interactive submissions). It is best to use the part(s) of the sequence that contain real DNA (good consensus regions) as the likelihood if these regions giving positive scores is higher than if they contained many 'N's which the programs will attempt to match to the entire database(s). If successful, these database scans will contain false positive hits, which are of limited use. For very short nucleic acid sequences the strategy recommended is to decrease the word length (ktup) from 6 to 1. In this way a significant increase in sensitivity is achieved without loss of biological significance. Fasta may not return as many hits as blast does in this type of searches but the relevance of fasta results is much higher than with blast very often.

Several other software available at web site for FASTA are given in Table 12.4.

Table 12.4: Software available at FASTA website

Program	*Description*
FASTA	Compares a protein sequence to another protein sequence or to a protein database, or a DNA sequence to another DNA sequence or a DNA library.
SSEARCH (Local only)	Performs a rigorous Smith-Waterman alignment between a protein sequence and another protein sequence or a protein database, or with DNA sequence to another DNA sequence or a DNA library (very slow).
FASTX/FASTY	Compares a DNA sequence to a protein sequence database, translating the DNA sequence in three forward (or reverse) frames and allowing frameshifts.
TFASTX/TFASTY	Compares a protein sequence to a DNA sequence or DNA sequence library. The DNA sequence is translated in three forward and three reverse frames, and the protein query sequence is compared to each of the six derived protein sequences. The DNA sequence is translated from one end to the other; no attempt is made to edit out intervening sequences. Termination codons are translated into unknown ('X') amino acids.
FASTF/TFASTF	Compares an ordered peptide mixture, as would be obtained by Edman degrdation of a CNBr cleavage of a protein, against a protein (fastf) or DNA (tfastf) database. A different format is required to specify the ordered peptide mixture.
FASTS/TFASTS	Compares set of short peptide fragments, as would be obtained from mass-spec. Analysis of a protein, against a protein (fasts) or DNA (tfasts) database. A different format is required to specify the ordered peptide mixture. Indicates three peptide fragments were found: MILG, MLLEYTD, and MGDAP. The commas (,) are required to indicate the number of fragments in the mixture, but there should be no comma after the last residue.
LALIGN/LFASTA PLALIGN	Compares two protein sequences to identify regions of sequence similarity. While FASTA and TFASTA report a single alignment between two sequences, LALIGN will report several sequence alignments if there are several similar regions. LALIGN can identify similarities due to internal repeats or similar regions that cannot be aligned by FASTA because of gaps. LALIGN reports sequence alignments and similarity scores. PLALIGN plots out a graph of the sequence alignments, which looks much like a "dot-plot".
PRSS/PRFX	Evaluates the significance of pairwise similarity scores using a Monte Carlo analysis. Similarity scores for the two sequences are calculated, and then the second sequence is shuffled 200 to 1000 times and compared with the first sequence. PRSS can use one of two shuffling strategies. One strategy simply keeps the amino acid composition of the entire shuffled sequence identical to the unshuffled sequence. The second, local shuffle, destroys the order but preserves the composition of small (10 - 25 residue) segments of the shuffled sequence. PRFX does a similar shuffle, but compares a translated DNA sequence to a protein sequence using the FASTX algorithm.
GREASE	Kyte-Doolittle hydropathy plot. A modern transmembrane prediction program is available from tempered.
GARNIER/CHOFAS	Protein secondary structure prediction.

EXPERIMENT 12B

To Perform Similarity Searches for a given Amino Acid Sequence with FASTA

About FASTA: A detailed description of FASTA is already given above.

About the sequence: is an amino acid sequence given Data Box 12.3. It is 338 amino acid in length and has a molecular weight of 35976. The sequence is saved in a notepad called experiment - 12.2 doc.

Hypothesis: It would be possible to extract amino acid sequence similar to the given sequence.

TESTING THE HYPOTHESIS

Computer set up: A windows environment with internet connection.

1. Log into FASTA web site. A web view is shown in Fig. 12.8
2. Paste the sequence in sequence text box.

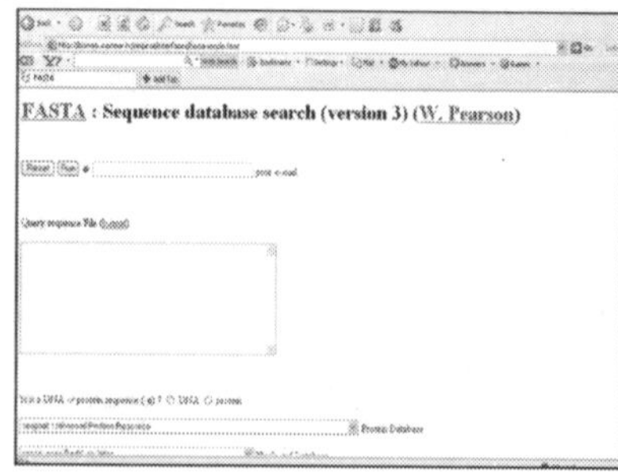

Fig. 12.8: Web view of FASTA

DATA

Data Box 12.3: Input sequence

```
>SQ   SEQUENCE   338 AA;  35976 MW;  E9D031A4BEAE48EC CRC64;
MFQAFPGDYD SGSRCSSSPS AESQYLSSVD SFGSPPTAAA SQECAGLGEM PGSFVPTVTA
ITTSQDLQWL VQPTLISSMA QSQGQPLASQ PPAVDPYDMP GTSYSTPGLS AYSTGGASGS
GGPSTSTTTS GPVSARPARA RPRRPREETL TPEEEEKRRV RRERNKLAAA KCRNRRRELT
DRLQAETDQL EEEKAELESE IAELQKEKER LEFVLVAHKP GCKIPYEEGP GPGPLAEVRD
LPGSTSAKED GFGWLLPPPP PPPLPFQSSR DAPPNLTASL FTHSEVQVLG DPFPVVSPSY
TSSFVLTCPE VSAFAGAQRT SGSEQPSDPL NSPSLLAL
//
```

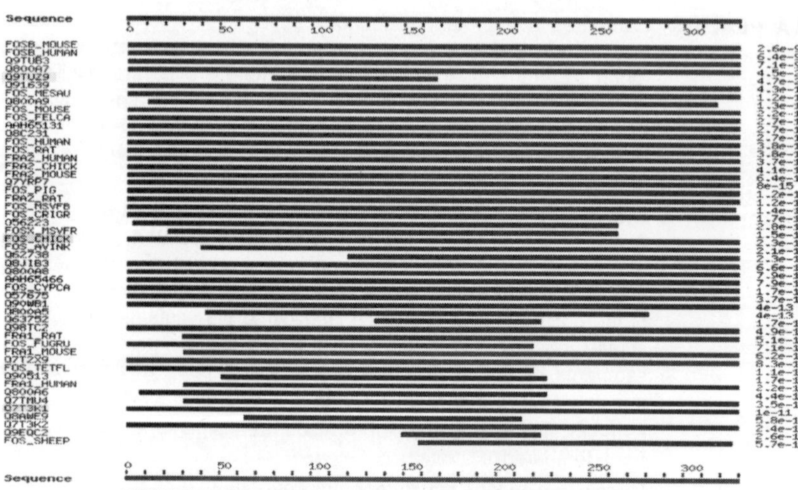

Fig. 12.9: Result of FASTA similarity research

Data Box 12.4

The textural results (score list) was as follows:

The best scores are:		*opt*	*bits*	*E(1414785)*
UNIPROT:FOSB_MOUSE P13346 Protein fosB.	(338)	2268	361.4	2.6e-98
UNIPROT:FOSB_HUMAN P53539 Protein fosB (G0/G1 swi	(338)	2195	350.1	6.4e-95
UNIPROT:Q9TUB3 Q9TUB3 FosB protein.	(338)	2151	343.3	7.1e-93
UNIPROT:Q800A7 Q800A7 FosB protein.	(394)	759	128.3	4.5e-28
UNIPROT:Q9TUZ9 Q9TUZ9 FOSB protein (Fragment).	(97)	577	99.6	4.7e-20
UNIPROT:Q91639 Q91639 Fos-related antigen-2.	(319)	500	88.2	4.3e-16
UNIPROT:FOS_MESAU O88479 Proto-oncogene protein c	(381)	492	87.0	1.2e-15
UNIPROT:Q800A9 Q800A9 Fra1 protein.	(338)	490	86.7	1.3e-15
UNIPROT:FOS_MOUSE P01101 Proto-oncogene protein c	(380)	486	86.1	2.2e-15
UNIPROT:AAH65131 AAH65131 Fos-like antigen 2.	(326)	483	85.6	2.7e-15
UNIPROT:Q8C231 Q8C231 Fos-like antigen 2.	(326)	483	85.6	2.7e-15
UNIPROT:FOS_FELCA Q8HZP6 Proto-oncogene protein c	(381)	484	85.8	2.7e-15
UNIPROT:FRA2_HUMAN P15408 Fos-related antigen 2.	(326)	480	85.1	3.7e-15
UNIPROT:FOS_RAT P12841 Proto-oncogene protein c-f	(380)	481	85.3	3.8e-15
UNIPROT:FOS_HUMAN P01100 Proto-oncogene protein c	(380)	481	85.3	3.8e-15
UNIPROT:FRA2_CHICK P18625 Fos-related antigen 2.	(323)	479	84.9	4.1e-15
UNIPROT:FRA2_MOUSE P47930 Fos-related antigen 2.	(326)	475	84.3	6.4e-15
UNIPROT:Q7YRP7 Q7YRP7 C-Fos.	(380)	474	84.2	8e-15
UNIPROT:FRA2_RAT P51145 Fos-related antigen 2.	(327)	469	83.4	1.2e-14
UNIPROT:FOS_PIG O97930 Proto-oncogene protein c-f	(380)	470	83.6	1.2e-14
UNIPROT:FOS_MSVFB P01102 p55-v-fos transforming p	(381)	469	83.5	1.4e-14
UNIPROT:FOSX_MSVFR P29176 Transforming protein v-	(244)	465	82.7	1.5e-14

```
>>UNIPROT:FOSB_MOUSE P13346 Protein fosB.                    (338 aa)
initn: 2268 init1: 2268 opt: 2268  Z-score: 1866.3  bits: 353.7 E(): 3.3e-96
Smith-Waterman score: 2268;  100.000% identity (100.000% ungapped) in 338 aa overlap (1-338:1-338)
                10        20        30        40        50        60
FOSB_M MFQAFPGDYDSGSRCSSSPSAESQYLSSVDSFGSPPTAAASQECAGLGEMPGSFVPTVTA
       ::::::::::::::::::::::::::::::::::::::::::::::::::::::::::::
UNIPRO MFQAFPGDYDSGSRCSSSPSAESQYLSSVDSFGSPPTAAASQECAGLGEMPGSFVPTVTA
                10        20        30        40        50        60
                70        80        90       100       110       120
FOSB_M ITTSQDLQWLVQPTLISSMAQSQGQPLASQPPAVDPYDMPGTSYSTPGLSAYSTGGASGS
       ::::::::::::::::::::::::::::::::::::::::::::::::::::::::::::
UNIPRO ITTSQDLQWLVQPTLISSMAQSQGQPLASQPPAVDPYDMPGTSYSTPGLSAYSTGGASGS
                70        80        90       100       110       120
               130       140       150       160       170       180
FOSB_M GGPSTSTTTSGPVSARPARARPRRPREETLTPEEEEKRRVRRERNKLAAAKCRNRRRELT
       ::::::::::::::::::::::::::::::::::::::::::::::::::::::::::::
UNIPRO GGPSTSTTTSGPVSARPARARPRRPREETLTPEEEEKRRVRRERNKLAAAKCRNRRRELT
               130       140       150       160       170       180
               190       200       210       220       230       240
FOSB_M DRLQAETDQLEEEKAELESEIAELQKEKERLEFVLVAHKPGCKIPYEEGPGPGPLAEVRD
       ::::::::::::::::::::::::::::::::::::::::::::::::::::::::::::
UNIPRO DRLQAETDQLEEEKAELESEIAELQKEKERLEFVLVAHKPGCKIPYEEGPGPGPLAEVRD
               190       200       210       220       230       240
               250       260       270       280       290       300
FOSB_M LPGSTSAKEDGFGWLLPPPPPPPLPFQSSRDAPPNLTASLFTHSEVQVLGDPFPVVSPSY
       ::::::::::::::::::::::::::::::::::::::::::::::::::::::::::::
UNIPRO LPGSTSAKEDGFGWLLPPPPPPPLPFQSSRDAPPNLTASLFTHSEVQVLGDPFPVVSPSY
               250       260       270       280       290       300

               310       320       330
FOSB_M TSSFVLTCPEVSAFAGAQRTSGSEQPSDPLNSPSLLAL
       ::::::::::::::::::::::::::::::::::::::
UNIPRO TSSFVLTCPEVSAFAGAQRTSGSEQPSDPLNSPSLLAL
               310       320       330
```

Fig. 12.10: A portion of pairwise alignment (only a portion is shown)

INTERPRETATION OF DATA

The top bar in Fig. 12.9 with the text "Sequence" represents the query sequence, which are 338 amino acids in this case. The bars underneath represent the sequences the query has been compared to. The part of the sequence that was aligned to the query is shown below this top bar, looking at the score list, the numbers in brackets are the length of the sequence. In the case of Q9TUZ9, which is 97 amino acids long, this is considerably shorter than the query sequence, and in other cases which is 394 amino acids long and thus longer than the query sequence, not the full sequence is plotted graphically as it runs off the edge of the plot. Also in another cases for example Q7T3K1, which is 369 amino acids long and thus longer than the query sequence, not the full sequence is plotted graphically as it runs off the edge of the plot as well as the alignment not spanning the full length of the query sequence.

The results in the graphical view in Fig. 12.10 are sorted by Smith and Waterman score (bits) and are coloured according to E() value (expectation value). The red colour represents (10^{-12}). We mainly concentrate on bars for further discussion. Considering the top (best) match with our query sequence:

UNIPROT: FOSB_MOUSE P13346 Protein fosB. (338) 2268 361.4 2.6e-98

Here Uniprot is the name of database and FOSB_MOUSE is Uniprot ID. P13346 is interpro ID and Protein fosB is protein description with length of 338 base pair. Opt here is 2268 and bits 354 and E-value is 3.3e –96 or 3.3×10^{-96}. Score list is textually given in data box 12.4. Underneath the score list, the pairwise alignments of sequences from the score list against your query sequence is given in Fig. 12.10. It would have been worthwhile to compare the results of similarity rearches for a DNA sequences and a amino acid sequence using FASTA.

12.3.2 Dynamic Programming Methods

The Smith Waterman evaluation finds an optimal, local alignment of nucleotide or peptide sequences and is typically used when low to moderate sequence identity is expected. Alignments here are optimal because it considers all possible ways that two sequences can be matched up and reports the one having the best score. There are three significant differences that make Smith Waterman more sensitive recently:

1. The BLAST algorithm begins by finding short regions of exact match between pairs of sequences. These seed regions are then extended, without gaps until the extension causes the score to drop below a threshold. Extended regions are then joined with gaps through a dynamic program. The effectiveness of a BLAST search is heavily dependent on the size of the initial seed region. Smaller regions produce better results because they are less likely to miss a true alignment but the computations are very expensive. On the other hand, longer initial word sizes make for faster searches but the results are not as good. The more evolutionary distance there is between two sequences, the greater the chance that BLAST will miss a real hit. Smith Waterman, by contrast, examines all alignments and reports an alignment having the highest score. This assures that an optimal alignment (there may be more than one) is always reported.
2. BLAST randomly substitutes one of the nucleic acids for "N" in DNA sequence searches. Many sequences such as genomic scaffolds or sequences masked by low-complexity filtering can contain long strings of "N". By randomly substituting nucleic acids for "N", BLAST may fail to create a seed and therefore miss a hit. Smith Waterman algorithm does not need to make this substitution because it uses explicitly specified scoring matrices. The reward for matches with "N" may be explicitly specified to represent the biology, chemistry, or quality of the sequences involved.

3. In DNA searches: BLAST provides a match and mismatch score parameter that is applied to all nucleic acids. This model is appropriate for evaluating differences due to sequencing error in which all modifications are equally likely, i.e., in contextual evaluations. Evaluating mismatches due to evolution requires more complicated scoring models that take into account actual differences between related sequences. This leads to matrices that may penalize certain substitutions differently than others, and additionally, may be asymmetric to penalize changes in one direction more heavily than another. Because Smith Waterman implementation uses explicitly specified substitution matrices for both DNA and protein searches, users can select the scoring matrix that best models the situation.

   Several Smith Waterman nucleotide and peptide variants available are given below:

   (*a*) SWN - Smith Waterman nucleotide (SWN) compares nucleic acid sequences.

   (*b*) SWP - Smith Waterman peptide (SWP) compares peptide sequences. Generally SWP is used for homology analysis and one of the evolutionary matrices, e.g., BLOSUM, is used.

   (*c*) Smith Waterman frame variants.

   (*d*) SWX - compares nucleic acid query sequences to protein data. This search is used to find putative homologous proteins for newly sequenced ESTs, RNAs and cDNAs.

   (*e*) TSWN - Searching a peptide sequence against nucleic acid coding regions is performed with reverse frame algorithm. This comparison allows a user to annotate unknown peptide sequences by comparing them to databases of nucleic acid coding regions or to locate putative genes with known proteins.

   (*f*) TSWX - offers a double frame nucleic acid to nucleic acid comparison at the protein level.

12.3.3 Profile Variants

A profile is a mathematical summary of the alignment of multiple, related nucleic acid or protein sequences. A detailed discussion on profile and profile database is given in chapter 4. A profile is built by scanning down each column of the multiple sequence alignment to determine the frequency of each character in the column. These position-dependent character frequencies are normalized by the frequency of each character in the multiple alignment to produce an odds ratio for the substitution of one character for another. Profile searches enable a comparison of database sequences with the family represented by the profile. The position-specific nature of these searches facilitates a comparison of the critical features characteristic of the family. For example, a multiple alignment of a protein or nucleotide family will have highly conserved regions interspersed with non-conserved regions. The profile gives high scores to matches of a query within the conserved areas and correspondingly very low scores for mismatches. In non-conserved regions match and mismatch scores have lower positive or negative impact to reflect the reduced contribution of those regions to the function or structure of the protein or nucleic acid sequence.

Profile similarity searches have several applications:

1. They permit greater accuracy in alignments of distantly related sequences.
2. Set of residues that are highly conserved are likely to part of active site and gives clues about function
3. The conservation pattern facilitates identification of other homologus sequences.
4. Patterns from other sequences are useful in identification of subfamily within a set of homologues.

Position specific iterative BLAST (PSI-BLAST) refers to a feature of BLAST 2.0 in which a profile or position specific scoring matrix, PSSM is constructed (automatically) from a multiple alignment of the highest scoring hits in an initial BLAST search. Calculating position-specific scores for each position in the alignment generates the PSSM. Highly conserved positions receive high scores and weakly conserved positions receive scores near zero. BLAST search and the results of each "iteration" used to refine the profile. This iterative searching strategy results in increased sensitivity.

EXPERIMENT 12C

Perform PSI-BLAST Similarity Searches and Compare it with Ordinary BLAST Search

About PSI-BLAST: A description of PSI-BLAST is given above.

About the sequence: This experiment uses as an example, the uncharacterized archaebacterial protein, MJ0577, from *Methanococcus jannaschii*.

Hypothesis: It would be possible to perform psiBlast similarity search for the uncharacterized archaebacterial protein, MJ0577, from *Methanococcus jannaschii.* The PSI-BLAST search in this experiment has two purposes:

1. To identify distant relatives of the MJ0577 family and
2. To gain insight into the function of this family of proteins.

TESTING THE HYPOTHESIS

Computer set up: A windows environment with internet connection.

Starting Materials: It is an amino acid sequence given in Data Box 12.5 is stored as expt. 12.3.doc

1. log in to web site of PSI-blast as shown in Fig. 12.11.
2. Wad the sequence in sequence text box.
3. Press Browse. Save Results.
4. Press "Run PSI-BLASTS iteration". Run PSI-BLAST till there is no new information.

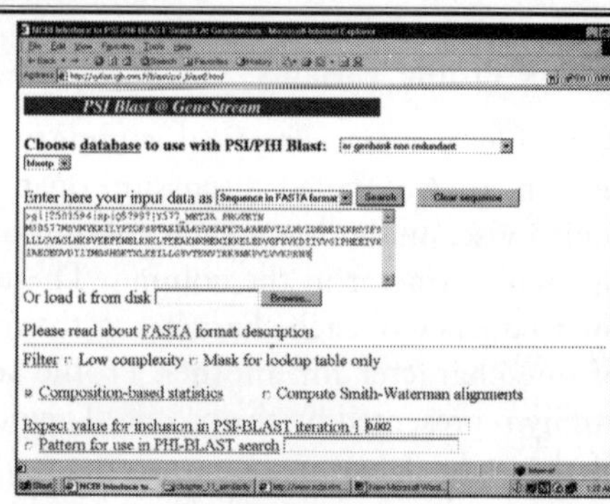

Fig. 12.11

DATA

Data Box 12.5: The given sequence

>gi|2501594|sp|Q57997|Y577_METJA PROTEIN MJ0577
MSVMYKKILYPTDFSETAEIALKHVKAFKTLKAEEVILLHVIDEREIKKRDIFSLLLGVA
GLNKSVEEFENELKNKLTEEAKNKMENIKKELEDVGFKVKDIIVVGIPHEEIVKIAED
EGVDIIIMGSHGKTNLKEILLGSVTENVIKKSNKPVLVVKRKNS

1.	sp|Q57997|Y577_METJA	PROTEIN MJ0577 >gi|2128018|pir||A64372...	314	2e-85
2.	pdb|1MJH|	Structure-Based Assignment Of The Biochemical F...	272	1e-72
3.	dbj|BAA29916|	(AP000003) 170aa long hypothetical protein [P...	107	6e-23
4.	sp|Q57951|Y531_METJA	HYPOTHETICAL PROTEIN MJ0531 >gi|212801...	91	4e-18
5.	gi|2622094	(AE000872) conserved protein [Methanobacterium t...	85	4e-16

6.	gi|2621993 (AE000865) conserved protein [Methanobacterium t...	81	4e-15
7.	gi|2621194 (AE000803) conserved protein [Methanobacterium t...	80	7e-15
8.	gi|2622163 (AE000877) conserved protein [Methanobacterium t...	79	2e-14
9.	sp|P42297|YXIE_BACSU HYPOTHETICAL 15.9 KD PROTEIN IN BGLH-W...	76	1e-13
10.	sp|Q50777|YB54_METTM HYPOTHETICAL 16.1 KD PROTEIN IN MTR RE...	66	2e-10
11.	gi|2648791 (AE000981) conserved hypothetical protein [Archa...	65	3e-10
12.	gi|2648610 (AE000970) conserved hypothetical protein [Archa...	64	5e-10
13.	gi|2983400 (AE000710) hypothetical protein [Aquifex aeolicus]	64	6e-10
14.	sp|P73475|YC30_SYNY3 HYPOTHETICAL 31.2 KD PROTEIN SLR1230 >...	63	1e-09
15.	gi|2983527 (AE000719) hypothetical protein [Aquifex aeolicus]	61	4e-09
16.	sp|O27222|YB54_METTH HYPOTHETICAL PROTEIN MTH1154 >gi|26222...	59	2e-08
17.	dbj|BAA30650| (AP000006) 208aa long hypothetical protein [P...	57	6e-08
18.	dbj|BAA31039.1| (AP000007) 149aa long hypothetical protein ...	56	1e-07
19.	emb|CAB50594.1| (AJ248288) hypothetical protein [Pyrococcus...	55	2e-07
20.	sp|P39177|UP12_ECOLI UNKNOWN PROTEIN FROM 2D-PAGE (SPOTS PR...	55	2e-07
21.	sp|P74148|YD88_SYNY3 HYPOTHETICAL 17.3 KD PROTEIN SLL1388 >...	52	2e-06

Fig. 12.12

INTERPRETATION OF OUTPUT

The experiment illustrates the potential for PSI-BLAST searches to identify even weak (subtle) homologies to annotated entries in the database. It demonstrates that PSI-BLAST is an important tool for predicting both biochemical activities and function from sequence relationships.

A BLAST search of this sequence revealed a number of probable homologs but no hits that provided useful information about the function or biochemical activities of this protein. From the graphical overview in Fig. 12.12, it is apparent that there are several high-scoring sequences that are highly related to the MJ0577 query. The top two bars are matches to MJ0577 itself. The next five hits are probable homologs. The other full-length bars towards the bottom may represent more distantly related homologs. Partial length bars reflect similarity between a portion of the query and a database entry. These regions of similarity may be a shared domain or motif. The nature of these lower scoring hits can be further explored by examining the corresponding alignment. To facilitate viewing any alignment of interest, each bar in the graphic and each E value in a description line (which are found below the graphic) is linked to the corresponding alignment. The description (also called definition) lines are listed below under the heading "Sequences producing significant alignments". The term "significant" simply refers to all those hits whose E value was less than the threshold. It does not imply biologica significance.

The description lines reveal that the sequence in the database with greatest similarity to MJ0577 is MJ0577 itself. The second hit is to the PDB database entry associated with the determination of the MJ0577 structure. The score for the structure entry is somewhat lower and E value somewhat higher because (see pairwise alignment - the missing residues will appear as dashes) a certain number of residues were omitted from this database entry because they were disordered in the structure.

The set of entries corresponding to the bars number 3 to 7 in the overview Fig. 12.12 are two orthologous sequences in two other Archaeal species, *Pyrococcus horikoshii* (#3) and *Methanobacterium thermoautotrophicum* (#5,6,7)). Interestingly, the top hits also include sets of paralogs in *Methanococcus jannaschii* (#1,4) and *Methanobacterium thermoautotrophicum* (#5,6,7). To examine the relationship among these sequences, they have been subject to multiple alignments using the multiple alignment algorithm, ClustalW. A multiple alignment can also be generated using the "flat query-anchored with identities" formatting option in BLAST.

Since all of the closest homologs of MJ0577 are unannotated, it is of interest to examine a few of the highest scoring annotated sequence in the BLAST hit list. The first sequences in the hit list (scroll down to find these) with positive identification are gi|2648945, whose annotation indicates that this sequence encodes a cationic amino acid transporter and gi|1787640, a putative filament protein. The scores and E values of these hits put the significance of the alignments in the "twilight zone" with respect to significance. Since the transporter entry is 780 aa long, it is apparent that MJ0577 is not itself a homolog of a transporter. It may however, share a domain with this family of proteins. The results of this BLAST search have revealed that MJ0577 is one member of a moderate size family of proteins that are found in Archaea and in Eubacteria. To gain insight into the function of the MJ0577 family of proteins, a more sensitive search with the profile based tool, PSI-BLAST, would be one way to proceed.

The search specified in this PSI-BLAST query is a comparison of the 168 amino acid MJ0577 sequence with all the sequences in the non-redundant (nr) protein database, which consisted of 440,253 different entries at the time this search was executed. The summary also shows that an E value threshold of 0.001 has been specified for inclusion of sequence hits into a profile to be used in the first iterations. The E value threshold for inclusion in subsequent PSI-BLAST iterations has been left at 0.001.

The output from the first stage PSI-BLAST search is identical to the BLAST search results with the exception that in this query the E value threshold was set at 0.001. The results thus far have revealed that MJ0577 is one member of a moderate size family of proteins that are found in Archaea and in Eubacteria. To gain insight into the function of the MJ0577 family of proteins, PSI-BLAST iterations may be launched. The power of PSI-BLAST is in its ability to generate a profile from the initial BLAST results and to use that profile to research the same database. This procedure is initiated by the user upon clicking the "Run PSI-BLAST iteration" button. The number of iterations that should be performed will depend on the goal of the experiment and the nature of the output. The results of the first iteration include numerous new hits that are better than the specified E value threshold. Although the majority of these hits are unannotated, several have known functions or resemble proteins with known functions. Annotated hits are shown in bold. To further evaluate the annotated hits, one may examine each alignment. Click on the corresponding score to jump down to the alignment for each annotated hit in the output of the first iteration. Is it a filament protein? Maybe so because the E value for the alignment of the 168 aa MJ0577 with the 168 aa *E. coli* filament protein suggests that the alignment is highly significant.

The alignment of the relatively short MJ0577 with the 780 aa cationic amino acid transporter indicates that the MJ0577 is unlikely to be a cationic amino acid transporter itself, but that it may share a domain in common with this protein. One way to evaluate whether the shared domain is informative with respect to the function of the uncharacterized MJ0577 protein is to perform a reverse PSI-BLAST search. Copy the gi number for the cationic amino acid transporter and paste it into a fresh PSI-BLAST query page. The results of this search can be easily grasped by examining the graphic. The large number of highly significant hits are to amino acid transporters of various sorts. The similarity to amino acid transporters resides between amino acid 45 and amino acid 440, for the most part. Modest similarity to MJ0577 is localized to a separate region of the protein, aa 550 - aa 720. Therefore, MJ0577 is not an amino acid transporter.

SUMMARY

Similarity research of biological sequences involves two types of analysis: homology evaluation and contextual analysis Homology evaluation looks for evidence that biological sequences are related but have been altered by evolution. Context analysis however just compare the text without any consideration to evolution. Orthologs sequences in different organism have same function. Similarity searches are assessed by Z-score, P-value and E-value scores.

Similarity research programs are classified into 3 types: fast research methods e.g. BLAST and FASTA; dynamic programming searching methods and profile based. However, FASTA and BLAST have become standard in the field of similarity searches and are being extensively used by biologists over entire world. In Experiment 12A similarity research for a given sequence to done using Weblastn. In Experiment 12C a PSI-blast similarity research is done for given sequence and it is compared with BLAST researches.

EXERCISE

1. Align the two histone H2b polypeptide sequences from yeast (CAA81268) and human (CAB11426). The sequences can be downloaded from protein sequence database UniProt. The number in the bracket refers to UniProt ID. Use the BLAST to align the polypeptides over their length (all defaults, BLASTP program). Report per cent identity, length of match. A high per cent identity is expected for histones. Use an identical procedure to align the two polypeptides from yeast (P03962) and human (BAA19923) that have orotodine decarboxylase activity. Use BLASTN with all defaults to align the two DNA sequences corresponding to the orotodine decarboxylases from yeast (gi 172061) and human (gi 2081624). Report % identity, and length of match. You should note that % identity in case of polypeptide is lower than DNA and that a large portion of the human orotodine decarboxylase protein is not aligned with any part of the yeast protein. Now answer the following questions:
   (*a*) Write a one sentence explanation of why two same proteins from the same organisms differ to different extents. If yeast and human diverged from a common ancestor, then shouldn't all genes within them have diverged at that time, and if mutation is a random process, then the genes would all be expected to differ by equal amounts, wouldn't they?
   (*b*) Write one sentence in which you rationalize why histones appear to be more conserved than a basic metabolic enzyme.
   (*c*) Write one sentence that describes why the large portion of the orotodine decarboxylase human sequence may be that is not aligned with any part of the yeast protein.
   (*d*) Write one sentence that describes why alingment protein appears to be a much less extensive identity than that in DNA. This illustrates a very important point about the relatedness of protein sequences versus that of DNA sequences.
2. Use the sequence of a metabolic enzyme (5,10-methylene tetrahydrofolate dehydrogenase—the *folD* gene product) from *E. coli* to find polypeptides in the non-redundant protein database that are close relatives.
   (*a*) Give the name of the database search program and the name of the database most appropriate for this task.
   (*b*) List the gi number(s) for the most similar polypeptide found in *Arabidopsis thaliana* and at least one gi number for the most similar polypeptide found in *Saccharomyces cerevisiae*.
3. What cutoff value of E would you use a PSI-BLAST research of all you want to know is whether your sequence is already present in a Databank.
4. You have sequenced a random segment of genomic DNA from *Oryza sativa*.
   (*a*) You would like to know whether a segment of DNA overlapping yours (identical match) has been sequenced. Which combinations of NCBI BLAST programs and databases would you use for this determination. Please be exhaustive and give your answer like:

      LIST the program and databases like this: blastp,nr and blastp,swissprot...etc.
   (*b*) You would like to know whether it is possible that this segment of DNA is part of an exon that encodes a protein b to any already-sequenced, publicly accessible sequence. List the blast programs and databases that you feel would exhaustively address the question of relatedness.
5. Explain why homologus genes have sequences that are similar and not identical.

13

GENE PREDICTION

A gene is a section of the DNA strand that carries the instructions for a specific function. Each gene is a sequence within a DNA that codes for single protein. For example, the globin genes contain instructions for making the hemoglobin protein, which is the protein, that allows our blood to carry oxygen throughout the body. The genomes contain the complete set of genes of an organism. Humans have about 30,000 different genes, which work together in complex ways to control much of what our bodies do.

There are two general types of gene in the human genome: non-coding RNA genes and protein-coding genes. Non-coding RNA genes represent 2-5 per cent of the total and encode functional RNA molecules. Many of these non-coding RNAs are involved in the control of gene expression, particularly protein synthesis hence are not translated. Protein-coding genes represent the majority of the total and are expressed in two stages: transcription and translation. They show incredible diversity in size and organisation and have no typical structure. There are, however, several conserved features of protein coding genes.

1. The boundaries of a protein-encoding gene are defined as the points at which transcription begins and ends. The core of the gene is the coding region referred to as CDS, which contains the nucleotide sequence that is eventually translated into the sequence of amino acids in the protein.
2. The coding region begins with the initiation codon, which is normally AUG. It ends with one of three termination codons: TAA, TAG or TGA.
3. On either side of the coding region are DNA sequences that are transcribed but are not translated. These untranslated regions or non-coding regions referred to as UTR often contain regulatory elements that control protein synthesis.
4. The coding region in the sequence in case of prokaryotes is continuous i.e. there are no divisions into exons and

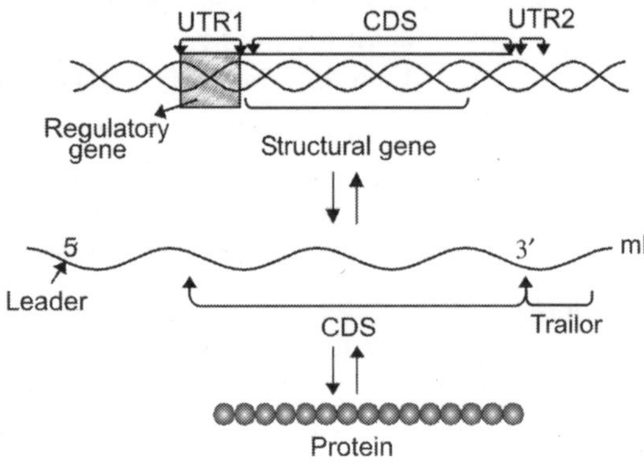

Fig. 13.1: Structure of a prokaryotic gene

introns in case of prokaryotes. Genes, mRNA and protein all are collinear in prokaryotes i.e. physical map of DNA matches an amino acid map of a protein. Structure of gene shown in Fig. 13.1 refers to structure of a typical prokaryotic gene.

5. In typical eukaryotes however, the region of the DNA coding for a protein is usually not continuous (Fig. 13.2). This region is composed of alternating stretches of exons and introns. It starts with the promoter region, which is followed by a transcribed but non-coding region called 5′ untranslated region (5′ UTR). Then follows the initial exon which contains the start codon. Following the initial exon, there is an alternating series of introns and internal exons, followed by the terminating exon, which contains the stop codon. It is followed by another non-coding region called the 3′ UTR. Ending the eukaryotic gene, there is a polyadenylation (poly A) signal the nucleotide Adenine repeating several times. The exon-intron boundaries (i.e., the splice sites) are signalled by specific short (2bp long) sequences. The 5′(3′) end of an intron (exon) is called the donor site, and the 3′(5′) end of an intron (exon) is called the acceptor site.

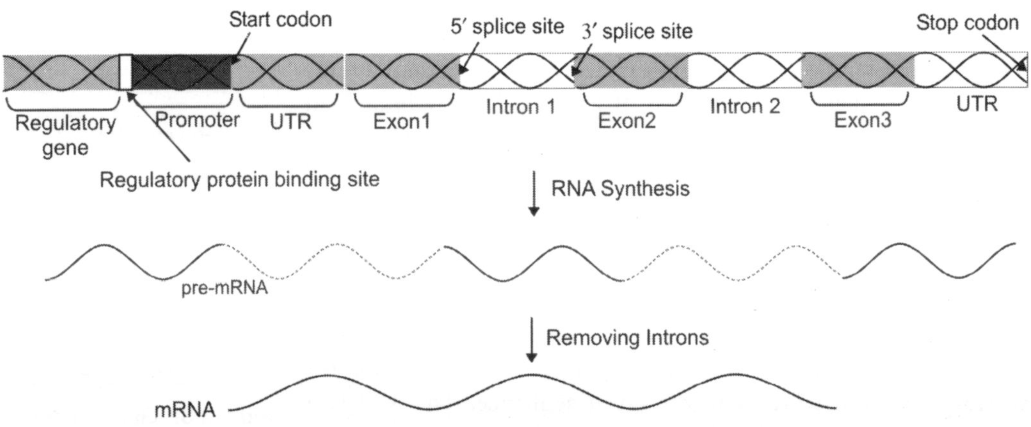

Fig. 13.2: Structure of typical eukaryotic gene

6. During transcription, both exons and introns are transcribed onto the RNA, in their linear order. Thereafter, a process called splicing takes place, in which the intron sequences are excised and discarded from the RNA sequence. The smallest protein-coding gene in the human genome is only 500 nucleotides long and has no introns. It encodes a histone protein. The largest human gene encodes the protein dystrophin, which is missing or non-functional in the disease muscular dystrophy. This gene is 2.5 million nucleotides in length and it takes over 16 hours to produce a single transcript. However, more than 99 per cent of the genes made up of its 79 introns.

7. In prokaryotes, the predicted genes may have conserved sequence patterns such as those for promoter recognition by RNA polymerases and transcription factors for ribosomal binding to mRNAs or for termination of transcription. Similarly in eukaryotes, the region at the 5¢ end of the gene may also have characteristic sequence patterns such as transcription factor binding sites and sequence patterns characteristics of RNA polymerase II promoters. Eukaryotic genes also have repeat sequence elements that probably reflect nucleosome structure. Loops of chromation are attached to the nuclear matrix by relatively short (100–1000 bp) sequence called matrix attachment region (MARS) or scaffold associated regions (SARS). These regions are considered to be indicator of the presence of expressed genes.

Because of presence of these conserved features it has been possible to make prediction about occurrence of a gene in a given sequence. Gene prediction based on sequence similarity has following limitations:

1. Analysis by sequence similarity can only reliably identify about 30% of the protein coding genes in a genome.
2. 50-80% of new genes identified have a partial, marginal, or unidentified homology.
3. Frequently expressed genes tend to be more easily identifiable by homology than rarely expressed genes.

13.1 FEATURES OF GENE FINDING

Process of identifying potential coding regions in an uncharacterized region of the genome is called as gene finding. Important facts about gene finding are highlighted as under:

1. **Computational gene finding is very different from biological gene finding:** Computational gene finding is the process of identifying potential coding region in an uncharacterized regions of a DNA sequence. The process of gene finding by biological method involves extracting mRNAs in the cell and identifying the location of DNA corresponding to this on the gene (Fig. 13.3). In computational gene finding, gene expression is not the only thing that we are looking for. Other biologically significant regions in a sequence which are identified computationally, which hint towards presence of gene in the given sequence. In a lot of cases, we don't even know what constitute one of these sites, so all we can do is to look for several conserved features in the given sequence. This still is a subject of active research. There are many different gene finding software packages and no one program is capable of finding everything. The sequence elements we look for finding a gene are ORF, repeat elements, splice sites, protein binding sites, MARS and SARS etc., promoter region and transportable elements characteristic features of some of these one discussed already discussed in chapter 9, Table 9.1. These sequence elements are great aid in gene identification.

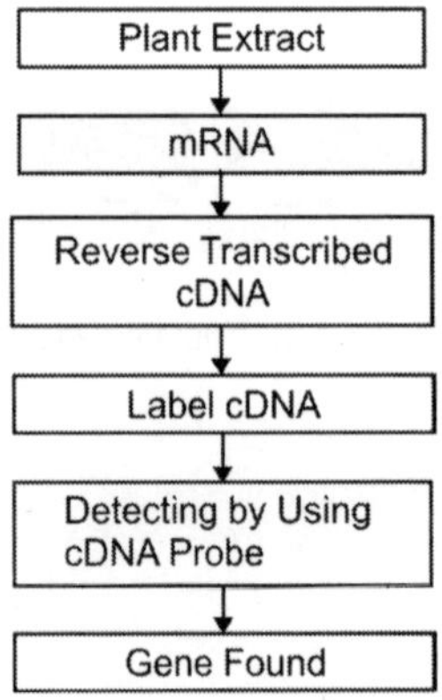

Fig. 13.3: Gene finding by biological method

2. **Gene finding is approached in two ways:** Starting from CDS to protein (Fig. 13.1) or from protein to cDNA (Fig. 13.2).
3. **Gene finding for eukaryotic or prokaryotes organisms is different:** Gene finding in prokaryotes is much easier than eukaryotes because of following reasons:
   (*a*) Prokaryotes have no introns.
   (*b*) Promotor regions and start sites more highly conserved in prokaryotes.
   (c) Eukaryotic DNA are wrapped around histones that might result in repeated patterns for histone binding. The promoter regions might be near these sites so that they remain hidden.
   (*d*) Abundant comparative information is available in prokaryotes.
4. **Gene finding is species-specific:** General characters for this species specificity are:
   (*a*) Codon usage patterns vary with species.

(*b*) Functional regions (promoters, splice sites, translation initiation sites, termination signals) vary with species.
(*c*) Common repeat sequences are species specific.
(*d*) Gene finding programs rely on this information to identify coding regions.

Table 13.1: Common terms used in gene finding

Term	*Description*
1. ORF	A open reading frame is one of the three possible ways of reading a nucleotide sequence as a series of triplets. An open reading frame contains no termination codons and is thus potentially translatable into protein.
2. Bio-dictionary	A exhaustive collection of seqlets that cover essentially all known samples of naturally occurring proteins.
3. Seqlets	Regular expressions to map to combinations amino acids that occur 2 or more times.
4. Putative gene	Gene proposed to exist.
5. (L, W) Pattern	Begins and ends with a literal character, W, is the length of the pattern and L is number of positions occupied by literals.
6. Pattern	A sequence of amino acids that contain optional position.

13.2 GENE FINDING METHODS

Gene finding strategy for beginners begins with choice of the appropriate type of gene finder. It is very important to be made sure that you're using gene finders for microbial (intronless) sequences only to analyze bacteria and archaea. If there is no organism-specific gene finder for your system, at least use one that makes sense like use an *Arabidopsis* gene finder for other plants. General protocol for gene identification in a sequence is given in Fig. 13.4.

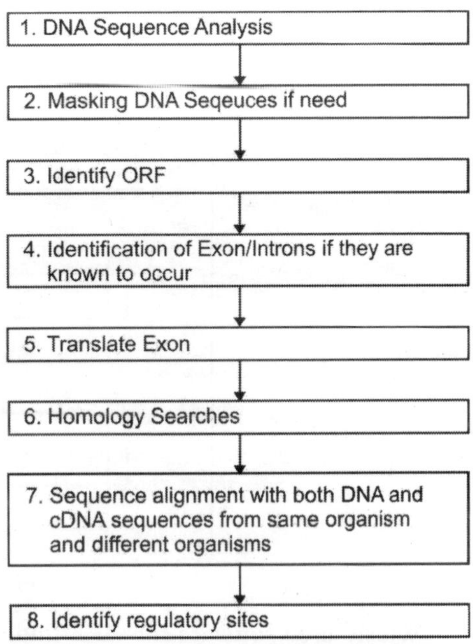

Fig. 13.4: Protocol for gene identification

13.2.1 Masking DNA Sequences

Before matching any new sequence with the various nucleotide databases (or protein databases), one must clean up the data from repeat sequences or vector sequences. There are several programs that will mask a nucleotide sequence from repeat and vector sequences by replacing the nucleotide position with an "X" or similar character. Ideally, this replacement doesn't change the original numbering of the query sequence. Several softwares used for vector masking are given in Table 13.2. If one is performing genomic sequence, further annotation of the sequence is possible besides sequence matching. One would also like to map the gene structure of their genomic sequence by noting the exon/intron boundaries, the promoter region and other sequence signals. There are several programs that will predict the location of these nucleotide sequence patterns and boundaries. However, because these programs rely on different algorithms, one should use the results of more than one program to determine the genomic structure of a DNA sequence.

Table 13.2: Repeat/Vector masking software

Site Name	*Description of Site*
Mask	This searches for vector sequences (e.g. plasmids, *E. coli*, etc.) in the query sequence and "masks" or "Xs" out that vector sequence from the query sequence.
CENSOR	The CENSOR web server allows users to have query sequences aligned against a reference collection of human, rodent, or plant repeats.
DNA Repeats	This is a collection of DNA repeat-finding sites.
Repeat Masker	Repeat Masker screens DNA sequences in fasta format against a library of repetitive elements and returns a masked query sequence ready for database searches as well as a table annotating the masked regions.
RepMask	Repeat masking using Repeat Masker, XBlast, XNU and SEG at TigemNet.
VecScreen	VecScreen is a system for quickly identifying segments of a nucleic acid sequence that may be of vector origin.
XBLAST	Reads BLAST output and masks query.

13.2.2 ORF Finding

Identifying ORFs is the simple first step in gene finding. Concept of ORFs is explained in Fig. 13.5. ORF refers to open reading frame. Every DNA sequence can be read in six ways i.e. 5′ first nucleotide, 5′ second nucleotide, 5′ third nucleotide, 3′ first nucleotide, 3′ second nucleotide and 3′ third nucleotide. By looking at the start and stop codons usually is possible to find out which ORF is present in a gene and what are UTR and CDS in this sequences. These are very useful in gene identification and gene annotation. There are 4 main steps in finding the ORF are summarised in Fig. 13.6.

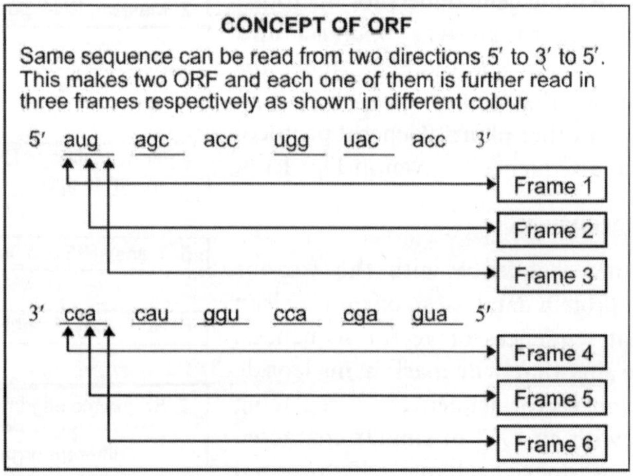

Fig. 13.5: Concept of OR

Widely used software for ORF finding is NCBI ORF Finder. In Fig. 13.7 an interface of NCBI ORF finder is shown. The ORF finder is a graphical analysis tool, which finds all open reading frames of a selectable minimum size in a user's sequence or in a sequence already in the database. In Fig. 13.7 a view of ORF finder showing result of a search sequence is shown. This tool identifies all open reading frames using the standard or alternative genetic codes. The deduced amino acid sequence can be saved in various formats and searched against the sequence database using the WWW BLAST server. The ORF finder should be helpful in preparing complete and accurate sequence submissions. It is also packaged with the Sequin sequence submission software.

Another popularly used ORF finding tool is retranslate at Expasy site is used for ORF finding in experiment 13A and details of ORF and ORF finding is given there.

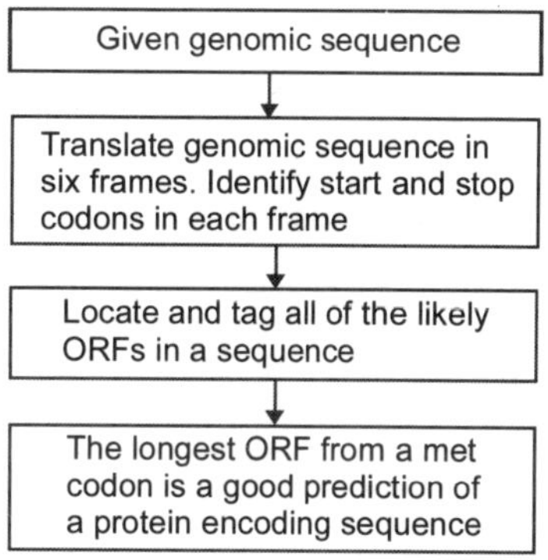

Fig. 13.6: Protocol for finding ORF

Once the ORF are predicted with any of the softwares it is an important task to choose the coding ORF from the six ORFs. Each species has a characteristic pattern of use of synonymous codons i.e. codons that stand for the same amino acid. There are also different patterns of use of codons in strongly us. Weakly expressed genes as, for example in *E. coli*, there is a strong preference for certain codon pairs within a coding region and for certain codons to be next to the termination codon. There is also a strong preference for codon pairs in eukaryotic exons that has been very useful for distinguishing exons and introns in eukaryotic genomes. Similarly organisms with a high genome content of GC have a strong bias of G and C in the third codon position. Based on these characteristics three tests have been devised to verify that a predicted ORF is in fact likely to encode a protein.

ORF finding based on start and stop codon frequency is a pattern-based procedure. Other pattern-based procedures recognize characteristic sequences associated with known features and genes, such as ribosome binding sites, promoter sites, histone binding sites, etc. These all are statistically based methods. Content-based gene finding methods rely on statistical information derived from known sequences to predict unknown genes. Some evaluative measures include: "coding potential" (based on codon bias), periodicity in the sequence, sequence homogeneity, etc. A standard content based gene finding procedure is depicted in Fig. 13.8.

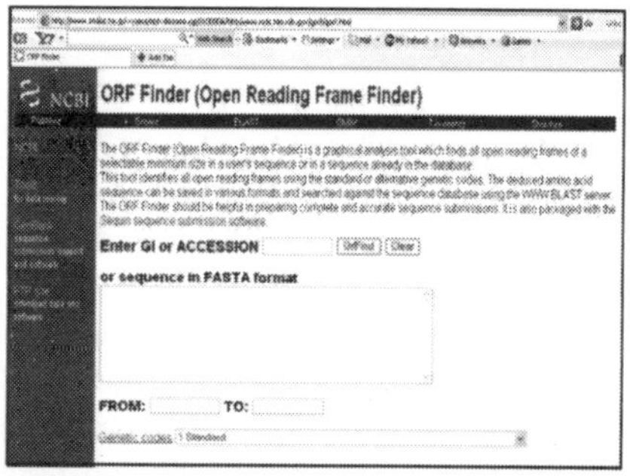

Fig.13.7: Web view of NCBI ORF finder

1. Check if the third base in the codons tends to be the same one more often than by chance alone.
2. Are the codons used in the ORF the same as those used in other genes (need codon usage frequency)?
3. Compare the amino acid sequence for similarity with other known amino acid sequences.

For the ORF to contain a genes it must be long enough (roughly 300 bp or more), should have average amino-acid composition specific for a given organism and should have codon use specific for the given organism. Still it is difficult to conclude that the given ORF is coding for a gene. Several common problems associated with ORF finding are:

1. A single-character sequencing error can hide a stop codon or insert a false stop codon, preventing accurate identification of ORFs.
2. Short exons can be overlooked.
3. Multiple transcripts or ORFs on complementary strand can confuse results.

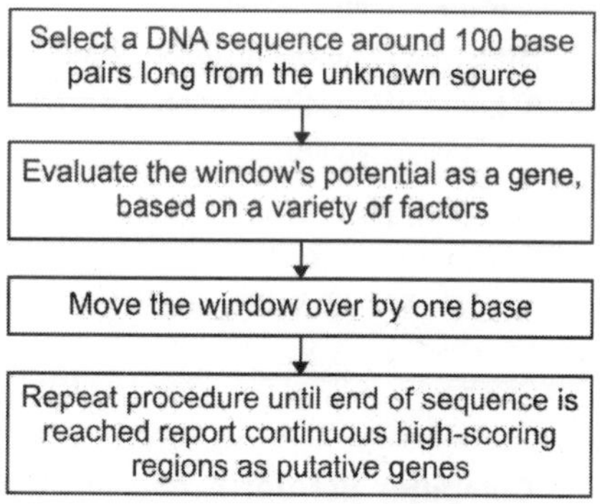

Fig. 13.8: A standard content-based alignment procedure

EXPERIMENT 13A

To study various functional aspects of given DNA sequence. It is assumed that the given sequence is isolated from a prokaryotic cell. Determine the most probable ORF, UTR and CDS in this sequence and finally translate using translate software at expasy site.

About CDS, ORF and UTR: CDS refers to the coding sequence that is actually expressed as protein i.e. they code for a amino acid. UTR refers to untranslated region of sequence, which usually has a role in regulation of expression and are not coding for an amino acid usually (see Fig. 13.13). ORF refers to oral reading frame. Every DNA sequence can be read in six ways i.e. 5′ first nucleotide, 5′ second nucleotide, 5′ third nucleotide, 3′ first nucleotide, 3′ second nucleotide and 3' third nucleotide. By looking at the start and stop codons usually is possible to find out which ORF is present in a gene and what are UTR and CDS in this sequences. These are very useful in gene identification and gene annotation.
Start and stop codon in a protein coding gene.

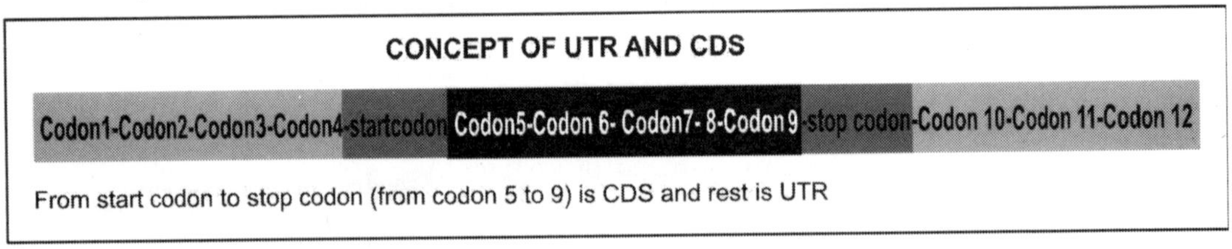

Fig. 13.9: Diagrammatic representation of occurrence CDS, UTR, start and stop codon in a protein coding gene

About Expasy Poteomics Tools: It provides a long list of proteomics tools. Many of these tools are local to the ExPASy server. Translate tool is local to expasy and translates a nucleotide sequence (DNA/RNA) to a protein sequence. IT is very easy to access translate tool and provides easy and clearer presentation comparatively. List of other tools available there is given below.

1. Translate Translates a nucleotide sequence to a protein sequence.
2. Transeq Nucleotide to protein translation from the EMBOSS package.
3. Graphical Codon Usage Analyser: Displays the codon bias in a graphical manner.
4. BCM search launcher: Six frame translation of nucleotide sequence(s).
5. Backtranslation: Translates a protein sequence back to a nucleotide sequence.
6. Genewise: Compares a protein sequence to a genomic DNA sequence, allowing for introns and frameshifting errors.
7. FSED: Frameshift error detection.
8. LabOnWeb: Elongation, expression profiles and sequence analysis of ESTs using Compugen LEADS clusters.

Hypothesis: It would be possible to determine the most probable ORF, UTR and CDS in the given DNA sequence. It is assumed that this sequence is from a prokaryotic cell. The translate tool at expasy server would be used for this purpose.

TESTING THE HYPOTHESIS

Computer Set up: A normal PC with internet connection in windows environment

1. Log on to web site of Expasy proteomic tools. This takes to the welcome page of the website (Fig. 13.10).

2. Copy the sequence (as given in Data Box 13.1) to a notepad (START<ACCESSARY<NOTEPAD) and save it as experiment_13.doc. Translate it manually to mRNA by writing complementary sequences. Save the mRNA sequences as Expt13_mRNA.doc (Data Box 13.1).
3. Copy the mRNA sequences to the text box of the web site and press translate.
4. Repeat the same with the starting DNA sequence. Save results as in Fig. 13.12.
5. Analyze the results.

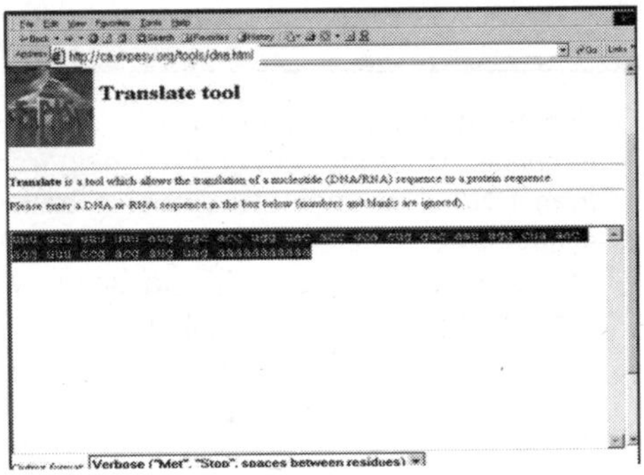

Fig. 13.10: Webview of expasy website

DATA

Data Box 13.1: The given sequence and the product of transcription

The given DNA sequence
aaa aaa aaa aaa tac tcg tgg acc atg tgg agt gac ctg tta acc gat ttg tcc aaa ggc tgc tac atc tttttttttt

↓

Product of manual transcription mRNA
uuu uuu uuu uuu aug agc acc ugg uac acc uca cug gac aau ugg cua aac agg uuu ccg acg aug uag aaaaaaaaaa

Data Box 13.2: Result of translation from mRNA

Translate Tool - Results of translation
Please select one of the following frames:

5′3′ Frame 1
F F F F Met S T W Y T S L D N W L N R F P T Met Stop K K K

5′3′ Frame 2
F F F L Stop A P G T P H W T I G Stop T G F R R C R K K K

5′3′ Frame 3
F F F Y E H L V H L T G Q L A K Q V S D D V E K K K

3′5′ Frame 1
F F F F Y I V G N L F S Q L S S E V Y Q V L I K K K

3′5′ Frame 2
F F F S T S S E T C L A N C P V R C T R C S Stop K K K

3′5′ Frame 3
F F F L H R R K P V Stop P I V Q Stop G V P G A H K K K K

Data Box 13.3: Result of translation from DNA directly, however DNA is not being transcribed rather straight serving as codons

Translate Tool — Results of translation
Please select one of the following frames:

5′3′ Frame 1
K K K K Y S W T Met W S D L L T D L S K G C Y I F F F

5′3′ Frame 2
K K K N T R G P C G V T C Stop P I C P K A A T S F F F

5′3′ Frame 3
K K K I L V D H V E Stop P V N R F V Q R L L H L F F F

3′5′ Frame 1
K K K K Met Stop Q P L D K S V N R S L H Met V H E Y F F F

3′5′ Frame 2
K K K R C S S L W T N R L T G H S T W S T S I F F F

3′5′ Frame 3
K K K D V A A F G Q I G Stop Q V T P H G P R V F F F F

INTERPRETATION OF THE DATA

The most modern definition of translation is synthesis of amino acid sequence on the mRNA template. Translate tool is also doing translation but it is starting at both DNA and mRNA. Starting from DNA it is understood that a transcription would follow and the resultant mRNA would be synthesized by the program and then it would be used for amino acid synthesis. However on comparison the results of Data Boxes 13.2 and 13.3 reveals that it does not happen in translate tool. When results are compared using the genetic code table, resultant amino acid chain as a result of mRNA template synthesis appears to be correct. Starting with DNA it only considers every thymine base to be uracil and do translation considering this DNA same as RNA. This is correction needs to be done in the algorithm of the program. The correct steps followed broadly should be as shown in Fig. 13.11.

Concept of ORF is explained once again in Fig. 13.12 with the help of experimental sequence. By looking at the sequence of amino acid synthesized it is not difficult which one of the ORF

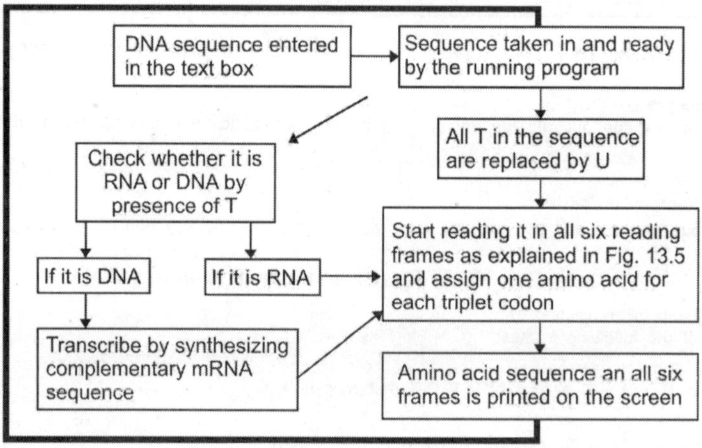

Fig. 13.11: Step involved and amino acid synthesis from a DNA sequence

Sequence to be analyzed

5′uuu uuu uuu uuu aug agc acc ugg uac acc uca cug gac aau ugg cua aac agg uuu ccg acg aug uag aaaaaaaaaa 3′

Sis ORF starting from six reading frames respectively are:

First ORF 5′-3′
5¢uuu uuu uuu uuu aug agc acc ugg uac acc uca cug gac aau ugg cua aac agg uuu ccg acg aug uag aaaaaaaaaa 3′
Second ORF 5′-3′
5¢uuu uuu uuu uua uga gca ccu ggu aca ccu cac ugg aca auu ggc uaa aca ggu uuc cga cga ugu aga aaa aaa aaa a 3′
Third ORF 5′-3′
5¢uuu uuu uuu uau gag cac cug gua cac cuc acu gga caa uug gcu aaa cag guu ucc gac gau gua gaa aaa aaa aaa 3′
First ORF 3′-5′
3¢ aaa aaa aaa aag aug uag cag ccu uug gac aaa ucg guu aac agg uca cuc cac aug guc cac gag uau uuu uuu uuu uu 5′
Second ORF 3′ - 5′
3 ′aaa aaa aaa aga ugu agc agc cuu ugg aca aau cgg uua aca ggu cac ucc aca ugg ucc acg agu auu uuu uuu uuu u 5′
Third ORF 3′– 5′
3′ aaa aaa aaa gau gua gca gcc uuu gga caa auc ggu uaa cag guc acu cca cau ggu cca cga gua uuu uuu uuu uuu

Fig. 13.12: Six ORFs of the given sequence

sequence is coding for a gene. For example a careful examination of amino acid sequence in six ORFs reveal that the gene encoding sequence is read in frame 5′ 3′ starting with first amino acid. The main reason for such a conclusion is the presence of start codon in the beginning and stop codon at the end. In the beginning of the experiment it was given that this sequence is from prokaryotes this leaves no scope for presence of introns and exons in the sequence. After assuming that it the first sequence that is coding for a gene, it is very easy to conclude that the region from start to stop codon refers to CDS and the sequence on both sides of CDS is the UTR in this sequence. All this information helps in gene identification.

13.3 GENE-FINDING SOFTWARE

Different gene finding software and their characteristic features is given in Table 13.2. Programs rarely use one measure to predict genes. Different programs use different algorithms and measures to make a gene prediction. For example, the popular gene finding program Grail looks for exon boundaries based on start and stop codons, splice site sequences, GC content, potential polyadenylation sites, and repetitive sequence elements to make prediction for occurrence of a gene. On the other hand Glimmer uses an interpolated Markov model approach. Markov model is a model for computing probabilities in the context of sequential events. Different values are combined using different methods like (probabilistic methods, discriminant analysis, neural net methods, etc.) to produce one score for the entire window. Detailed explanation about Markov model is beyond the scope of this book.

GRAIL is a suite of tools designed to provide analysis and putative annotation of DNA sequences both interactively and through the use of automated computation. GRAIL was developed by Uberbacher et al. of DOE. GRAIL is a program that looks for exon boundaries based on start and stop codons, splice site sequences, GC content, potential polyadenylation sites, and repetitive sequence elements. The coding recognition portion of the system uses a neural network, which combines a series of coding prediction algorithms. There are three basic versions of this neural network, GRAIL 1, GRAIL 1a and GRAIL 2. GRAIL 1 has been in place for about three years. It uses a neural network. GRAIL 1a is an updated version of GRAIL 1. GRAIL 2 uses a scheme that facilitates the use of more genomic context information like splice junctions, translation starts, non-coding scores of 60-base regions on either side of a putative exon in the exon recognition process. GRAIL 2, is therefore, not appropriate for sequences without genomic context. These changes have improved the overall performance compared to GRAIL 1, particularly for short exons. Description about other common gene prediction softwares is summarized in Table 13.3.

Most gene finding software are web-based, but they differ in terms of submit sequence; input sequence length may be limited. For example three versions of Grail use different size window for submit sequence. GRAIL 1 recognizes coding potential within a fixed size (100 base) window. It evaluates coding potential without looking for additional features (information such as splice junctions, etc). GRAIL 1a also uses a fixed-length window to locate the potential coding regions and then evaluates a number of discrete candidates of different lengths around each potential coding region, using information from potential exon candidate, defined as an open reading frame bounded by a pair of start/donor, acceptor/donor or acceptor/stop sites.

13.4 MODELS IN GENE PREDICTION

13.4.1 Neural Network Technology

Neural network techniques involve a set of many simple units that hold symbolic data, which are interconnected by a network of links associated with numeric weights. Neural network structure is reminiscent of the organisation of neurons in brain. Neural networks can adjust the parameters that define the underconnections among its units in response to data; thus can be used to identify very complex patterns coming from the input data. Units operate only on their symbolic data and on the

HISTORICAL PROFILE 15

Edward C. Uberbacher for Gene Finding Software Grail

Edward C. Uberbacher received his B.A. degree in chemistry from the Johns Hopkins University in 1974, and PhD in physical chemistry from the University of Pennsylvania in 1979, with a thesis in the area of macromolecular crystallography. Starting in 1980, he did post-doctoral studies at the University of Pennsylvania Department of Biophysics (Johnson Foundation), and the Biology Division of Oak Ridge National Laboratory - University of Tennessee Graduate School of Biomedical Sciences, investigating the structure and function of genetic materials using crystallography, electron microscopy, and computational modeling. In 1985 he became an investigator at the Center for Small-Angle Scattering Research at ORNL, pursuing structural and dynamic studies of macromolecules in solution using neutron and X-ray scattering techniques and molecular modeling techniques. In 1987, he also became a Research Assistant Professor at the University of Tennessee Graduate School of Biomedical Sciences, and an investigator in the ORNL Biology Division, focusing on X-ray and neutron crystallography, scattering, and other biophysical methods and solved a crystal structure of the nucleosome core particle. In 1988 he became a consultant at the ORNL Engineering Physics and Mathematics Division to develop AI and high-performance computing methods for genomic DNA sequence analysis, and in 1991 joined the staff of the Computer Science and Mathematics Division as the Informatics Group leader, where he received an R&D 100 award for the development of the GRAIL DNA sequence analysis system. In 1997, he became the head of the Computational Biology Section in Life Sciences, and became a co-developer of the PROSPECT computational protein fold prediction system, which received an R&D 100 award in 1998. In 2003-2004, Dr. Uberbacher performed part-time duties as an IPA at the DOE Office of Biological and Environmental Research contributing extensively to the GTL computing roadmap. He is currently the Program Leader for Computational Biology and an Adjunct Professor in the Genome Science and Technology Program at the University of Tennessee. His scientific interests include the application of pattern recognition, artificial intelligence, concurrent processing techniques for computational biology, algorithm development for computational biology, computational genome sequence analysis, mass spectrometry analysis, and macromolecular structure, dynamics and docking.

inputs that they receive through their connection. Neural network topology is depicted in Fig. 13.13. Neural network methods are applied to gene finding and also used for predicting secondary structure of proteins.

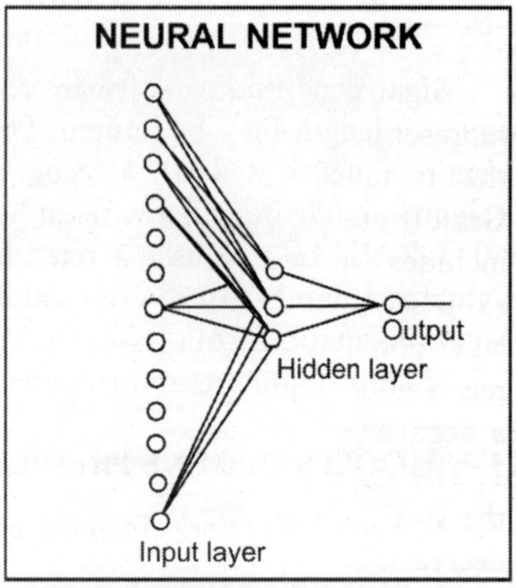

Fig. 13.13: Grail II neural net

The model in neural network comprises of three layers of processing units the input layer, the output layer, and the hidden layer between these layers. Data are sent from input layer to the hidden layer and from the hidden layer to the output layer through junctions between the units. The input layer of units reads the sequence, one unit per amino acid residue and transmits information on amino acid that location. A small window of sequence is read at a time and information is sent as signals through junctions to a number of sequential units in the hidden layer by all of the input units within the window as shown by the lines jouning units in Fig. 13.13. These signals are each individually modified by weighing factor and then added together to give a total input signal into each hidden unit. The resulting signal is then transformed by the hidden unit into number that is very close either to one or a zero.

Table 13.3: GENE-finding software

Name of Software	*Description*
ORNL Grail (v1.3)	Gene finder for human, mouse, *Arabidopsis*, *Drosophila* and *E. coli*. It is based on neural networks and masks human and mouse repetitive elements. It incorporates pattern-based searches for several types of promoters and simple repeats. It has accuracy in 75-95% range.
GLIMMER	Glimmer (Gene Locator and Interpolated Markov Modeler) uses interpolated Markov models (IMMs) to identify the coding regions and distinguish them from non-coding DNA. It is gene finder for bacterial and archaebacterial genomes. It predicts genes with around 98% accuracy when compared with published annotations. TIGR has developed this software.
GENSCAN	Gene finder for human and vertebrate sequences. Maize and Arabidopsis-optimized versions are available now. Probabilistic method based on known genome structure and composition: number of exons per gene, exon size distributions, hexamer composition, etc. Only protein coding genes predicted. Accuracy in 50-95% range. Burge et al., of Stanford and MIT developed this program. This is designed to predict complete gene structures—including exons, introns, promoter and polyadenylation signals in genomic sequences. It differs from the majority of existing gene finding algorithms in that it allows for partial as well as complete genes and for the occurrence of multiple genes in a single sequence, on either or both DNA strands.
Gene Mark	Gene finder for bacterial and archaebacterial sequences. Gene Mark and Gene Mark HMM available as web servers. Accuracy is in 90-99% range. It is provided by Georgia Tech. This program is used for gene prediction in sequences.
CRITICA	Gene finder for bacterial and archaebacterial genomes. Combines sequence homology-based prediction with content-based statistical (dicodon probability) analysis. Accuracy is in 90-99% range. No web server.
Gene Parser	Predicts the most likely combination of exons and introns using dynamic programming. The intron and exon positions are aligned subject to the constraint that they alternate. A neural network is used to adjust the weights given to the sequence indicators of know exon and intron regions such as codon usage, information content, length distribution, hexamer frequencies, and scoring matrices. University of Colorado developed this software.
TWINSCAN	TWINSCAN is a modern gene prediction system designed for the analysis of eukaryotic genomic sequences. TWINSCAN is one of the most accurate gene prediction systems today. It achieves this by simultaneously modeling both gene structure and evolutionary conservation. TWINSCAN was designed with high-throughput analysis in mind.

Signals from hidden units are then sent to three individual output units, each output unit representing one type character gene finding. Grail II neural Net finds exons in eukaryotic genes, that is, takes inputs and predicts if a gene is present. Grail II neural net is shown in Fig. 13.12. Grail II uses a neural network to identify patterns characteristics of coding sequences. This method includes the use of three layers: an input layer for the data with the data coming from a candidate exon sequence, and a hidden layer for discerning relationships among the input data. An output layer comprising one neuron indicates whether or not the region is likely to be exon. Each neuron receives information from a set in the layer above, some with a positive value and others with a negative value, sums these values; and then convert them to an output of approximately 0 or 1. The system is trained using a set of known coding sequences, and as each sequence is utilized, the strengths and types of connection (position or negative) between the neurons are adjusted, decreasing or increasing the signal to the next neuron in a manner that produces the correct output in Grail II a candidate sequence is evaluated by calculating pattern frequencies in the sequence and applying these values to the neural network. If output is close to a value of 1, then the region is predicted to be an exon. Neural net with reference to secondary structure prediction is discussed in chapter 15.

13.4.2 Hidden Markov Models (HMM)

HMM is a statistical model for an ordered sequence of symbols, creating as a stochastic state machine that generates a symbol each time a transition is made from one state to the next. In sequence analysis, a HMM is usually a probabilistic model of periodic patterns in a sequence for example patterns found in exons of a gene. It is also a model of multiple sequence alignment.

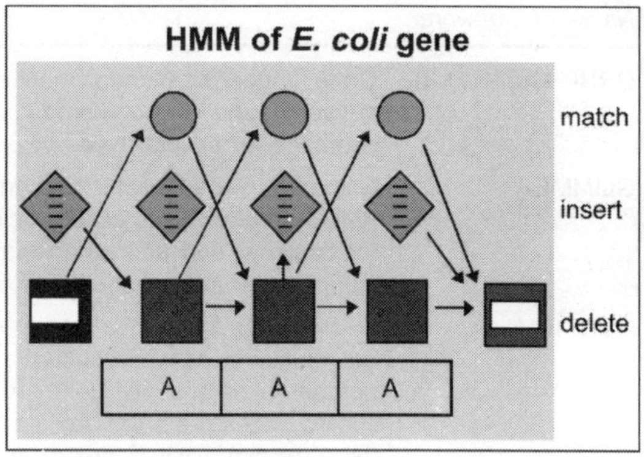

Fig. 13.14: HMM of *E. coli* gene

HMM is applied to model *E. coli* genome. This model will read through a sequence of unknown gene composition and find the genes i.e. a series of codons that specify amino acids flanked by a start and stop codons have been used to train the model. HMM for finding most probable set of genes in *E. coli* sequences of unknown base composition is shown in Fig. 13.14. This model is designated to generate a sequence of amino acid encoding codons of appropriate length of an *E. coli* gene starting with an AUG codon and ending with a stop codon. In Fig. 13.3 the model for each codon is respresented by a set round, diagonal on square boxes representing match, insert and delete states respectively. The model is shown for AAG codon. Each of the 61 codons has a similar structure. If a sequence were extremely accurate only match state would be needed in the model. Further discussion on HMM is beyond the scope of this book. The insert and delete states allow an ORF with an extra or missing base to be recognised directional arrows (transmissions) give the probability of passing from one state to another state in the model. Further discussion on HMM is beyond the scope of book. A version of GeneMark called GENEMARK-HMM (Table 13.2) uses HMM of this type to search for *E. coli* genes.

EXPERIMENT 13B

Identify the Given Gene Sequence using Genebuilder

About Genebuilder: Genebuilder is an integrated computing system for protein-coding gene prediction at Webgene—Tools for prediction and analysis of protein-coding gene structure. The GeneBuilder system is based on prediction of functional signals and coding regions by different approaches in combination with similarity searches in proteins and EST databases. The potential gene structure models are obtained by using a dynamic programming method. The program permits the use of several parameters for gene structure prediction and refinement. During gene model construction selecting different exon homology levels with a protein sequence selected from a list of homologous proteins can improve the accuracy of the gene structure prediction. In case of low homology GeneBuilder is still able to predict the gene structure. GeneBuilder system has been tested by using the standard set and the performances are: 0.89 Sensitivity and 0.91 Specificity at nucleotide level. The total Correlation Coefficient is 0.88.

GeneBuilder is composed of many modules. Each module is executed independently. The GeneBuilder results output is based on results obtained by the automatic execution of different programs and it is shown in a separate window of the browser. GeneBuilder system can be used for human, mouse, *Drosophila*, C. *elegans*, *Arabidopsis* and *Aspergillus* sequences. The set of Potential Coding Fragments (PCFs) is used to construct the gene models with the maximum coding potential by using the dynamic programming technique. It is also to use one homologous protein selected from the list of potential homologous proteins to refine the predicted gene model.

About the sequence: Our aim is to identify the gene for which given sequence codes it really does not make any sense to give the identity of the given sequence. For those who are trying to repeat the present experiment the sequence in the present experiment can be extracted from...and is saved as 'sequence.txt'. Contents of sequence file are shown in data box 13.4.

Hypothesis: it is assumed that the given sequence is coding for an important protein in human being and it contain only one complete gene. This DNA sequence is directly coding for protein i.e. not a complementary strand is used. We would be able find the gene coded by this protein using genebuilder software.

TESTING THE HYPOTHESIS

Computer set up: A normal PC with internet connection in windows environment:

1. Log on to website of GeneBuilder at http://www.itba.mi.cnr.it/webgene/. This takes us to the welcome page of the web site with interface shown in Fig. 13.15. Paste the sequence given in Data Box 13.4.
2. Select the following options:
   (*a*) Organism. We would select human here because our sequence is hypothesized to be belonging to human beings.
   (*b*) Mode (Gene, Exon). The GENE option is used for predicting the full gene model. The EXON option is used for only selecting the exons with the best scores. We would select gene here.
   (*c*) Strand (Direct, Complement). By default the analysis is performed on the direct strand. If the complement option is selected, the analysis will be executed on the complementary DNA strand. We would select direct here.

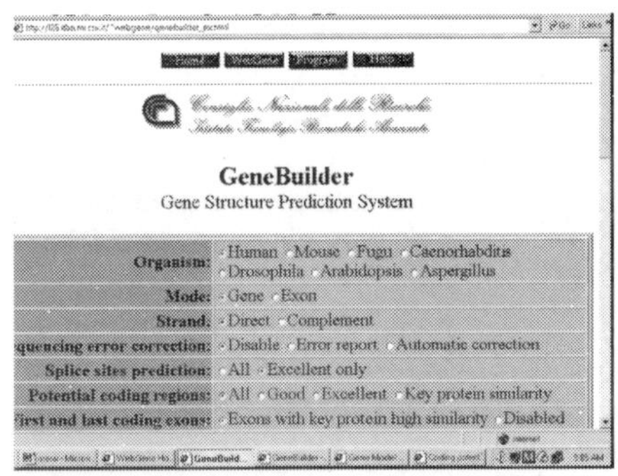

Fig. 13.15: Web view of gene builder

   (*d*) Sequencing error correction (Disable, Error report, Automatic correction). By selecting the automatic correction option, it is possible to find and automatically correct potential sequencing errors due to the frame-shifts and substitutions in the stop codons. The gene model predicted can be substantially improved if these errors are eliminated. The Error report option is used to generate only the report without correcting the sequence under analysis. We select automated here.
   (*e*) Splice sites prediction (All, Excellent only). For the splice site prediction we use the classification analysis combined with the weight matrix technique. When the excellent only option is selected, the program is able to find 95% of real splice sites. About 15% of pseudosites will be predicted as splicing signals. When the all option is selected, the program is able to find 98% of used for gene reconstruction. In this mode and with GENE option GeneBuilder will try to reconstruct potential gene. This mode is very useful as the first step of sequence analysis when no information about gene content is available for a query sequence. As a second step, proteins found by the searches of the predicted peptides against the protein database can be used to obtain a more accurate gene structure. With the Good option selected, only the exons having "good" quality will be used for gene reconstruction. When the Excellent exons option is used only the exons having "excellent" quality will be used for gene reconstruction. Finally when the Protein similarity

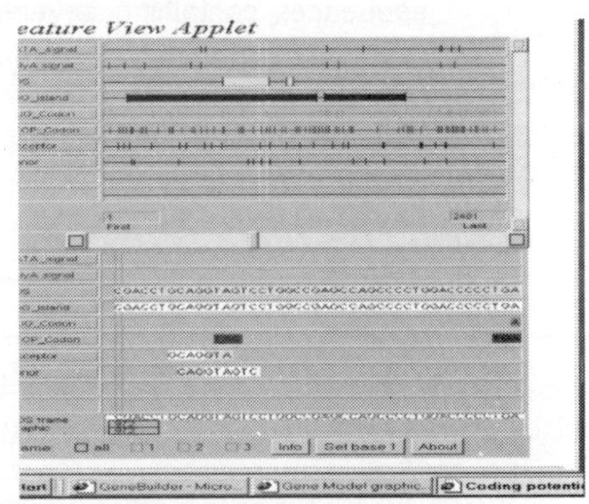

Fig. 13.16: Web view of gene builder

option is selected, only the exons with similarity to a selected homologous protein will be used for gene reconstruction. We select all here.

(*f*) First and last coding exons (Disable, Exons with high protein homology). With the Exons having high protein homology option selected it is possible to require more accuracy in the determination of the first and last exons (potential genes must start and finish with well confirmed coding exons.) This option is very important where several genes are present in a query sequence and the gene localization can be confirmed by using homology with a chosen protein.

(*g*) Sequence segment for coding region predictions (Start, End). The first and last segment positions for coding region prediction: This option is particularly the real splice sites, but 30-35% of all predicted sites will be false. We select excellent here.

(*h*) Potential coding regions (All, Good, Excellent, Key protein similarity). Potential coding regions are found by combining the protein coding potential calculated by using the dicodon statistic and the splicing signals. With the All option selected, all potential coding exons (with and without similarity to key proteins) will be Yes option selected the program is able to predict the repeated elements present in the sequence and mask them before searching the ESTs homology search in GeneBuilder. We select no here.

(*i*) EST mapping (Yes, No). With the Yes option selected the GeneBuilder performs a homology search against the EST database and the position of the homologous EST sequences is reported in the output in relation to a query sequence. The repeated elements can be automatically masked before the homology search. This information is used only when the similarity between the EST sequences and the query sequence is greater than 95. We select no here, 12. TATA box prediction (Good, Marginal). With the option Good only the better matches are presented as potential TATA-box. With the option Marginal all potential TATA-box are described. We select good here.

(*j*) POLY-A site prediction (Poly A pattern length). We select 10 here.

(*k*) Search for potential binding sites of transcription factors (Vertebrates, Fungi, Insects, Plants, Miscellaneous, All). By selecting the Vertebrates, Fungi, Insects, Plants, Miscellaneous options it is possible to search the input sequence by using the appropriate group of weight matrices. The group matrix for vertebrates' genomes is used as default. It is also possible to choose the matrices to be used individually.

(*l*) Mail to (Yes, No) Results of GeneBuilder in text format can be sent by e_mail. This is useful when the analysis is of long sequences or is over poor network connections. In this case the e_mail address is mandatory. We select no here useful when analyzing very long sequences containing several genes.

(*m*) Complete gene model (Yes, No). With the Yes option selected the program is able to reveal only the models with complete potential gene structure including the first and last exons. With no selected any gene models, including partials are reported. We select yes here.

(*n*) Result obtained is shown in Fig. 13.16. Details of information gathered is given in Data boxes 13.5 to 13.8.

DATA

Data Box 13.4: The given sequence

AGCTTTCTTC TTTTCCCTGT TGCTCAAATA AATAGTGTTC TTTGCTCAAA
CCCCCTTTCC CTCCTCCTTC TGCAATCTCA GCGCCTAGCG AAATCTGTTT
TCTTCATTGT AACCTCAGCT TCACCGCAAT TAATTTTTTT TCCCTCTGGT
CACAAGATAA TTCCTGACGC CAGTGAGTCT GGAGGTCAGA CGAACAGCAA
ATTGGGGAAC AAGGCGGCAC TAATTCCTTA CAAGTTCCTT GAAAAATCTT
TCGCTTAAAA AAAACGGGGG GTGGGGGGAG CTTCTTTGCT GTTCAGGGAT
TTATGCCTCG CGGAGCTGTG GCTCGAACCA GTGTTGGCTA AGGCGGACTG
GCAGGGGCAG GGAAGCTCAA AGATCTGGGG TGCTGCCAGG AAAAAGCAAA

```
TTCTGGAAGT TAATGGTTTT GAGTGATTTT TAAATCCTTG CTGGCGGAGA
GGCCCGCCTC TCCCCGGTAT CAGCGCTTCC TCATTCTTTG AATCCGCGGC
TCCGCGGTCT TCGGCGTCAG ACCAGCCGGA GGAAGCCTGT TTGCAATTTA
AGCGGGCTGT GAACGCCCAG GGCCGGCGGG GGCAGGGCCG AGGCGGGCCA
TTTTGAATAA AGAGGCGTGC CTTCCAGGCA GGCTCTATAA GTGACCGCCG
CGGCGAGCGT GCGCGCGTTG CAGGTCACTG TAGCGGACTT CTTTTGGTTT
TCTTTCTCTT TGGGGCACCT CTGGACTCAC TCCCCAGCAT GAAGGCGCTG
AGCCCGGTGC GCGGCTGCTA CGAGGCGGTG TGCTGCCTGT CGGAACGCAG
TCTGGCCATC GCCCGGGGCC GAGGGAAGGG CCCGGCAGCT GAGGAGCCGC
TGAGCTTGCT GGACGACATG AACCACTGCT ACTCCCGCCT GCGGGAACTG
GTACCCGGAG TCCCGAGAGG CACTCAGCTT AGCCAGGTGG AAATCCTACA
GCGCGTCATC GACTACATTC TCGACCTGCA GGTAGTCCTG GCCGAGCCAG
CCCCTGGACC CCCTGATGGC CCCCACCTTC CCATCCAGGT AAGCCTCGAA
GTCGGGACAG GGCTGAACAC CCAGGCAAGG ATGCTGCGGG ACCCTCGGAG
CTCCCGATTG CCTCGCGTAA CTCTTCCCTC TTTTCCTCTA ATCAGACAGC
CGAGCTCGCT CCGGAACTTG TCATCTCCAA CGACAAAAGG AGCTTTTGCC
ACTGACTCGG CCGTGTCCTG ACACCTCCAG GTGAGTATCT CCTCTCTTGG
AGAGGGAGGT TTAAACGGCA AGTCCTGGAG TTGGCAGACG TTTTGAAAAA
TTGCCACTCA CTCGGTTTAG GGAAACTGAG GCCAGAGAGG GACAAGTGAC
TTGCCCATGG TTGCATCAAA TGAATGGCAG AGTCAGTTTC CATGTGATGT
GCATTTAAGC CTTAATGCGC CTGGCCCTGC CTCCGCAGTG GCCGAGGTCT
GGCAAGTAGA CATGGTCCGA CTAAATACAA GTCTTTCTGT TCCATGTTGT
ATAGGAGCTG TCTTCGGCAG CCCCCTCCCA GCTAGTGTCA ATTCCAAGTA
GGAGGGGTAG CGCAACGTCC GCCTGTGGTC TTTGGCGCCA ACTGGGTGGG
GGCAGCGTGG GGGGCGGAGT TATCAGGCTG GAGGTACAGA CCAAGTTTCC
TCCCTGGCGC CGGCCAGTCT GCGGACGGCC CCCGCCTCGG CACGCTCGGC
GGAAACTGAC TGCTCCTTGG TCTTCTTTCC TCCCCCGCCC AGAACGCAGG
TGCTGGCGCC CGTTCTGCCT GGGACCCCGG GAACCTCTCC TGCCGGAAGC
CGGACGGCAG GGATGGGCCC CAACTTCGCC CTGCCCACTT GACTTCACCA
AATCCCTTCC TGGAGACTAA ACCTGGTGCT CAGGAGCGAA GGACTGTGAA
CTTGTGGCCT GAAGAGCCAG AGCTAGCTCT GGCCACCAGC TGGGCGACGT
CACCCTGCTC CCACCCCACC CCCAAGTTCT AAGGTCTTTT CAGAGCGTGG
AGGTGTGGAA GGAGTGGCTG CTCTCCAAAC TATGCCAAGG CGGCGGCAGA
GCTGGTCTTC TGGTCTCCTT GGAGAAAGGT TCTGTTGCCC TGATTTATGA
ACTCTATAAT AGAGTATATA GGTTTTGTAC CTTTTTTACA GGAAGGTGAC
TTTCTGTAAC AATGCGATGT ATATTAAACT TTTTATAAAA GTTAACATTT
TGCATAATAA ACGATTTTTA AACACTTGTG TATATGATGA CACCCGTCTC
CATTAAGTAC TAATGATGCT TTCTCGCACA TGGCCGAATT TTGGGAGCTT
TGGGAAAGTG AACTTGCTTA TTCTACGAGA GGGAAATGAA AAACTGCCTG
GTTGAGAGGG GATGGGGTGG AGAGAGAAGG GTTCATGATG GGAGTCTCAT
GTCCATTGAG GGATGGGTGC AGAGAAAAGT TCTGGCTCTG CCTCATTATT
TCAGAGATGA AACCAGAGAC TGGTGCAAGC T
```

Data Box 13.5: Portion of output details of sequence removed to avoid repetition

```
ID HS1D3HIH
XX
OS Homo sapiens
XX
CC Search in direct strand
XX
CC Results of gene prediction by GeneBuilder system
CC using ESTs, coding potential and similarity with key protein(s)
CC URL: www.itba.mi.cnr.it/webgene
XX
CC Best splice signals mode
XX
CC First and last CDSs have expressed similarity
XX
CC All potential CDS were used for prediction
XX
FH Key Location/Qualifiers
FH
FT CDS 739..1038
FT /note=potential CDS predicted by GeneBuilder system
FT /note=first CDS
```

```
FT CDS 1146..1202
FT /note=potential CDS predicted by GeneBuilder system
FT /note=last CDS
XX
FT CpG_island 139..1350
FT /note=detected CpG island
FT CpG_island 1388..1896
FT /note=detected CpG island
XX
CC Translated peptide
CC MKALSPVRGCYEAVCCLSERSLAIARGRGKGPAAEEPLSLLDDMNHCYSRLRELVPGVPR
CC GTQLSQVEILQRVIDYILDLQVVLAEPAPGPPDGPHLPIQTAELAPELVISNDKRSFCH
XX
SQ 2481 BP; 521 A; 662 C; 699 G; 599 T; 0 other;
```

Data Box 13.6: Poly site predicted

```
POLY-A site prediction
Position Pattern
______________
27 CAAATAAATA
128 GCAATTAATT
254 CGCTTAAAAA
545 GCAATTTAAG
606 TGAATAAAGA
1403 GCATTTAAGC
1474 TAAATACAAG
2108 ATAATAGAGT
2206 ATAATAAACG
```

Data Box 13.7: Tata box

```
TATA box prediction
Position Pattern
______________________
606 TGAATAAAGA
636 TCTATAAGTG
636 TCTATAAGTG
636 TCTATAAGTG
1402 TGCATTTAAG
1621 GTTATCAGGC
2105 TCTATAATAG
2105 TCTATAATAG
2115 AGTATATAGG
2184 TTTATAAAAG
2184 TTTATAAAAG
2184 TTTATAAAAG
2229 TGTGTATATG
```

Data Box 13.8: Transcription factors vew

```
MatInspector Release 2.0 August 1996 Fri Oct 29 10:22:46 2004
Solution parameters:
~~~~~~~~~~~~~~~~~~~~
sequence file: /usr/local/webgene/htdocs/tmp/12054.matinsp.inp
core sim:1.00
 matrix sim: 0.85
Explanation for column output:
~~~~~~~~~~~~~~~~~~~~~~~~~~~~~~

->  Matrix positions correspond to sense strand numbering, but all sequences are given in 5'-3' direction.
->  n/a in column 'core simil.' indicates, that no core search was conducted.
->  Capital letters within the sequence indicate the core string.
```

```
matrix | matrix | core | matrix | sequence
name | position(str)| simil. | simil. |
________________________________________________
Inspecting sequence (1 - 2481):
V$AHRARNT_01 | 607 (+) | 1.000 | 0.890 | ataaagaggCGTGcct
V$AHRARNT_01 | 649 (+) | 1.000 | 0.885 | cgcggcgagCGTGcgc
V$AHRARNT_01 | 1597 (+) | 1.000 | 0.872 | tgggggcagCGTGggg
V$AHRARNT_01 | 1688 (-) | 1.000 | 0.929 | tccgccgagCGTGccg
V$AHRARNT_01 | 1987 (+) | 1.000 | 0.887 | ttttcagagCGTGgag
V$AP1FJ_Q2 | 144 (-) | 1.000 | 0.884 | tgTGACcagag
V$AP1FJ_Q2 | 163 (+) | 1.000 | 0.902 | ccTGACgccag
V$AP1FJ_Q2 | 180 (-) | 1.000 | 0.901 | tcTGACctcca
V$AP1FJ_Q2 | 511 (-) | 1.000 | 0.921 | tcTGACgccga
V$AP1FJ_Q2 | 640 (+) | 1.000 | 0.918 | agTGACcgccg
V$AP1FJ_Q2 | 669 (-) | 1.000 | 0.916 | agTGACctgca
V$AP1FJ_Q2 | 950 (-) | 1.000 | 0.909 | gaTGACgcgct
V$AP1FJ_Q2 | 1165 (-) | 1.000 | 0.901 | gaTGACaagtt
V$AP1FJ_Q2 | 1201 (+) | 1.000 | 0.926 | acTGACtcggc
V$AP1FJ_Q2 | 1217 (+) | 1.000 | 0.906 | ccTGACacctc
V$AP1FJ_Q2 | 1345 (+) | 1.000 | 0.929 | agTGACttgcc
V$AP1FJ_Q2 | 1377 (-) | 1.000 | 0.926 | acTGACtctgc
V$AP1FJ_Q2 | 1532 (-) | 1.000 | 0.871 | atTGACactag
V$AP1FJ_Q2 | 1705 (+) | 1.000 | 0.927 | acTGACtgctc
V$AP1FJ_Q2 | 1838 (+) | 1.000 | 0.896 | ctTGACttcac
V$AP1FJ_Q2 | 1944 (-) | 1.000 | 0.948 | ggTGACgtcgc
V$AP1FJ_Q2 | 2145 (+) | 1.000 | 0.948 | ggTGACtttct
V$AP1FJ_Q2 | 2236 (+) | 1.000 | 0.912 | gaTGACacccg
V$AP1_Q2 | 163 (+) | 1.000 | 0.860 | ccTGACgccag
V$AP1_Q2 | 180 (-) | 1.000 | 0.859 | tcTGACctcca
V$AP1_Q2 | 511 (-) | 1.000 | 0.893 | tcTGACgccga
In 0 seq. 0 matches to the matrix V$SEF1_C (re: < 0.01) were found.
In 1 seq. 2 matches to the matrix V$SOX5_01 (re: 1.10) were found.
In 1 seq. 8 matches to the matrix V$SP1_01 (re: 7.52) were found.
In 1 seq. 5 matches to the matrix V$SP1_Q6 (re: 1.70) were found.
In 1 seq. 1 matches to the matrix V$SREBP1_01 (re: 0.21) were found.
In 0 seq. 0 matches to the matrix V$SREBP1_02 (re: < 0.01) were found.
In 0 seq. 0 matches to the matrix V$SRF_01 (re: < 0.01) were found.
In 0 seq. 0 matches to the matrix V$SRF_C (re: 0.08) were found.
In 0 seq. 0 matches to the matrix V$SRF_Q6 (re: 0.16) were found.
In 1 seq. 12 matches to the matrix V$SRY_01 (re: 30.95) were found.
In 1 seq. 3 matches to the matrix V$SRY_02 (re: 3.94) were found.
In 0 seq. 0 matches to the matrix V$STAT1_01 (re: < 0.01) were found.
In 0 seq. 0 matches to the matrix V$STAT3_01 (re: < 0.01) were found.
In 1 seq. 4 matches to the matrix V$STAT_01 (re: 0.48) were found.
In 0 seq. 0 matches to the matrix V$T3R_01 (re: 0.03) were found.
In 0 seq. 0 matches to the matrix V$TAL1ALPHAE47_01 (re: 0.14) were found.
In 0 seq. 0 matches to the matrix V$TAL1BETAE47_01 (re: 0.11) were found.
In 0 seq. 0 matches to the matrix V$TAL1BETAITF2_01 (re: 0.06) were found.
In 0 seq. 0 matches to the matrix V$TATA_C (re: 0.30) were found.
In 1 seq. 1 matches to the matrix V$TAXCREB_01 (re: 0.06) were found.
In 0 seq. 0 matches to the matrix V$TAXCREB_02 (re: < 0.01) were found.
In 1 seq. 9 matches to the matrix V$TH1E47_01 (re: 2.04) were found.
In 1 seq. 1 matches to the matrix V$TST1_01 (re: 5.66) were found.
In 0 seq. 0 matches to the matrix V$USF_01 (re: 1.09) were found.
In 0 seq. 0 matches to the matrix V$USF_02 (re: 43.36) were found.
In 0 seq. 0 matches to the matrix V$USF_C (re: 2.19) were found.
In 0 seq. 0 matches to the matrix V$USF_Q6 (re: 2.52) were found.
In 1 seq. 6 matches to the matrix V$VBP_01 (re: 10.64) were found.
In 0 seq. 0 matches to the matrix V$VJUN_01 (re: < 0.01) were found.
In 0 seq. 0 matches to the matrix V$VMAF_01 (re: 0.99) were found.
In 1 seq. 3 matches to the matrix V$VMYB_01 (re: 2.29) were found.
In 1 seq. 3 matches to the matrix V$VMYB_02 (re: 0.82) were found.
In 1 seq. 13 matches to the matrix V$YY1_01 (re: 12.71) were found.
```

In 1 seq. 1 matches to the matrix V$YY1_02 (re: 0.13) were found.
In 1 seq. 1 matches to the matrix V$ZID_01 (re: 0.33) were found.

Data Box 13.9: BLAST search results for the gene

```
BLAST search results
The BLAST similar 5 sequences are:
Score          E                                                    Sequences     producing
significant alignments:              (bits) Value
ID3_HUMAN (Q02533) DNA-binding protein inhibitor ID-3       (ID-like p...  244     2e-65
ID3_RAT  (P41138) DNA-binding protein inhibitor ID-3.        236            4e-63
ID3_MOUSE (P41133) DNA-binding protein inhibitor ID-3       (ID-like p...  235     1e-62
ID2_HUMAN (Q02363) DNA-binding protein inhibitor ID-2.       84             3e-17
ID4_HUMAN (P47928) DNA-binding protein inhibitor ID-4.       84             4e-17
```

INTERPRETATION OF DATA

A portion of output details of sequence are given in Data Box 13.5. It gene occurrence of several coding sequences first at 739 .... 1038 and last at 1146 ... 1202. It gene the translated prepated of this sequence. Besides it detailed 2 CpG-island one at 139 ... 1350 and other at 1388 ... 1896. CpG island are the region of DNA with a high density of CpG dinucleolides. In Mammalian cells it is often associated with the promoter region of genes.

It also predicted the presence of poly A site (Data Box 13.6), TATA box (Data Box 13.7) and transcription factors (Data Box 13.8). It gene a poly A sites and position pattern of these is given in Data Box 13.6. If indicated the sites where poly A tail will be added to eukaryotic on RNA Molecules in processing. Similarly TATA box sequence at different positions indicates the pressure of promoter. It also gave a long list of transcription factor binding site. Based on all this information and BLAST search, genebinder finds this sequence as a human DNA-binding protein inhibitor ID-1.

SUMMARY

A gene is a section of the DNA strand that carries the instructions for a specific function. One gene is a sequence within DNA that codes for single protein. Every gene typically contain several conserved features. Eukaryotic gene in comparison to prokaryotic gene contains several exons (expressed region) and introns (intervening regions). Besides it has a 5′ untranslated region (UTR) and 3′ UTR. Computational gene finding is very different from biological gene finding. Gene finding can be approached in two ways and gene finding for eukaryotic organisms is very different from prokaryotic organisms. Methods for gene finding involve DNA sequence analysis, making DNA sequences, identifying ORF exons or introns, translating exon, homology searches and sequence alignment and identification of regulatory sites. Most widely used software for ORF finding is NCBI ORF finder. Models in gene prediction involve Neural Networks and Hidden Markov Models (HMM). Experiment 13A is designed to study various functional aspects of a given DNA sequence. In experiment 13B usage of software genebuilder is studied.

EXERCISE

1. Use the sequence saved in sequence.txt (Experiment 13B, Data Box 13.4). Pick three other gene finding programs and submit the sequence. Answer the following:
   (*a*) How many exons are there in the unknown sequence?
   (*b*) What are the start and end points for each exon?

(*c*) Do the three gene finding programs agree on the above answers or are there discrepancies?

(*d*) What other elements could you identify with these programs? Look for Poly A sites, GC content, etc.

(*e*) Can you translate the sequence into a protein? What is the length of the protein sequence?

(*f*) What else can you say about the putative protein sequence? (Molecular weight, other characteristic attributes, matches in a database, structure.)

2. Let us suppose a gene has three exons and two introns. What would be the consequence if one of the introns is accidentally not spiced off? How would it effect translation?

14

PHYLOGENETIC ANALYSIS

Phylogeny is the study of the evolution of life forms from more than one ancestor. Evolution is called as phyletic evolution if the study involves changes within a single lineage. Phylogenetic analysis thus is the means of explaining species multiplication, splits and diversions within ancestral lines that causes appearance of more than one species or a cluster of species is called as clade. Biological systematists who wanted to reconstruct evolutionary genealogies of species based on morphological similarities originally developed phylogenetic analysis. The German entomologist Willi Hennig was the first author to propose an explicit method of phylogenetic analysis, and the publication of his work in English quickly led to the widespread use of his approach. Phylogenetic methods used to reconstruct the relationships between macromolecular sequences also involve the application of Hennigian principles.

The evolutionary history inferred from phylogenetic analysis is usually depicted as branching tree like diagram that represent the evolutionary relationship among different species. An example of a phylogenetic tree, also called a cladogram or a dendrogram, is displayed in Fig. 14.1. This shows that archaebacteria and eubacteria are more closely related to each other than to bacteria. Similarly in eukarya, fungi and plants are more related to each other than to slime molds.

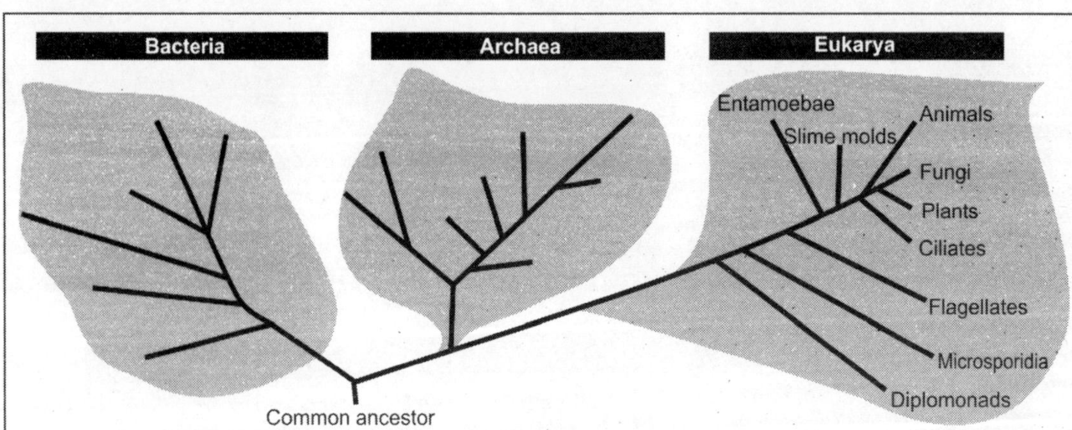

Fig. 14.1: Three domain phylogenetic tree of life on all earth

Interpretation of phylogenetic tree is very simple. Lines with names associated with them refers to species or group of species all more closely related to each others than to other groups on trees. Following these lines back to the points where they join other lines leads to ancestral species to the names groups. Lines that lead back to the same line represent species that share a common ancestor. In the Fig. 14.2 phylogenetic trees are explained with the help of two simple diagrams. A small tree diagram (Fig. 14.2a) shows that descendant species 1 and 2 are relatives since they recently evolved from same ancestral species. We consider more species. The larger tree diagram (Fig. 14.2b) shows relationships among four descendant species. Here, species 1 and 2 are each others close relatives and species 3 and 4 are each others close relatives. However all did have a common ancestor in the more distant part.

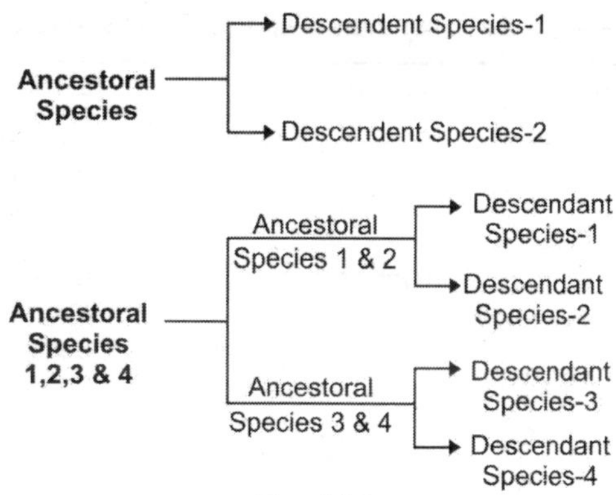

Fig. 14.2

Phylogenetic tree can be rooted or unrooted. The Fig. 14.3 shows that the four different representations of a phylogenetic tree among ten taxa in a unrooted and unrooted tree. The root represents the common ancestor of the entire taxonomic units. The direction and the length of each branch correspond to evolution time. The phylogenetic trees are drawn with Treeview software. Fig. 14.3 shows typical rooted type of tree. The four types in Fig. 14.3a, b, c and d are called radial, stanting cladogram phylogram and rectangular respectively. An unrooted tree is a tree that only specifies relationships among taxonomic unit but not the evolution path. It does not identify common ancestors. A rooted tree is more commonly constructed to study evolutionary relationships. Rooting a tree can be accomplished by adding an outgroup, or species that is known to be more distantly related to each of the remaining species than they are to each other. The point in the tree where the edge to the outgroup join is therefore the best candidate for the root position.

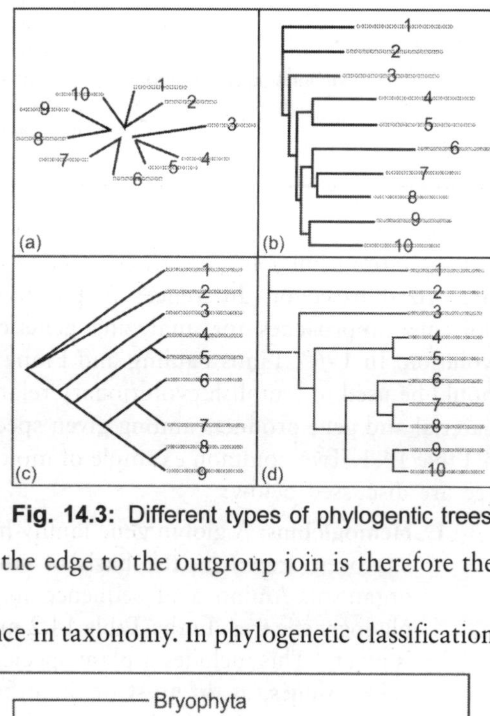

Fig. 14.3: Different types of phylogentic trees

Construction of phylogenetic is tree of key importance in taxonomy. In phylogenetic classification taxonomists name groups of organism that are all close relatives of one another. Classification schemes are hierarchal nature. Large groups containing many species are divided into smaller more specific groups, each with former species. Each of these is divided into still smaller groups with fewer species. These groupings reflect phylogenetic relationships. The life cycle, adaptation to life on land gametophyte phase or sporophyte phase; vascular tissue, pollen and seed etc. are certain characters based on which plants have been classified into six major groups as shown in Fig. 14.4.

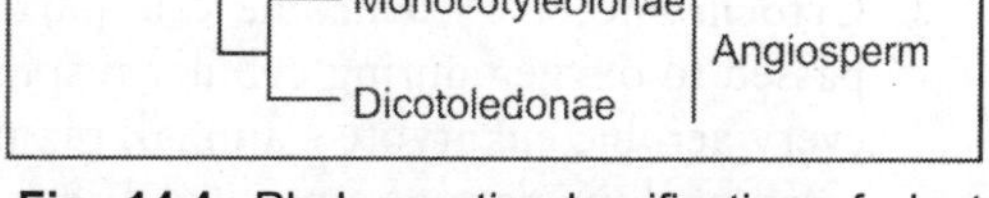

Fig. 14.4: Phylogenetic classification of plants

Table 14.1: Experimental approaches used to compare species at the molecular level technique explanation

Level	*Technique*	*Explanation*
DNA Level	DNA sequencing	The degree of similarities and differences between DNA sequences are used to establish evolutionary relationships. This approach, along with the analysis of the deduced amino acid sequences of structural genes, is the most common and reliable method in use.
	Hybridization	Genetic material from two different species is allowed to hybridize. The rate of hybridization will be faster for closely related species.
	Restriction mapping	Segments of DNA are isolated from different species and subjected to restriction mapping see. Closely related species will have more similar restriction maps.
	Cytogenetic analysis	The chromosomes of different species are analyzed microscopically. Closely related species will have identical or similar chromosome numbers and banding patterns.
Protein Level	Deduced amino acid	DNA sequencing is technically much easier to perform than amino acid sequence sequencing. Therefore, to determine an amino acid sequence of a protein, the coding region of a structural gene is determined and the amino acid sequence is then deduced using the genetic code. The amino acid sequences of a given protein will be more similar between closely related species.
	Immunological methods	Antibodies that recognize specific macromolecules, usually on the cell surface, are tested on different species. Antibodies that recognize macromolecules from one species will often recognize closely related species, but not from distantly related species.

Initially morphological differences were used to construct evolutionary trees in which species that are more similar in appearance tended to be placed closer on a tree. In addition species were categorized based on differences in physiology, biochemistry and even behavior. The advent of molecular approaches for analyzing genetic material and genes has revolutionized the field of evolution. In 1963, Linus Pauling and Emile Zukerkandh were first to suggest that molecular data should be used to establish evolutionary relationships. There are many ways to compare the genetic material and gene products among given species. Some of the molecular approaches are highlighted in Table 14.1. Two common example of molecular homology being used for drawing a phylogenetic tree are discussed below:

1. **Hemoglobins:** A globin gene family has evolved from a single anesstoral gene. This inference has been drawn on the basis of study of amino acid sequence of hemoglobin in several organism. Amino acid sequence in b-chain subunit of hemoglobin is used to draw the phylogenetic tree. In the Table 14.2 number of different amino acids in b-chain of hemoglobin is given. This includes a plant species soyabean. The human beta chain contains 146 amino acid residues, as do most of the others. The numbers represent the number of amino acid differences between the beta chain of humans and the hemoglobins of the other species. In general, the number is inversely proportional to the closeness of kinship. Smaller is the difference more closely are the species related. All the values listed are for the beta chain except for the last three, in which the distinction between alpha and beta chains does not occur.
2. **Cytochrome c:** Cytochrome c is part of the respiratory chain down which electrons are passed to oxygen during cellular respiration. Cytochrome c is found in the mitochondria of every aerobic eukaryote - animal, plant, and protist. In 1960 E.M. Fitch and E. Margoliash assembled the data of amino acid sequence for cytochrome c in a variety of organisms. The amino acid sequences of many of these have been determined, and comparing them shows

Table 14.2: Number of different amino acids in beta chains of Hemoglobins of human from different organism

Species	*No. of different amino acids in chain*
1. Human beta chain	0
2. Gorilla	1
3. Gibbon	2
4. Rhesus monkey	8
5. Dog	15
6. Horse, cow	25
7. Mouse	27
8. Gray kangaroo	38
9. Chicken	45
10. Frog	67
11. Lamprey	125
12. Sea slug (a mollusk)	127
13. Soyabean (leg hemoglobin)	124

that they are related. Human cytochrome c contains 104 amino acids, and 37 of these have been found at equivalent positions in every cytochrome-c that has been sequenced. It is assumed that each of these molecules has descended from a precursor cytochrome in a primitive microbe that existed over 2 billion years ago. Table 14.3 shows the N-terminal 22 amino acid residues of human cytochrome c with the corresponding sequences from six other organisms aligned beneath. All the vertebrate cytochromes (the first four) start with glycine (Gly). The *Drosophila*, wheat and yeast cytochromes have several amino acids that precede the sequence shown here (indicated by <<<). In every case, the heme group of the cytochrome is attached to Cys-14 and Cys-17 (human numbering). In addition to the two Cys residues, Gly-1, Gly-6, Phe-10, and His-18 are found at the equivalent positions in every cytochrome c that has been sequenced.

Table 14.3: Molecular homology of cytochrome-c

		1					6				10				14			17	18		20		
Human		Gly	Asp	Val	Glu	Lys	Gly	Lys	Lys	Ile	Phe	Ile	Met	Lys	Cys	Ser	Gln	Cys	His	Thr	Val	Glu	Lys
Pig		-	-	-	-	-	-	-	-	-	-	Val	Gln	- -	Ala	-	-	- -	-	-	-		
Chicken		-	-	Ile	-	-	-	-	-	-	-	Val	Gln	- -	-	-	-	- -	-	-	-		
Dogfish		-	-	-	-	-	-	-	-	Val	-	Val	Gln	- -	Ala	-	-	- -	-	-	Asn		
Drosophila	<<<	-	-	-	-	-	-	-	-	Leu		Val	Gln	Arg		Ala	-	- -	-	-	-	Ala	
Wheat	<<<	-	Asn	Pro	Asp	Ala	-	Ala	-	-	-	Lys	Thr	- -	Ala	-	-	- -	-	Asp	Ala		
Yeast	<<<	-	Ser	Ala	Lys	-	-	Ala	Thr	Leu	-	Lys	Thr	Arg	-	Glu	Leu	- -	-	-	-	-	

Phylogenetic tree shown in Fig. 14.5 depicts the relationship between these species of eukaryotes. It was assumed that the more identities are there between two molecules, the more recently they have evolved from a common ancestral molecule and thus the closer the kinship of their owners. Thus the cytochrome-c of the rhesus monkey is identical to that of humans except for one amino acid, whereas yeast cytochrome-c differs from that of humans at 44 positions. (There are no differences between the cytochrome-c of humans and that of chimpanzees.) With such information, one can reconstruct an evolutionary history of the molecule and thus of their respective owners. The numbers represent the minimum number of nucleotide substitutions called as minimal mutational distance in the gene for cytochrome-c needed to produce these 20 proteins from a series of hypothetical ancestral genes at the various branching points (nodes). The tree corresponds quite well to what we have long believed to be the evolutionary relationships among the vertebrates.

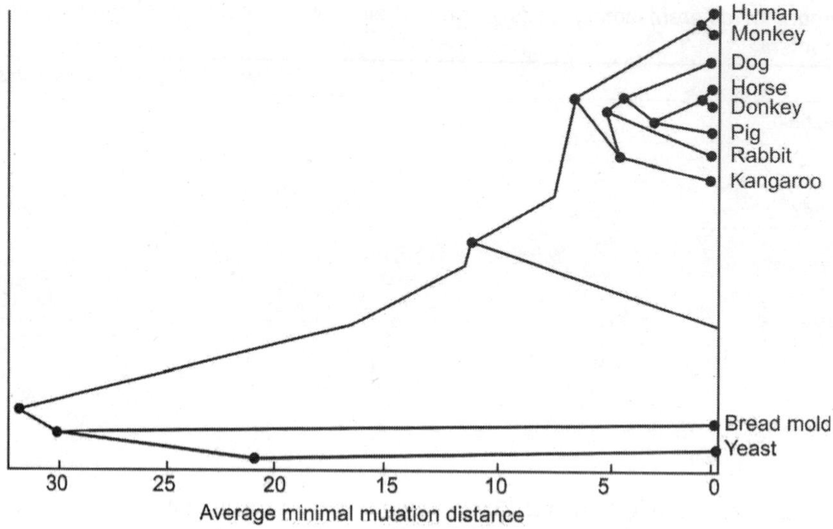

Fig. 14.5: Phylogenetic tree based on cytochrome-c (based on data taken from Fitch and Margoliash1960)

Tree of life represents the phylogenetic tree drawn for all living organisms on earth. In Fig. 14.1 tree of life showing lineage of three major domains of life *i.e.* eukarya, prokarya and bacteria are shown. The rooting of the Tree of Life, and the relationships of the major lineages, are controversial (see news corner). The monophyly of Archaea is uncertain, and recent evidence for ancient lateral transfers of genes indicates that a highly complex model is needed to adequately represent the phylogenetic relationships among the major lineages of Life. The Tree of Life web project (ToL) is a collaborative effort of biologists from around the world. On more than 3000 world wide web pages, the project provides information about the diversity of organisms on earth, their evolutionary history (phylogeny), and characteristics.

14.1 FEATURES OF A PHYLOGENETIC TREE

1. **Phylogenetic tree are useful in identification of orthologs, paralogs, analogs and xenologs:** A pair of genes found in two species are orthologues when encoded protein are 60-80% identical in alignment. These protein almost certainly have the same three-dimensional structure domain structure and biological function. The genes encoding these proteins are said to have originated from a common ancestor gene at an earlier evolutionary times. Orthology is predicted by a very close phylogenetic relationship between sequences. Paralogs are gene that are related through gene duplication events. Orthology and paralogy is explained in details with homology in chapter 12. These events may lead to production of a family of related protein with similar biological functions within a species.

   Analogous genes refers to a gene in species I and different gene species II which have converged to same function by different evolutionary path. Xenologous genes are genes in two species that are related through transfer of genetic material between species, even though the species are separated by a long evolutionary distance. Xenologous genes are also called horizontally transferred genes. Transposable elements transfer and symbiotic relationships between organisms may be prime event leading to such exchanges in xenologous genes. Analogous genes have similar active site in case of proteins, but rest of the structure of the protein is different.

2. **Differences among species are the key to reconstructing the phylogenetic tree:** Species differ in the characters. The characters may be observable and measurable properties of the individuals. If a species depends on a character for its continued survival, that character will not change and any mutations to it will be eliminated. Such characters are essential characters. And most visible characters are essential for the species. This means that if we choose essential characters, any differences should be counted as very significant. If one species evolved by changing an essential characteristic, whatever ecological forces supported that change may also apply to other species, and that could lead to parallel evolution. Thus, differences or similarities in essential characters are very relevant to the reconstruction of the general shape of the phylogenetic tree, but they really cannot be used to determine the relative lengths of the lines within the tree. Some species have been stable for millions of years. Others evolve very fast.
3. **A tree isn't always the best model:** For example:
   (*a*) *For individuals within a species*: The genetic material of an individual doesn't derive from a single earlier existing individual. Animals and plants that multiply by sexual reproduction receive half their genetic material from each of two parents, so a tree like this is inappropriate. For species that multiply asexually, a tree is appropriate. Even for species that usually multiply asexually—such as many one-celled creatures—the occasional exchange of genetic material through conjugation is so important that trees are inappropriate.
   (*b*) *For closely related species*: Individuals do occasionally mate between closely related species, and their progeny survive to contribute to the gene pool of one or both of the parent species. As the species diverge, such intermixing of genetic material becomes more rare. One solution is to treat closely related species as one larger variable species. Another is simply not to consider closely related species.
   (*c*) *Hybrid species*: In the plant world it occasionally happens that a new tetraploid species arises from two diploid species. The two parent species need to be somewhat related for this to happen.
   (*d*) *Distant interaction*: There are a couple of ways that genetic material from one species can find its way into a distantly related or unrelated species. Among bacteria, sometimes a bacterium of one species can inject the genetic material of a bacterium of another species and incorporate part of it into its own genetic material. Rare as this may be, the effects are significant. Sometimes viruses can inadvertently transport genetic material from one species to another. When some viruses break out of cells of one species, they may infect other species and carry that material to them.

   In spite of these exceptions, a tree model is usually a pretty good model to show the relations among species.
4. **The algorithm for drawing a phylogenetic tree is not based on any actual data or theory of species evolution:** In the next section methods of construction of phylogenetic trees is discussed. A general procedure for creation of a random phylogenetic tree with n number of leaves is as follows. First, there are n unrelated species. Usually, two are selected at random to have a common ancestor, but 1/10 of the time; three are selected (and 1/100 of the time four, etc.). An ancestral species is created for these species, and it is placed back in time before the older of the child species. The amount further back in time is 2/10 plus a random number between 0 and 1, which makes an average of 0.7 time units back. This process is repeated until all the species have a common ancestor. The vertical scale shown is arbitrary, except that it is the same scale used to generate the random tree.

5. All phylogenetic analysis is based on some assumptions:

(*a*) Each position in the sequence has evolved independently.

(*b*) Each position in the sequence has evolved according to same stochastic procedure.

(*c*) Each sequence in the sample has evolved according to same stochastic process.

14.2 METHODS FOR CONSTRUCTION OF PHYLOGENETIC TREES

Inferring a phylogeny is an estimation procedure, in which a best estimate of the evolutionary history is made on the basis of incomplete information. Indeed, we usually do not have information about the past, but have access only to contemporary sequences. Because many different phylogenetic trees can be produced from a set of data we must specify criteria for selecting one or a few trees as representing our best estimate of the true evolutionary history. Most phylogenetic methods seek to accomplish this goal by defining a criterion for comparing alternative phylogenies and deciding which one is the best.

A phylogenetic reconstruction, consists of two steps:

1. Defining an optimality criterion, or objective function, i.e., the value that is assigned to a tree and is subsequently used for comparing one tree to another;
2. Developing specific algorithms to compute the value of the objective function, thereby to identify the tree or a set of trees that have the best values according to this criterion.

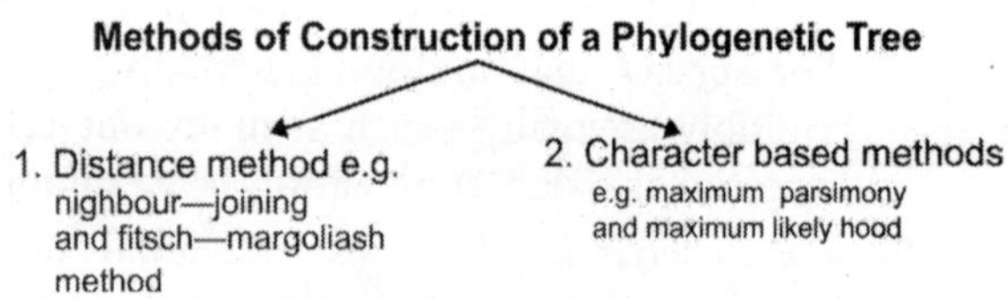

Fig. 14.6: Method of construction of a phylogenic tree

There are many statistical methods that can be used to reconstruct phylogenetic trees from molecular data. Methods of construction of a phylogenetic tree are classified into distance methods and character based methods (Fig. 14.6) depending upon the types of data used in phylogenetic inference or types of computational methods. Distance methods compute pairwise distances according to some measure and then discard the actual data and use only fixed distances to derive a tree. Character based method on the other hand derives the tree based on distribution of actual data patterns for each character. Pairwise distances are therefore not fixed as they are determined by tree topology. Character based methods are further classified into maximum parsimony and maximum likelihood method. Maximum parsimony method uses character state data with the global optimality criterion: the smaller the number of evolutionary changes required by a tree, the better the tree. Maximum likelihood method uses both character data and evolutionary distances and chooses the tree with the highest maximum likelihood value as the best tree.

In distance methods, evolutionary distance or the dissimilarity is computed for all pairs of taxa, and a phylogenetic tree is constructed by considering the relationships among these distance values. Distance methods are much less computationally intensive than other methods. Commonly used distance method is UPGMA-method (unweighted pair-group method using arithmetic averages). UPGMA is not the most powerful method for estimating phylogenies from geneticdata, but is quite straightforward. Furthermore UPGMA method works reasonably well under many circumstances. The starting point for UPGMA is a table of pairwise genetic distances among a given group of species data used in explanation of UPGMA method is a hypothetical data and the values in Fig. 14.7 represents the genetic distances between the species. The evolutionary tree constructed indicates A and B are one another's closest relative and they share a common ancestor. Detailed algorithm of UPGMA method is given Fig. 14.8.

	A	B	C	D	E
A	–				
B	(1.62)	–			
C	2.26	2.21	–		
D	4.70	5.133	4.543	–	
E	4.77	4.76	4.753	1.95	–

(a)

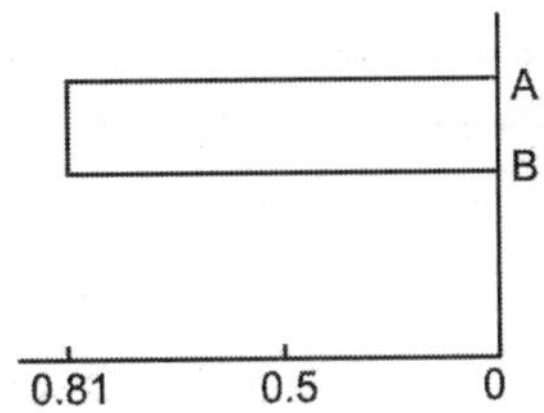

↓

Search for the table for the smallest genetic distance between any pair of species in our table it is 1.62 between species A and B.

↓

Place these species on neighbouring branches of an evolutionary tree. Define the length of these two branches to half the genetic distance between the species it is 0.81 in our case

↓

Recalculate the genetic distances between the cluster and the other species as the average of distances between each cluster members and the other species. In our case genetic distance between A–B cluster and C is the average of A–C and B and C genetic distance. This is equal to (2.26+2.21)/2 = 2.23. Repeat this step for all the species. The smallest distance in the table now is 1.95 D and E on neighbouring bronchus of our evolutionary tree with branch length connecting their concern ancestor equal to 0.98

↓

Redraw the table as given below:

	A–B	C	D	E
A–B	–			
C	2.23	–		
D	4.9	4.53		
E	4.76	4.75	(1.95)	

(b)

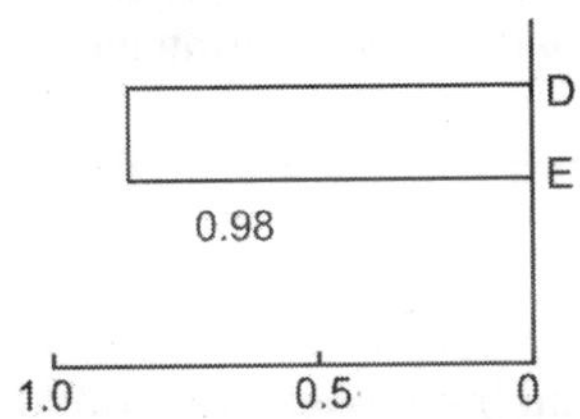

↓

Repeat the step as earlier. Smallest genetic distance now 2.23, between A–B–C. The genetic distance for A–B and C is 1.11

↓

Redraw the table as given below:

–	A–B	C	D–E
A–B			
C	(2.23)		
DE	4.84	4.64	

(c)

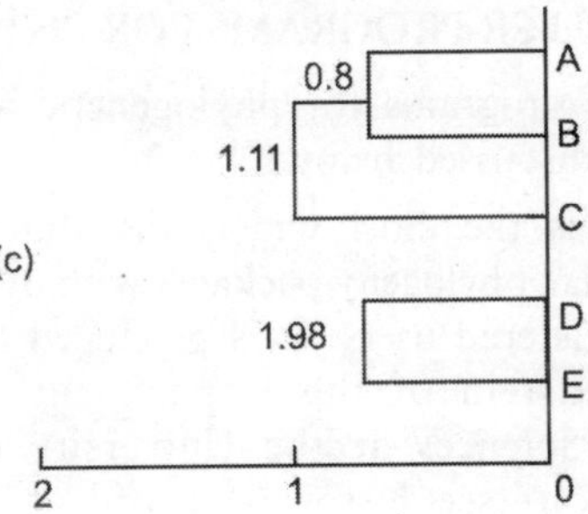

↓

Reducing the table final time leaves us with distance of 4.77 and with this the evolutionary tree is complete. The genetic distance for A–B–C and D–E is 2.38

↓

A–B–C	A–B–C	D–E
D–E	4.77	

(d)

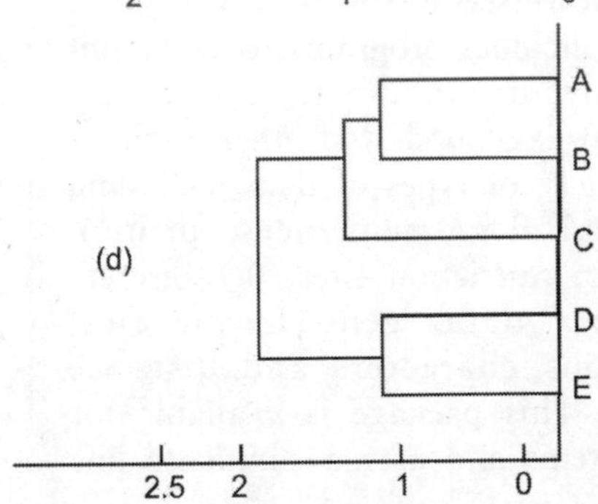

Fig.14.7: Simple graphical working in UPGMA method

HISTORICAL PROFILE 16

Dr. Joseph Felsenstein for Developing Popular Phylogentic Analysis Software

Dr. Joseph Felsenstein

Joseph Felsenstein is Professor in the Deparment of Genome Science and Biology and Adjunct Professor in the Departments of Computer Science and Statistics at the University of Washington in Seattle. He is best known for his work on phylogentic analysis and principal author of phylogentic software Phylip. He has written book Inferring Phytogeny.

His group lately been working on methods for estimating population parameters (such as effective population size, mutation rate, and so on) from population samples of molecular sequences. He has been using a computationally intensive method known as Markov Chain Monte Carlo Integration to make approximate calculations of the statistical likelihoods for different values of the population parameters and are now distributing a free package of computer programs, LAMARC, to do these calculations.

Since a phylogenetic tree is constructed from a limited amount of data, it is important to examine the reliability of the tree obtained. Program dnadist, protdist, neighbour and genedist used from the PHYLIP package to implement distance method.

14.3 COMPUTER PROGRAMS FOR PHYLOGENETIC TREES

A number of programs for phylogenetic analysis are available. Some of the common programs however are discussed below:

1. **PHYLIP:** is the most widely distributed and popular phylogeny package, with over 7,000 registered users. It is produced by Joe Felsenstein of the Department of Genome Sciences at the University of Washington (see historical profile 17). PHYLIP includes programs to carry out parsimony, distance matrix methods, maximum likelihood, and other methods on a variety of types of data, including DNA and RNA sequences, protein sequences, restriction sites, 0/1 discrete characters data, gene frequencies, continuous characters and distance matrices. This package is available for most systems and uses a command line interface.

INITIALIZATION

- Initialize n clusters with the given species, one species per cluster
- Set the size of each cluster to 1
- In the output tree T, assign a leaf for each species

ITERATION

- Create a new cluster-(*ij*), which has n (*ij*) = *ni* + *nj* members
- Connect *i* and *j* on the tree to a new node, which corresponds to the new cluster (*ij*), and give the two branches connecting *i* and *j* to (*ij*) length D*ij*/2 each
- Compute the distance from the new cluster to all other cluster (except for *i* and *j*, which are no longer relevant) as a weighted average of the distance.
- Delete the columns and rows in D that correspond to clusters *i* and *j*, and add a column and row for cluster (*ij*)

Fig. 14.8: Algorithm: UPGMA method

2. **PAUP:** Package is produced by David Swofford of the Laboratory of Molecular Systematics, National Museum of Natural History, Smithsonian Institution, Washington, D.C. and is a commercial software package. PAUP is the most sophisticated parsimony program, with many options. It includes parsimony, distance matrix, invariants, and maximum likelihood methods and many indices and statistical tests. The package has a graphical user interface, but most operations still have to be done on the command line, and similar to PHYLIP, intensive reference to the manual is necessary for successful use of this package.
3. **MEGA2:** Though PHYLLIP and PAUP cover almost all needs of modern phylogenetic analysis, in many cases extensive study is required to learn the use of command line features of these packages. The package MEGA2 (Molecular Evolutionary Genetic Analysis) form the Microsoft Windows platform offers a convenient and easy-to-use graphical user interface, which facilitates the use of most popular methods of phylogenetic analysis. For MEGA2 users, PAUP and PHYLIP remain as indispensable, highly sophisticated tools for special cases. MEGA2 carries out parsimony, distance matrix and likelihood methods with branch-and-bound search available for parsimony and bootstrapping available for all methods and has a reasonably good set of incorporated statistical tests and sequence evolution models. The GUI allows sophisticated tree manipulating and output of trees in emf or bmp formats. MEGA2 is produced by Sudhir Kumar, Koichiro Tamura, Ingrid Jakobsen and Masatoshi Nei.

EXPERIMENT 14A

Use the Multiple Aligned Sequence of Experiment 11A and Draw a Phylogenetic Tree of these Sequences using TREEVIEW Software

About the software: **TREEVIEW** is a program for displaying and printing phylogenies. The program reads most NEXUS tree files (such as those produced by PAUP and COMPONENT) and PHYLIP style tree files (including those produced by fast DNAml and CLUSTALW). It is available for both Apple Macintosh and Microsoft Windows PCs. TREEVIEW is free and can be obtained on the Internet from the world wide web site.

The Hypothesis: It is possible to draw phylogenetic tree using DNA sequences for gene rbcL for ribulose-1, 5-bisphosphate carboxylase/oxygenase (RUBpcase), large subunit, from 10 different sources. Four types of phylogenetic trees is radial, cladogram, phylogram, and rectangular cladogram would be drawn. The phylogenetic tree may reveal important features concerning similarities and differences within the gene family.

TESTING THE HYPOTHESIS

Computer set up: Windows environment with treeview installed in it.

Starting material: Nucleotide sequence of gene encoding for the larger subunit of enzyme rubpcase as collected in Experiment - 11A is used and a complete alignment is done in the same way. The structure of input file is shown in Data Box 11.1 and classification of these organisms is given in Data Box 11.2. The output file for clustal tree is saved as Experiment 11A. dnd.

1. Log into the treeview software. An interface like Fig. 14.11 is seen.
2. Open the file experiment_10A.ph.
3. In the tree menu chose all the four types of trees viz. radial, slanting cladogram, phylogram and rectangular one by one as graphic and save as graphic image.

DATA

The output file and different types of trees produced by treeview software are shown in Fig. 14.4.

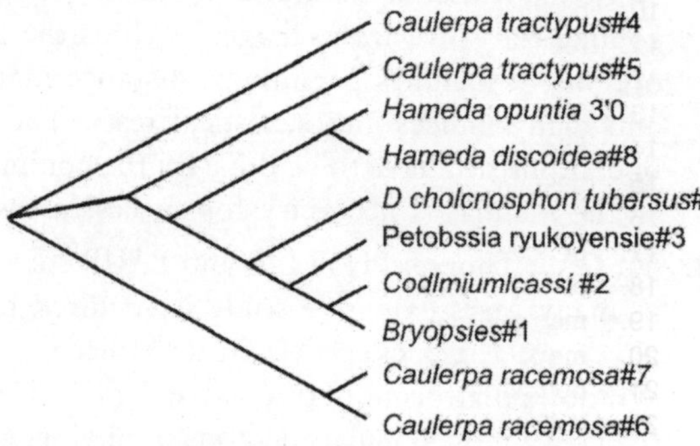

Fig. 14.9: Phylogenetic tree drawn with tree view

INTERPRETATION OF DATA

Radial slanting cladogram, phylogram, and rectangular phylogenetic tree are shown in Fig.14.3a, b, c and d respectively. Tree view in combination with clustal is very easy to use and depicts the results reasonably well. Typical phylogram rooted type is enlarged in Fig. 14.9. All the species based on enzyme Rubcase have been depicted to have common single ancestry.

EXPERIMENT 14B

To Draw a Phylogenetic Tree using the Phylip Software and Compare the Results Obtained with ClustalX

About study sequence: The acquired immunodeficiency syndrome (AIDS) is caused by the human immunodeficiency virus (HIV). HIV is known to be the worst since the black plague. Phylogenetics is becoming increasingly more important in the analysis of HIV sequence data. Such analyses have been used to classify the diversity of HIV sequences into major groups and subtypes within these groups. Phylogenetics has been used to track epidemiological change over time and to detect epidemiological linkage among patients. In the specific case of HIV it has been proved that phylogenetic analysis of HIV sequences has led scientists to better understand the origin and evolution of the virus and therefore estimate its impact on human health. A set consisting of 14 different gp120 protein sequences from isolates of HIV-1, HIV-2 and chimpanzee SIV was seleted for the present study. Only partial length i.e. 180 AA was selected for the present study. (Data Box 14.1). The gp120 protein is crucial for binding of the virus particle to target cells. It is the specific affinity of gp120 for the CD4 protein that targets HIV to those cells of the immune system that express CD4 on their surface (e.g., T-helper lymphocytes, monocytes, and macrophages).

About phylip: PHYLIP (Phylogeny Inference Package) Version 3.5c executables for Windows 95, 98, NT systems is selected for phylogenetic analysis in the present study. This package includes 3 programs. These programs are for specific purposes of being used for DNA or proteins or for a different process of phylogenetic analysis. Detailed contents of programs included in this package and their brief description is given in Table 14.4. A brief description about the programs used in present study is given below:

Table 14.4: Content of phylip package

1.	protpars	parsimony for protein sequence data
2.	dnapars	DNA parsimony program
3.	dnamove	interactive DNA parsimony
4.	dnapenny	branch and bound method for DNA
5.	dnacomp	DNA compatibility program
6.	dnainvar	computation of Lake's and Cavender's invariants
7.	dnaml	DNA maximum likelihood program

8.	dnamlk	DNA maximum likelihood with molecular clock
9.	dnadist	computes distance matrix from sequences
10.	protdist	computes distance matrix from sequences
11.	restml	maximum likelihood for restriction sites
12.	seqboot	makes multiple data sets by bootstrap resampling
13.	fitch	Fitch-Margoliash and least-squares methods
14.	kitsch	F-M, L-S methods with evolutionary clock
15.	neighbor	neighbour-joining and UPGMA methods
16.	contml	maximum likelihood program
17.	gendist	computes genetic distances
18.	contrast	contrasts etc. for comparative method studies
19.	mix	Wagner, Camin-Sokal parsimony and mixtures
20.	move	interactive Wagner, Camin-Sokal and mixed parsimony
21.	penny	finds all most parsimonious trees
22.	dollop	Dollo and polymorphism parsimony methods
23.	dolmove	interactive Dollo and polymorphism parsimony
24.	dolpenny	branch and bound for Dollo, polymorphism
25.	clique	compatibility program
26.	factor	recode multistate to binary characters
27.	drawgram	makes plots of cladograms, phenograms
28.	drawtree	makes plots of unrooted phylogenies
29.	consense	majority-rule and strict consensus trees
30.	retree	reroots, rearranges and changes lengths on trees

Hypothesis: that HIV-1 arises from cross-species transmission from chimpanzees and HIV-2 arises from cross-species transmission from macaque monkeys and test it using phylogenetic analysis on a chosen data set using phylogenetic tools.

TESTING THE HYPOTHESIS

Computer setup: A window environment with clustalX and phylip installed in it.

1. Run clustalX like experiment and do multiple sequence alignment (Fig. 14.10) save the tree file within dnd and phylip format.
2. Run the dnd file in tree view the save phylogenetic tree (Fig. 14.11).
3. Ordinarily, PHYLIP programs are run in the foreground by executing them at a >GCG> prompt (i.e., starting gcg first to initialize the gcg shell). Programs can be run in the background by altering the input file to include the responses to a program's pre-run interactive menu. An interface of protpars webview running is shown in Fig.14.12.
4. Input data file shown in Data Box 14.1 is saved as sequences.dat. It is run in the background, taking its input data from sequences.dat, putting its interactive output to file called "screenout", and using a file called "input" as the place to store the interactive input. The file "input" need only contain two lines: sequences.dat Y which is what you would have typed to run the program interactively, in response to the program's request for an input file name if it did not find a file named "infile", response the menu.

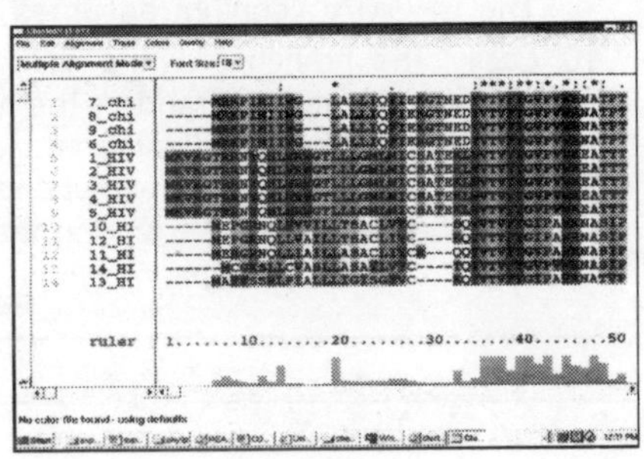

Fig. 14.10

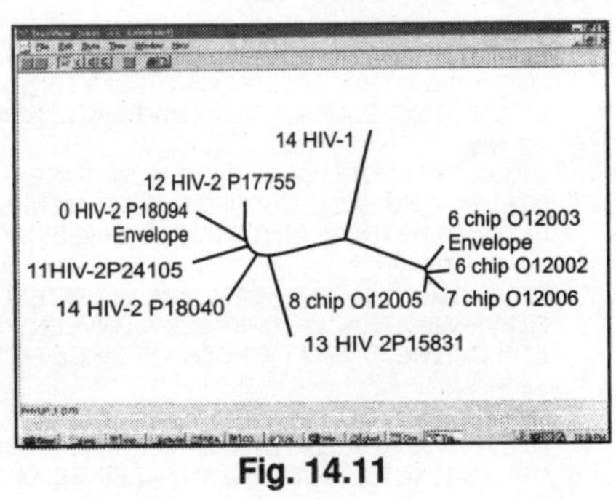

Fig. 14.11

5. To run the program in background, you would simply give the command: protopars < input > screenout and which runs the program with input responses coming from "input" and interactive output being put into file "screenout". The usual output file and tree file will also be created by this run (keep that in mind as if you run any other PHYLIP program from the same directory while this one is running in background you may overwrite the output file from one program with that from the other).

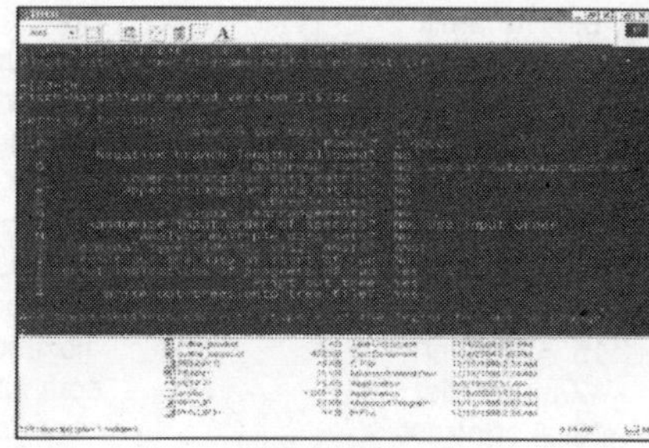

Fig. 14.12

6. Analyses in the present interface are rendered according to the distance method. Four programs within PHYLIP are employed here. They are, SEQBOOT, PROTDIST, NEIGHBOUR, and CONSENSE.
7. Once multiple alignment has been completed, the data set is transmitted to SEQBOOT. SEQBOOT generates multiple possible arrangements of the alignment (reflecting the number of conceivable evolutionary paths).
8. PROTDIST reads in the data from SEQBOOT and computes a distance score for protein sequences. This step is most critical, since no subsequent analysis can be made without a measure of sequence divergence or similarity. A Dayhoff PAM matrix is used for computation of distance scores between pairs of sequences. A distance score reflects the number of single amino acid alterations required in order generate an identity sequence from a second sequence.
9. Neighbour implements the Neighbor-joining method to determine the most reasonable positioning of branches. Two sequences having the smallest distance scores are joined as "neighbours" and will share a node below them (or to their left) in the final tree. (Fig 14.14). The user option "Tree" affects the execution of NEIGHBOUR. The default setting gives rise to unrooted trees. An unrooted or rooted tree is produced depending upon certain assumptions of the two alternative algorithms that can be used in NEIGHBOUR. The Neighbour-joining method does not assume a constant rate of mutation, known as the molecular clock hypothesis, and as such this algorithm always produces unrooted trees. Alternatively, if the user specifies a rooted tree, then NEIGHBOUR implements another algorithm, the unweighted pair group method with arithmetic mean (UPGMA). The UPGMA algorithm assumes a molecular clock and generates rooted trees.
10. Lastly, the branch ordering data is passed to CONSENSE for resampling computations. Any phylogenetic method renders the most likely tree, i.e., those relationships that are most reasonable given the sequence alignments. As such, any single tree is only one of many possible trees that could have arisen over evolutionary time. Resampling methods, therefore, are designed to find the most probable tree among the many possible evolutionary paths that could have generated a given set of proteins.

DATA

Data Box 14.1

```
The sequences
>1_HIV-1
MKVKGTRRNYQHLWRWGTLLLGMLMICSATEKLWVTVYYGVPVWKEATTTLFCASDARAY
DTEVHNVWATHACVPTDPNPQEVVLGNVTENFNMWKNNMVEQMQEDIISLWDQSLKPCVK
LTPLCVTLNCTDLGKATNTNSSNWKEEIKGEIKNCSFNITTSIRDKIQKENALFRNLDVV
>2_HIV-1
MKVKGTRRNYQHLWRWGTLLLGMLMICSATEKLWVTVYYGVPVWKEATTTLFCASDARAY
DTEVHNVWATHACVPTDPNPQEVVLGNVTENFNMWKNNMVEQMQEDIISLWDQSLKPCVK 120
LTPLCVTLNCTDLGKATNTNSSNWKEEIKGEIKNCSFNITTSIRDKIQKENALFRNLDVV 180
>3_HIV-1
MKVKGTRRNYQHLWRWGTLLLGMLMICSATEKLWVTVYYGVPVWKEATTTLFCASDARAY 60
DTEVHNVWATHACVPTDPNPQEVVLGNVTENFNMWKNNMVEQMQEDIISLWDQSLKPCVK 120
LTPLCVTLNCTDLGKATNTNSSNWKEEIKGEIKNCSFNITTSIRDKIQKENALFRNLDVV 180
>4_HIV-1
MKVKGTRRNYQHLWRWGTLLLGMLMICSATEKLWVTVYYGVPVWKEATTTLFCASDARAY 60
DTEVHNVWATHACVPTDPNPQEVVLGNVTENFNMWKNNMVEQMQEDIISLWDQSLKPCVK 120
LTPLCVTLNCTDLGKATNTNSSNWKEEIKGEIKNCSFNITTSIRDKIQKENALFRNLDVV 180
```

```
>5_HIV-1
MKVKGTRRNYQHLWRWGTLLLGMLMICSATEKLWVTVYYGVPVWKEATTTLFCASDARAY 60
DTEVHNVWATHACVPTDPNPQEVVLGNVTENFNMWKNNMVEQMQEDIISLWDQSLKPCVK 120
LTPLCVTLNCTD——LGKATNTNSSNWKEEI-KGEIKNCSFNITTSIRDKIQKENALF 174
>6_chip_O12003_Envelope glycoprotein gp160 Chimpanzee immunodeficiency virus
(SIV(cpz)) (CIV)
MRKPIHIIWG LALLIQFIEK GTNEDYVTVF YGVPVWRNAT PTLFCATNAS MTSTEVHNVW
ATTSCVPIDP DPIVVRLNTS VWFNAYKNYM VESMTEDMLQ LFQQSHKPCV KLTPMCVRMN
CTEYNGITTN TTSSTAKTTT SPVAGMQLQK CQFNQSTGFK CQFNQSTGFK DKKQEMNAIF
>7_chip_O12006
MRKPIRIIWG LALLIQFIEK GTNEDYVTVF YGVPVWRNAT PTLFCATNAS MTSTEVHNVW
ATTSCVPIDP DPIVVRLNTS VWFNAYENYM VESMTEDMLQ LFQQSHKPCV KLTPMCVRMD
CTEYNEKSTT TTATTTTKGT VTIVDGMQLQ ECKFNQSTGF KDKKQSMKAI FYKGDLMECQ
>8_chip_O12005
MRKPIHIIWG LALLIQFIEK GTNEDYVTVF YGVPVWRNAT PTLFCATNAS MTSTEVHNVW
ATTSCVPIDP DPIVVRLNTS VWFNAYENYM VESMTEDMLQ LFQQSHKPCV KLTPMCVKMN
CTEYNGENTT ISTVTAKPTT SPVAGMQLQK CQFNQSTGFK DKKQEMNAIF YKGDLMECQE
>9_chip_O12002
MRKPIHIIWG LALLIQFIEK GTNEDYVTVF YGVPVWRNAT PTLFCATNAS MTSTEVHNVW
ATTSCVPIDP DPIVVRLNTS VWFNAYKNYM VESMTEDMXQ LFQQSHKPCV KLTPMCIKMN
CTGYNGTPTT PSTTTSTVTP KTTTPIVDGM KLQECNFNQS TGFKDKKQKM KAIFYKGDLM
>10_HIV-2_P18094Envelope polyprotein GP160 [Precursor] Human immunodeficiency virus
type 2 (isolate BEN) (HIV-2)
MEPGRNQLFV VILLTSACLV YCSQYVTVFY GIPAWKNASI PLFCATKNRD TWGTIQCLPD
NDDYQEIILN VTEAFDAWNN TVTEQAVEDV WHLFETSIKP CVKLTPLCVA MNCSRVQGNT
TTPNPRTSSS TTSRPPTSAA SIINETSNCI ENNTCAGLGY EEMMQCEFNM KGLEQDKKRR
>11_HIV-2_P24105
MERGRNQLLI AILLASACLI YCRQQYVTVF YGVPAWKNAS IPLFCATKNR DTWGTIQCLP
DNDDYQEIPL NVTEAFDAWD NTITEQAIED VWNLFETSIK PCVKLTPLCV AMKCNISTSD
TTMIRTTTPS TAKEAPISDN SPCIRTNNCS GLEEEKIVKC HFNMTGLERD KKKQYNETWY
>12_HIV-2_P17755
MEPGRNQLLV AILLTSACLI YCKQYVTVFY GIPAWRNASI PLFCATKNRD TWGTIQCLPD
NDDYQEITLN VTEAFDAWDN TVTEQAIEDV WRLFETSIKP CVKLTPLCVA MNCNITSGTT
ATPSPPNITI IDENSTCIGD NNCTGLGKEE VVECEFNMTG LEQDKKRKYN DAWYSRDVVC
>13_HIV_2_P15831
MAYFSSRLPI ALLLIGISGF VCKQYVTVFY GIPAWRNATV PLICATTNRD TWGTVQCLPD
NGDYTEIRLN ITEAFDAWDN TVTQQAVDDV WRLFETSIKP CVKLTPLCVA MNCSKTETNP
GNASSTTTTK PTTTSRGLKT INETDPCIKN DSCTGLGEEE IMQCNFSMTG LRRDELKQYK
>14_HIV-2_P18040
MCGKSLLCVA SLLASAYLVY CTQYVTVFYG VPVWRNASIP LFCATKNRDT WGTIQCKPDN
DDYQEITLNV TEAFDAWDNT VTEQAVEDVW SLFETSIKPC VKLTPLCVAM SCNSTTNNTT
TTGSTTGMSE INETSPSYSD NCTGLGKEEI VNCQFYMTGL ERDKKKQYNE TWYSKDVVCE
```

INTERPRETATION OF DATA

Protdist program in the PHYLIP package uses protein sequences to compute a distance matrix under four different models of amino acid replacement.. The distance for each pair of species estimates the total branch length between the two species. The program reads in protein sequences and writes an output file containing the distance matrix.

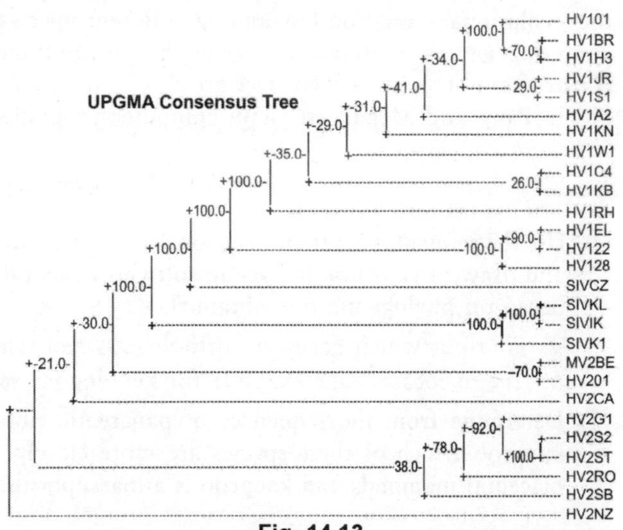

Fig. 14.13

Neighbour program implements the UPGMA method of clustering. Neighbour constructs a tree by successive clustering of lineages, setting branch lengths as the lineages join. The tree is not rearranged thereafter. The tree does not assume an evolutionary clock, so that it is in effect an unrooted tree. The Bootstrap Consensus Tree for this method is shown in Fig. 14.13. The unrooted tree drawn by Phylip is shown in Fig. 14.14. It is seen that all HIV-1 and the HIV-2 sequences cluster out separately. It can be seen that HIV-1 arises from cross-species transmission from chimpanzees.

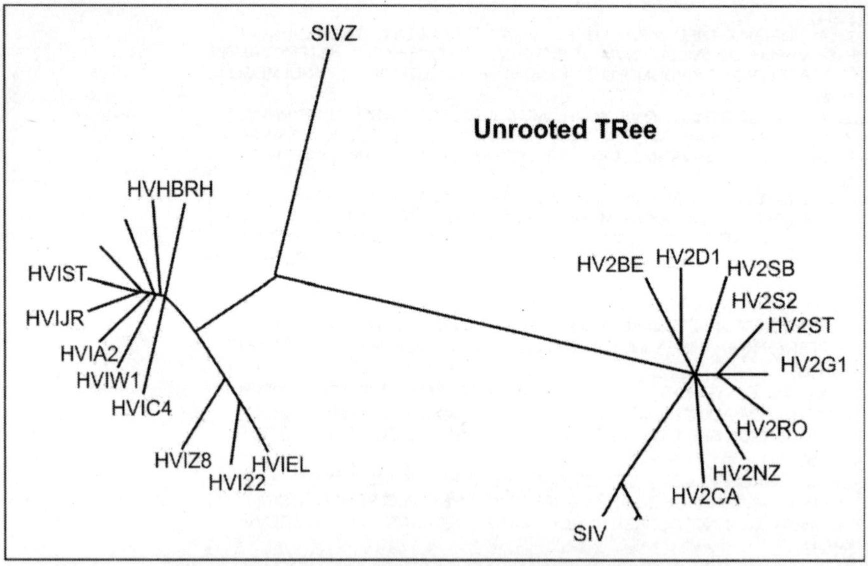

Fig. 14.14: The unrooted tree drawn by Phylip

SUMMARY

Phylogeny is the study of the evolution of life forms from more than one ancestor. Evolution is called as phyletic evolution if the study involves changes within a single lineage. The evolutionary history inferred from phylogenetic analysis is usually depicted as branching tree like diagram that represent the evolutionary relationship among different species. The advent of molecular approaches for analyzing genetic material and genes has revolutionized the field of evolution. The method of construction of a phylogenetic tree are classified into distance methods and character based methods. Phylip Paup and Mega2 are main computer programs for drawings phylogenetic trees.

EXERCISE

1. Use the sequences given in question 1 in exercise of chapter 11 for enzyme glutamate synthetase and draw phylogentic tree using softwares clustral and phylip. Answer the following questions based on phylogentic tree obtained.
   (*a*) Find out which genes are orthologous and which are paralogous?
   (*b*) Try to locate some example for xenologous sequences also?
2. Determine from the sequences of pancreatic ribonuclease from horse, minke whale and red kangroo which of these species are more closely related. We know that horse and whale are placental mammals and kangroo is a massuptical, we expect horse and whale to be the closest

pair. Select a suitable database to retrieve the sequences. Design a strategy for construction of alignment and phylogentic tree.

3. Two diploid species of closely related maize plant (called A and B here) were analysed with regard to genes that encode for enzyme hexokinase. Species A has two distinct copies of this gene: A1 and A2 i.e. deploid species is A1A1 A2A2. The other species has three copies of the hexokinase gene, which we call as B1, B2 and B3. A deploid member of the species would have B1B1 B2B2 and B3B3#. These hexokinase genes from the two species were subjected to DNA sequencing and percentage of sequence identify was compared and is gives below:

% Sequence Identify

A1	*A1*	*A2*	*B1*	*B2*	*B3*
	A1	A2	B1	B2	B3
A1	100	62	54	94	53
B1	62	100	91	49	92
B2	54	91	100	6T	90
B3	52	92	90	64	100

If we assume that hexokinase genes were never lost in evolution, how many distinct hexokinase genes do you think were there in most recent ancestor that preceded the divergence of these two species. Explain your answer? Explain why species of B has three distinct copies of this gene while species A has only 2. Relate this to orthologus and paralogous genes.

(**Hint:** Please note A1 is very similar to B2 and A2 is very similar to B1 and B3)

15

PROTEIN STRUCTURE PREDICTION

A protein is a polymer comprised of amino acid residues. Two amino acids can be joined together so that there is an amide bond between them. The phrase amide bond is a general term for a –C(=O) NH-group; in the specific context of proteins and related molecules we also use peptide bond to describe this kind of covalent link. Structure of a peptide bond between lysine and phenyl alanine is shown in Fig. 3.3. We can join together more and more amino acids, forming a polymer. A short polymer of amino acids is known as a polypeptide (or peptide), while a biosynthesised polymer of tens or hundreds of residues is a protein. In chapter 3 a detailed description about protein databases is given. Also an introduction to primary, secondary, tertiary and quartenary structures of proteins is given.

The structures of all molecules extend to three-dimensions. Various non-superimposable three dimensional arrangements of atoms that are interconvertible without breaking covalent bonds are generally described as conformations. The term conformation must not be confused with configuration which refers to chirality of atoms. Different conformations of a protein molecule must have the same configuration i.e. it should either be a D or L isomer of an amino acid. Different conformation of a molecule is important in study of its 3-D structure because bond lengths vary by ±0.05A° and bond angles by about ±5° at room temperature. For this reason only the energetically most stable arrangements are usually classified as individual conformations.

In Fig. 15.1 the structure of two amino acids in a polypeptide chain is shown. The R group is different for each of the 20 amino acids is shown in Fig. 3.2. Neighbouring amino acids are joined by a peptide bond between the C = O and NH groups already shown in Fig. 3.3. The N—Cα—C′ sequence is repeated throughout the protein, forming the backbone of the three-dimensional structure. The amino acid at one end of the chain has a free NH_2 group and the amino acid at the

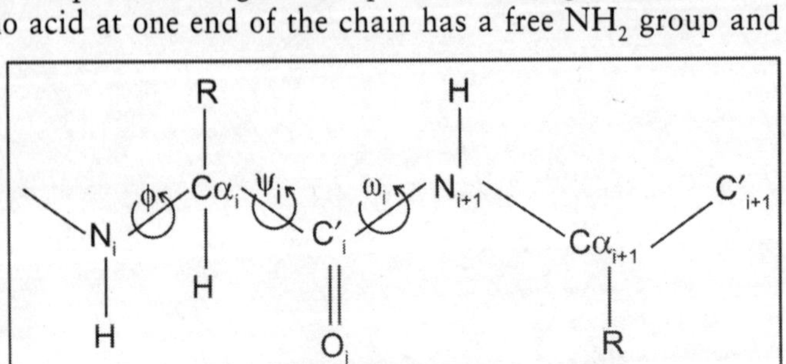

Fig. 15.1: Three dimensional structure of peptide bond

other end has a free COOH group. The bond on each side of the Cα atom are quite free to rotate, but many combinations of angles are not possible for most amino acids due to spatial constraints from the R group and neighboring positions in the chain. The comformation of the protein backbone in space is determined by rotation angles about the bond between atoms N and Cα (in ϕ angle) and the bond between atoms Cα in C (the ψ angle).

The rotation angle about the bond between atoms C and N (the ω-angle) always has a value of 180° since the peptide bond is always constrained in the planar trans conformation. The distribution of these two angles ϕ and Ψ for the amino acids in a particular protein is often plotted on a graph called a Ramachandran plot.

15.1 THREE-DIMENSIONAL CONFORMATION OF PROTEINS

In case of proteins each amino acid residue in a polypeptide chain contains three bonds of the polypeptide backbone plus those of the side chain. The peptide bond has a partial double bond character and is limited to cis and trans rotations. The other two bonds of backbone have single bond character as do many of the bonds of many side chain. Rotation about these bonds produce a different conformation. On an average each amino acid in a chain can exist in eight conformations. A small polypeptide of 10 residues might therefore be able to produce 8^{10} conformations. Because so many conformations are possible, the conformational properties of random polypeptides are best calculated statistically. Such calculations require detailed knowledge of the conformational properties of the monomeric unit of the polymer, i.e. relative energies of all its possible conformations. Rotations about bonds are described as torsion or dihedral angles, which usually are taken to lie in the range of –180° to +180°. ϕ, ψ and ω are three torsion angles, betwen N–Cα, C′ – C′ and C′ –N respectively.

One of the most useful ways of describing the conformation of a protein chain is in terms of the torsion angles, phi and psi of each residue. The permitted values of ϕ and ψ were first determined by Ramachandan and his collegues using hard-sphere model of atoms and fixed geometries of bonds. The permitted values of ϕ and ψ are usually indicated on a two dimensional map of ϕ–ψ plane called Ramachandran plot.

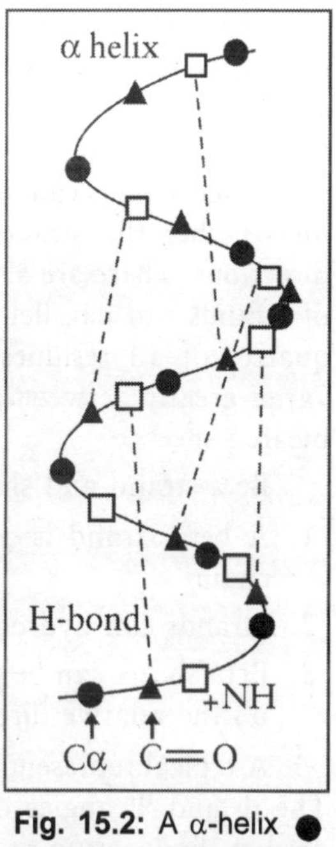

Fig. 15.2: A α-helix ● in the Fig. refers to Cα, ▲ to C = 0 and □ to NH. The dotted line connecting ▲ and □ of i and (i + 4)th residue is deplicting a hydrogen bond

15.2 REGULAR CONFORMATIONS OF POLYPEPTIDES

The phrase 'secondary structure' describes the regular three-dimensional shapes taken up by groups of residues close to one another. The common three-dimensional conformations are alpha helices, beta strands and beta sheets, turns and random coils.

15.2.1 Alpha Helix

The alpha helix is one of the most regularly observed and important secondary structure elements. About one-third of all residues in proteins are found in alpha-helical conformations, but the proportion varies enormously between different proteins. The alpha helix has the following properties:

1. It completes one turn every 3.6 residues
2. It rises approximately 5.4 Angstroms with each turn
3. It is held together by hydrogen bonds between the C = O of residue *i* and the NH of residue *i* + 4;
4. It is typically slightly curved

A typical representation of an alpha helix is shown in Fig. 15.2. In the α helix, note that each C = O group at amino acid position *n* in the sequence is hydrogen-bonded with the NH group at position *n* + 4. There are 3.6 residues per turn. The helix is usually right-handed, but short sections of 3 to 5 amino acids of left-handed helices occur occasionally. The average Φ and Ψ angles of the amino acids in the right-handed helix are approximately 60° and 40°, respectively. The R side chains of the amino acids are on the outside of the helix.

The commonest location of α-helices is at the surface of protein cores, where they provide an interface with the aqueous environment. The inner-facing side of the helix tends to have hydrophobic amino acids and the outer-facing side hydrophilic amino acids. Thus, every third of four amino acids along the chain will tend to be hydrophobic, a pattern that can be quite readily detected. Other α-helices buried in the protein core or in cellular membranes have a higher and more regular distribution of hydrophobic amino acids, and are highly predictive of such structures. Helices exposed on the surface have a lower proportion of hydrophobic amino acids. Amino acid content can be predictive of an α-helical region. Regions richer in alanine, glutamic acid, leucine, and methionine and poorer in proline, glycine, tyrosine, and serine tend to form an α helix. Proline destafilizes or breaks an α helix but can be present in longer helices, forming a bend. There are computer programs for predicting quite reliably the general location of α–helices in a new protein sequence.

15.2.2 β-strands and β-sheets

A beta-strand is a secondary structure element in which the protein chain is extended into an almost linear geometry. β-strands are formed by H bond between an average of 5-10 consecutive amimo acids in one portion of chain with another 5–10 for further down the chain. Beta-strands can form hydrogen bond together to form a beta sheet. This can be either parallel or anti-parallel depending on whether the strands are running in the same or opposite directions. There are also mixed beta sheets, in which some pairs of strands run parallel and some anti-parallel. Approximately a quarter of all residues are in beta-strands, though again this varies greatly between proteins. They are sometimes called beta-pleated sheets.

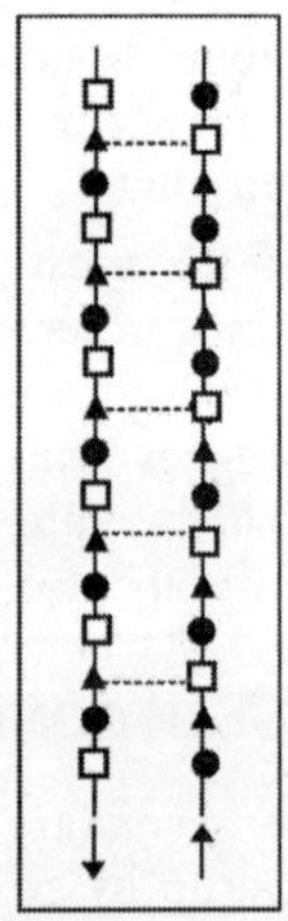

Fig. 15.3: The β-sheet▲ refers to C = 0 ● to Cα and □ refers to NH. The dotted line connecting ▲ to □ refers to a hydrogen bond

Beta-strand and sheet share the following properties:

1. A beta strand is an extended conformation of the protein chain
2. Strands can hydrogen bond together to form beta sheets
3. Beta sheets can be parallel, anti-parallel or mixed depending on the relative directions of the strands

A typical representation of a beta sheet is shown in Fig. 15.3. The Φ and Ψ angles of amino acids in β plated sheets very considerably. It is more difficult to predict the location of β-sheets than of α-helices.

15.2.3 Turns

Turns generally occur when the protein chain need to change direction in order to connect two other elements of secondary structure. The most common is the beta turn in which the change of direction is executed in the space of four residues. Some commonly observed features of beta turns are a hydrogen bond between the C = O of residue *i* and the N-H of residue *i* + 3 and a tendency to involve glycine and/or proline, but these are not universal. The term beta hairpin is often used to describe a beta turn joining two anti-parallel beta strands together. Beta turns are subdivided into numerous types on the basis of the details of their geometry. Gamma turns are three residue turns which often incorporate a hydrogen bond between the C = O of residue i and N-H of residue $i + 2$. Several other structures are important in regular conformation of polypeptides, for example two cystine residues, close in space but not necessarily in sequence, may be joined together by a disulfide bridge. This is much more common in the oxidising extracellular environment than for proteins in the more reducing intracellular environment.

15.2.4 Supersecondary Structure

One or more of secondary structural elements very often follows a regular pattern like α-β-α or β-α-β, β-hairpin, α-hairpin, etc. These structures are called as supersecondary structures. These structures are a higher level of structure than secondary. Structure but do not constitute entire structural domains. Different level of organisation of protein structure is summarised in Fig. 15.4.

15.2.5 Domains

A Domain is an independently folding unit of three-dimensional protein structure. In heirarchy domain falls between supersecondary structure and tertiary structures of a complete monomer. In general, a protein molecule consists of one or more chains, each of which consists of one or more domains. Domains are compact units within the folding pattern of a polypetide chain that looks as if it has an independent identity. Domain in context to structure of proteins refers to a segment of a polypeptide chain that can fold into a three-dimensional structure irrespective of the presence of other segments of the chain. The separate domains of a given protein may interact extensively or may be joined only by a length of polypeptide chain. A protein with several domains may use these domains for functional interactions with different molecules. The RNA-binding protein, L1 has features typical of multi domain proteins. The binding site appears in the left between two domains and the relative geometry of two domains is flexible. 3Dee, a database of protein domain definitions, is provided at http:/barton.ebi.ac.uk/servers/3Dee.htm1/. A structural classification of protein domains is maintained at http://www2.embl-ebi.ac.uk/dali/domain, and ddbase, a database of protein domains, may be found at http://www3.icgeb.trieste. Structural domains are also called as homology domains.

Domains are important because:

1. They usually correspond to the functional units of a protein. Often various combinations of the same set of domains are found in different arrangements in different proteins.
2. We believe them to be the key evolutionary units of protein structure. The essential structural features of domains sharing a common ancestor are often retained even when so much evolution has taken place that no sequence relationship can be seen.
3. They are the entities through which we categorise and understand tertiary structure.
4. To understand protein folding, we need to understand processes occurring at the domain level.

15.2.6 Folds

It refers to a larger combination of secondary structural units in the same configuration. Thus, proteins sharing the same fold have the same combination of secondary structures that are connected by similar loops. An example is the Rossman fold comprising several alternating α-helices and parallel β-strands. In the SCOP, CATH, and FSSP databases described later, the known protein structures have been classified into hierarchical levels of structural complexity with the fold as a basic level of classification. From a survey of the currently known proteins structures in the Brookhaven protein Data Bank approximately 500 independent folds have been identified. Foldon is a related

(a) Primary structure of protein

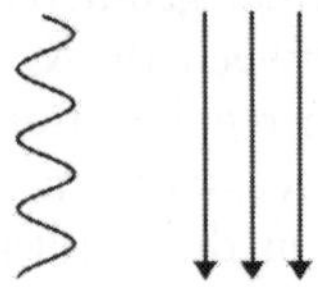

(b) Secondary structure

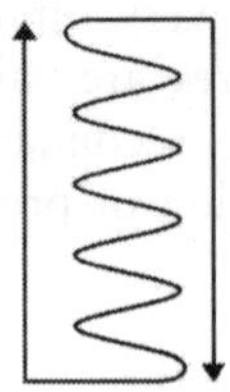

β-α-β-motif

(C) Super secondary structure

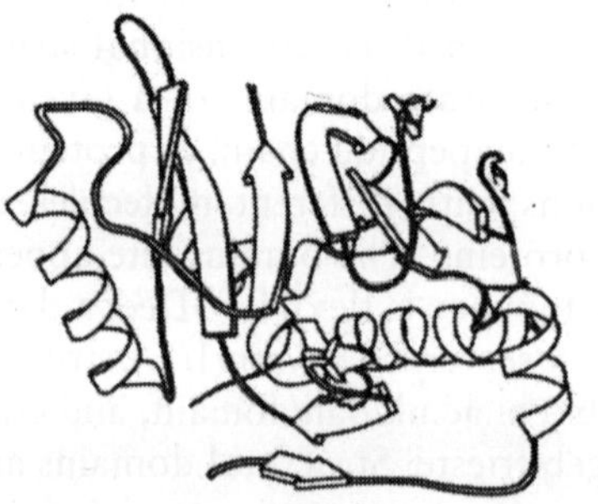

(d) Tertiary structure made up of α-helix and β-sheets

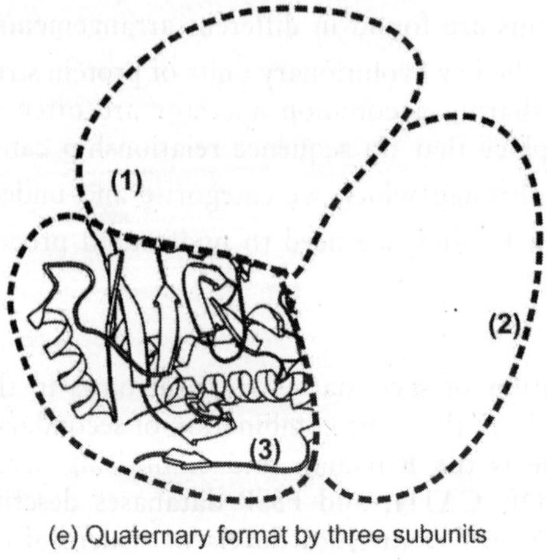

(e) Quaternary format by three subunits

Fig. 15.4: Different level of protein structure organisation term that has been used to describe an independently folding unit. Tim barrel or $(\alpha/\beta)_8$ another example of fold has been dissected from the structure of enzyme RuBpcase in Experiment 15A. This has eight parallel β-strands, with an α-helic on the outside connecting each pair of β-strands.

Tertiary structure describes the ways in which various elements of secondary structure are arranged to form the overall three-dimensional structure of a protein. A key difference between the three dimensional structures of proteins and those of other biological molecules, such as DNA, is the sheer variety of protein structures. It is fair to say that when Watson and Crick discovered the double helix they essentially solved the structure of DNA. For proteins, the situation is quite different. Kendrew published an atomic resolution model of myoglobin in 1960 and some thought that protein structure too had been solved. Yet even now, with almost 20000 protein structures known, we still find some totally new tertiary structures, though all are built from the same familiar secondary structure elements (helix, strand, turn, coil).

Module is a region of conserved amino acid patterns comprising one or more motifs and considered to be a fundamental unit of structure of function. The presence of a module has also been used to classify proteins into families.

15.2.7 Quaternary Structure

Quaternary structure is the formation of larger oligomeric protein molecules by the interaction of multiple protein chains. One well known examples is hemoglobin where four chains come together to form a tetrameric unit. Quaternary level of structure is often important from the points of view of the function and stability of the protein. For instance, an active site may be formed by residues from different chains. Another possibility is that the subunits can operate co-operatively, working together more efficiently than would be possible if they were spatially separate.

15.3 PROTEIN STRUCTURE EVALUATION

At present only two techniques can elucidate the dimensional structure of protein to high resolution: X-ray differaction analysis of single crystals and NMR analysis of small proteins in solution which are discussed in details (box for revolutionary technique). In 1953 Perutz showed that diffracted X-rays from protein crystals could be phased by comparing profile from crystals of the protein with and without heavy atoms attached. In 1959 he determined the molecular structure of hemoglobin and shared the Nobel prize for chemistry with John Kendrew (Historical Profile 17). As more and more protein structures have been solved by X-ray crystallographic and NMR methods, these structures have been classified by various means into structural databases. An introduction to these databases is already given in chapter 3. This classification is based on types, order, connections, and relative positions of secondary structures that are compared using the known atomic coordinates of atoms in each structure and method.

Once the three-dimensional coordinates of each atom in the protein molecule have been found, a table of these coordinates is deposited with the Protein Data Bank as a PDB entry. PDB is best established database for biological macromolecular stuctures. It contains structures of proteins, nucleic acids and a few carbohydrates. Started by late Water Hamilton at Brookhaven national Laboratories, New York in 1971, PDB is now managed by Research Collaboratory for Structural Bioinformatics (RSCB), and National Institute of Standards and technology and the San diego Supercomputer Center, all in USA. PDB entries such as shown in Data Box 16.1 give the atomic coordinates of the amino acids in proteins, protein fragments, or proteins bound to substrates or inhibitors. Pdb files may be easily retrieved from the PDB web site and displayed with a molecular viewer such as Rasmol. Structural information may also be stored in forms other than pdb, but pdb is the most accessible for the molecular biologist. This process has been discussed in details in Experiment 16A with enzyme rubpcase.

HISTORICAL PROFILE 17

Sir Max Ferdinand Perutz and Sir John Cowdery Kendrew

Max Ferdinand Perutz

John Cowdery Kendrew

Max Ferdinand Perutz, was an Austrian-British molecular biologist. He was born in Vienna in 1914. In 1936 he became a research student at the Cavendish Laboratory in a crystallography group directed by J.D. Bernal, and remained in Cambridge subsequently. In 1953 Perutz showed that the diffracted X-rays from protein crystals could be phased by comparing the patterns from crystals of the protein with and without heavy atoms attached. In 1959 he determined the molecular structure of the protein hemoglobin, which transports oxygen in the blood, using this method. In 1962 he received the Nobel Prize for Chemistry, with John Kendrew.

In the history of science, Perutz is also known as the supervisor of James D. Watson and Francis Crick during the early 1950's during which time Watson and Crick determined the structure of deoxyribonucleic acid (DNA). Perutz established the MRC Laboratory of Molecular Biology, Cambridge, England in 1962 and was chairman until 1979. He remained active in research to the end of his life. From the mid-1980s on he was a regular reviewer/essayist for The New York Review of Books on biomedical subjects.

An Italian biography of Max Perutz entitled "Piccole Visioni - La Grande Storia di una Molecola" was written by marta Paterlini and published by Codice Edizioni, Turin, in May 2006, as English biography by Georgina Ferry is due for publication by Chatto & Windus in July 2007. Mas Perutz's major contribution to molecular biology in Cambridge is also well documented in The History of the University of Cambridge: Volume 4 (1870 to 1990) published by CUP in 1992.

Sir John Cowdery Kendrew was an English biochemist and crystallographer who shared the 1962 Nobel Prize in Chemistry with Max Perutz; their group in the Cavendish Laboratory investigated the structure of heme-containing proteins.Their work was done at the MRC Laboratory of Molecular Biology in Cambridge. Kendrew determined the structure of the protein myoglobin, which stores oxygen in muscle cells.

In 1947 the MRC agreed to make a research unit for the Study of the Molecular Structure of Biological Systems. The original studies were on the structure of sheep hemoglobin, but when this work had progressed as far as was possible using the resources then available, kendrew embarked on the study of myoglobin, a molecule only a quarter the size of the hemoglobin molecule. His initial source of raw material was horse heart, but the crystals thus obtained were too small for X-ray analysis. Kendrew realized that the oxygen-conserving tissue of diving mammals could offer a better prospect, and a chance encounter led to his acquiring a large chunk of whale meat from Peru. The whale-sourced myoglobin gave large crystals with clean X-ray diffraction patterns. However, the problem still remained insoluble, until in 1953 Max Perutz discovered that the phase problem in analysis of the diffraction patterns could be solved by comparison of patterns from two crystals—one from the native protein, and one from the protein with heavy metals attached to it. An electron density at 6 angstrom (0.6 nanometre) resolution was obtained by 1957, and by 1959 an atomic model could be built at 2 angstrom (0.2 nm).

REVOLUTIONARY TECHNIQUE 1: X-ray Crystallography that Reveals Three-Dimensional Structure of Proteins in Atomic Detail

The understanding of protein structure and function has been greatly enriched by X-ray crystallography, a technique that can reveal the precise three-dimensional positions of most of the atoms in a protein molecule. Determination of atomic details by X-ray crystallography need crystals of the protein of interest. Crystals can be obtained by adding ammonium sulfate or another salt to a concentrated solution of protein to reduce its solubility.

The protein crystal is mounted in a capillary and position precise orientation with respect to the X-ray beam results in an X-ray photograph a regular array of spots called reflections. The intensity of each spot is noted. These intensities are the basic experimental data of an X-ray crystallographic analysis. The next step is to reconstruct an image of the protein from the observed intensities. In light microscopy or electron microscopy the diffracted beams are focused by lenses to directly form an image. However, lenses for focusing X-rays do not exist. Instead, the image is formed by applying a mathematical relation called a Fourier transform. For each spot, this operation yields a wave of electron density, whose amplitude is proportional to the square root of the observed intensity of the spot. Each wave also has a phase—that is, the timing of its crests and troughs relative to those of other waves. The phase of each wave determines whether it reinforces or cancels the waves contributed by the other spots. These phases can be deduced from the well-understood diffraction patterns produced by heavy-atom reference markers such as uranium or mercury at specific sites in the protein.

The stage is then set for the calculation of an electron-density map, which gives the density of electrons at a large number of regularly spaced points in the crystal. This three-dimensional electron-density distribution is represented by a series of parallel sections stacked on top of each other. Each section is a transparent plastic sheet (or a layer in a computer image) on which the electron-density distribution is represented by contour lines like the contour lines used in geological survey maps to depict altitude. The next step is to interpret the electron-density map. A critical factor is the resolution of the X-ray analysis, which is determined by the number of scattered intensities used in the Fourier synthesis. The fidelity of the image depends on the resolution of the Fourier synthesis. A resolution of 6 Å reveals the course of the polypeptide chain but few other structural details. The reason is that polypeptide chains pack together so that their centers are between 5 and 10 Å apart. Maps at higher resolution are needed to delineate groups of atoms, which lie from 2.8 to 4.0 Å apart. The ultimate resolution of an X-ray analysis is determined by the degree of perfection of the crystal. For proteins, this limiting resolution is usually about 2 Å.

REVOLUTIONARY TECHNIQUE 2: Nuclear Magnetic Resonance

Nuclear magnetic resonance (NMR) is a physical phenomenon based upon the quantum mechanical magnetic properties of an atom's nucleus. All nuclei that contain odd numbers of protons or neutrons have an intrinsic magnetic moment and angular momentum. The most commonly measured nuclei are hydrogen and carbon-13, although nuclei from isotopes of many other elements can also be observed. NMR studies magnetic nuclei by aligning them with a very powerful external magnetic field and perturbing this alignment using an electromagnetic field. The resulting response to the external perturbing electromagnetic field is the phenomenon that is exploited in NMR spectroscopy and magnetic resonance imaging.

Nuclear magnetic resonance was first described and measured in molecular beams by Isidor Rabi in 1938. Eight years later, in 1946, Felix Bloch and Edward Mills Purcell refined the technique for use on liquids and solids, for which they shared the Nobel Prize in physics in 1952. Purcell had worked on the development and application of RADAR during World War II at Massachusetts

Institute of Technology's Radiation Laboratory. His work during that project on the production and detection of radio frequency (RF) energy, and on the absorption of such energy by matter, preceded his discovery of NMR. They noticed that magnetic nuclei, like ^{1}H and ^{31}P, could absorb RF energy when placed in a magnetic field of a strength specific to the identity of the nuclei. When this absorption occurs, the nucleus is described as being in resonance. Interestingly, for analytical scientists, different atoms within a molecule *resonate* at different frequencies at a given field strength. The observation of the resonance frequencies of a molecule allows a user to discover structural information about the molecule.

The development of nuclear magnetic resonance as a technique of analytical chemistry and biochemistry parallels the development of electromagnetic technology and its introduction into civilian use. Nuclear magnetic resonance is extremely useful for analyzing samples non-destructively. Radio waves and static magnetic fields easily penetrate many types of matter and anything that is not inherently ferromagnetic. For example, various expensive biological samples, such as nucleic acids, including RNA and DNA, or proteins, can be studied using nuclear magnetic resonance for weeks or months before using destructive biochemical experiments. This also makes nuclear magnetic resonance a good choice for analyzing dangerous samples.

EXPERIMENT 15A

To Study the Primary, Secondary and Tertiary and Quaternary Structure of Enzyme Rubpcase from Spinach

About PDB: PDB refers to protein data bank. Protein data bank contain 3-D structure of protein that is obtained by making crystals of proteins and using X-ray differaction to determine the atomic coordinates of each molecule of protein. As on May 15, 2007 there are 43459 structures of proteins available a PDB. PDB structure entries give atomic co-ordinate of the amino acids in proteins, protein fragments or protein bound to substrates or inhibitors. The PDB structure entries, consisting of a collection of files having nondescript names, cannot be easily grasped in a biochemically meaningful context. The structure of a PDB entry can be displayed with a molecular viewer such as Rasnol, Chine CnED. In the present study RASMOL to be used for viewing the PDB files.

About Rasmol: Rasmol is a molecular graphics program intended for the visualization of proteins, nucleic acids and small molecules. The program is aimed at display, teaching and generation of publication quality images. The program reads in molecular co-ordinate files and interactively displays the molecule on the screen in a variety of representations and colour schemes. Supported input file formats include Brookhaven Protein Databank (PDB), Tripos Associates' Alchemy and Syby1 Mol1 formats, Molecular Design Limited's (MDL) Mol file format. The loaded molecule can be shown as wireframe bonds, cylinder 'Dreiding' stick bonds, alpha-carbon trace, space-filling (CPK) spheres, macromolecular ribbons (either smooth shaded solid ribbons or parallel strands), hydrogen bonding and dot surface representations. Different parts of the molecule may be represented and coloured independently of the rest of the molecule or displayed in several representations simultaneously. The displayed molecule may be rotated, translated, zoomed and z-clipped (slabbed) interactively using either the mouse, the scroll bars, the command line or an attached dial box. Rasmol can read a prepared list of connands from a 'script' file (or via inter process communication) to allow a given image or viewpoint to be restored quickly. Rasmol can also create a script file containing the commands required to regenerate the current image. Finally, the rendered image may be written out in a variety of formats including either raster or vector PostScript, GIF, PPM, BMP, PICT, Sun rasterfile or as a MolScript input script or a Kinemage. Rasmol has been developed at the University of Edinburgh's Biocomputing Research Unit and the BioMolecular Structure Department. Glaxo Research and Development, Greenford, U.K.

About Enzyme Rubpcase: The enzyme Rubpcase is the most abundant enzyme in the world, being responsible for carbon fixation in photosynthesis. There are several entries available for this at PDB, of these 8RUB.pds is selected for further study. The enzyme is a Hexadecamer L8S8 of Eight L (Large) Subunits and 8S (small) subunits. The L subunit has been assigned chain identifiers A, B, C, D, E, F, G and H. The S subunit has been assigned chain identifiers I, J, K, L, M, N, O and P. In the catalytically competent Molecule, dimers of the large subunits form functional units. In reference 1 these dimers are designated *AB*, *CD*, *EF*, and *GH*.

TESTING THE HYPOTHESIS

1. Click to open a new window for the PDB Home Page as seen in Fig. 15.5.
2. Slide the PDB window aside so that you can read these instructions and carry them out in the PDB window.
3. Find the Search button and type in 8RUC (the first character is the number eight) in the box beside it. Then click the Search button or press <return>. The Structure Summary information page for 8RUC appears as seen in Fig. 15.5.
4. At the top of the page shows the pdb file. Open and save it to your computer.
5. Open the enzymes pdb file in a note pad data as shown in box is seen study this thoroughly.
6. Dawnload rasmol binaries from the web page of rasmol.
7. Open the PDB file saved in rasmol and study all the views of two views are shown in Fig. 15.7.
8. Edit the pdb file to dissect the eight subunits. Reopen them separately in rasmol.
9. Dissect the pdb file to study the folds, motifs and alpha and beta subunit. Figures as shown in Fig. 15.7 needs to be drawn.

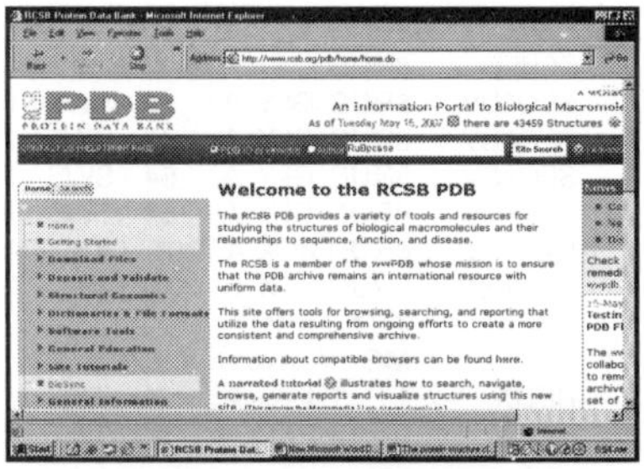

Fig. 15.5: Web view of PDB

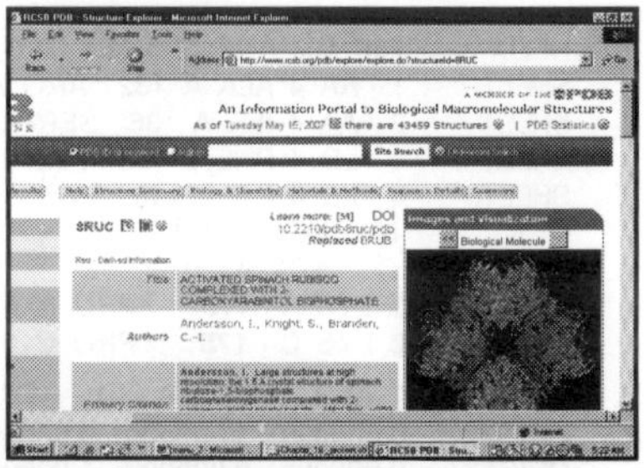

Fig. 15.6: Res ult obtained for 8RUB.pdb

DATA

Data Box 15.1: The Pdb file

```
HEADER    LYASE (CARBON-CARBON)                    22-FEB-96   8RUC
TITLE     ACTIVATED SPINACH RUBISCO COMPLEXED WITH 2-CARBOXYARABINITOL
TITLE     2 BISPHOSPHATE
COMPND    MOL_ID: 1;
COMPND    2 MOLECULE: RIBULOSE-1,5-BISPHOSPHATE CARBOXYLASE/OXYGENASE;
COMPND    3 CHAIN: A, C, E, G, I, J, K, L;
COMPND    4 SYNONYM: RUBISCO;
COMPND    5 EC: 4.1.1.39
SOURCE    MOL_ID: 1;
SOURCE    2 ORGANISM_SCIENTIFIC: SPINACIA OLERACEA;
SOURCE    3 ORGANISM_COMMON: SPINACH;
SOURCE    4 ORGAN: LEAF
KEYWDS    PHOTOSYNTHESIS, CARBON DIOXIDE FIXATION, PHOTORESPIRATION,
KEYWDS    2 LYASE, OXIDOREDUCTASE, MONOOXYGENASE, CHLOROPLAST,
KEYWDS    3 LYASE (CARBON-CARBON)
EXPDTA    X-RAY DIFFRACTION
AUTHOR    I.ANDERSSON,S.KNIGHT,C.-I.BRANDEN
REVDAT    1  01-AUG-96 8RUC     0
SPRSDE    01-AUG-96 8RUC       8RUB
..
.
```

```
.
.
REMARK   3   BOND ANGLES              (DEGREES) : 1.6
REMARK   3   DIHEDRAL ANGLES          (DEGREES) : 24.1
REMARK   3   IMPROPER ANGLES          (DEGREES) : 1.4
REMARK   999 THE ONLY PUBLISHED SEQUENCE FOR RBS (OBTAINED BY SEQUENCING
REMARK   999 THE PROTEIN) DIFFERS IN 30 OUT OF 123 RESIDUES. THIS
REMARK   999 ASPECT IS DISCUSSED IN DETAIL IN THE *JRNL* REFERENCE.
.
.
.
DBREF    8RUC A    9   475  SWS    P00875   RBL_SPIOL      9    475
DBREF    8RUC C    9   475  SWS    P00875   RBL_SPIOL      9    475
SEQADV   8RUC GLY A   16  SWS  P00875    GLY    16 CONFLICT
SEQADV   8RUC GLY A   47  SWS  P00875    GLY    47 CONFLICT
SEQADV   8RUC GLY A   54  SWS  P00875    GLY    54 CONFLICT
SEQADV   8RUC GLY A   64  SWS  P00875    GLY    64 CONFLICT
MODRES   8RUC KCX A  201  LYS  CARBAMYLATED AT THE EPSILON NITROGEN
MODRES   8RUC KCX C  201  LYS  CARBAMYLATED AT THE EPSILON NITROGEN
HET      KCX  A 201        12
HET      MG   A 476         1
HELIX    1    1 LYS A    21  TYR A   24  1                               4
HELIX    2    2 PRO A    50  GLU A   60  1
.
.
.
                                     11
HELIX    100 100 PRO L    80  GLU L   93  1                              14
SHEET    1   AA 4 ALA A 132  ARG A 139  0
SHEET    2   AA 4 ILE A  36  SER A  43 -1  N  SER A  43   O  ALA A 132
SHEET    3   LD 4 PHE L  98  PHE L 104 -1  N  PHE L 104   O  VAL L  39
SHEET    4   LD 4 ILE L 113  TYR L 118 -1  N  TYR L 118   O  ILE L  99
LINK       N    KCX A 201                C    THR A 200
LINK       C    KCX A 201                N    ASP A 202
CISPEP   1 LYS A  175      PRO A  176          0          -1.15
CISPEP   2 LYS C  175      PRO C  176          0           1.61
SITE     1 ACA  4 KCX A 201  ASP A 203  GLU A 204   MG A 476
ORIGX2   0.000000  1.000000  0.000000         0.00000
ORIGX3   0.000000  0.000000  1.000000         0.00000
SCALE1   0.006361  0.000000  0.000000         0.00000
ATOM     1  N    ALA A   9      -33.951  70.802  77.969  1.00 43.78          N
ATOM     2  CA   ALA A   9      -35.279  70.579  78.625  1.00 44.39          C
ATOM     3  C    ALA A   9      -35.736  69.151  78.331  1.00 44.68          C
ATOM     4  O    ALA A   9      -35.488  68.631  77.239  1.00 44.47          O
ATOM     5  CB   ALA A   9      -36.296  71.592  78.110  1.00 42.92          C
ATOM     6  N    SER A  10      -36.390  68.528  79.310  1.00 44.11          N
ATOM     7  CA   SER A  10      -36.888  67.161  79.181  1.00 43.13          C
ATOM     8  C    SER A  10      -37.909  66.906  80.290  1.00 42.32          C
ATOM     9  O    SER A  10      -37.921  67.614  81.304  1.00 41.14          O
```

INTERPRETATION OF DATA

Ribulose-1,5-bisphosphate carboxylase/oxygenase (rubisco) from spinach as studied from the summary info page at PDB website shows with entry 8ruc.pdb is a hexadecamer (L8S8, Mr = 550, 000) consisting of eight large (L, 475 residues) and eight small subunits (S, 123 residues). This stucture is based on study that describes the refined 1.6A crystal structure of activated rubisco complexed with a transition state analogue, 2-carboxyarabinitol-bisphosphate as submitted by Anderson et al in 1996 (Anderson, I. Large structures at high resolution: the 1.6A crystal structure of spinach ribulose-1,5-bisphosphate carboxylase/oxygenase complexed with 2-carboxyarabinitol bisphosphate. *J.Mol.Biol. v259,* pp. 160-74, 1996).

Summary page at PDB website also provide detailed information about CATH, SCOP and Pfam classification of the enzyme. Information about domain, class, fold superfamily in Scop classification is summarised in Table 15.1 This suggests that this protein is mainly belonging to the class of alpha and beta proteins. The fold expected in the structures of different subunits are TIM beta/alpha barrel, Ferredoxin like fold and Rubisco small subunit. Tim barrel is shown in Fig. 15.7(d). In the CATH classification in Table 15.2 folds are studied under architectire.

This entry presents the crystallographic asymmetric unit which contains half the molecule: Four L and Four S subunits related by a four–fold Axis through the molecular center at X - 0, Y = 1/4, Z = 1/4. The four I chains present are A, C, E, G and the four S chains present are I, J, K, L. Also included are four magnesium ions, four inhibitor molecules (CAP) and 1520 waters.

The Contents of PDB File: You can view PDB files can be studied when opened with a note pad. The view is shown in Data Box 16.1. The can also be directly studied on the web. The contents of the file, in order of appearance, are

- HEADER lines, containing the file name and date.
- COMPND lines, containing the name of the protein.
- SOURCE lines, giving the organism from which the protein was obtained.
- AUTHOR lines, listing the persons who placed this data in the protein Data Bank.
- REVDAT lines, listing all revision dates for data on this protein.
- REMARK lines, which contain (1) references to journal articles about the structure of this protein, and (2) general information about the contents of this file.
- SPRSDE lines, which list older coordinate files of this same structure.
- SEQRES lines, which give the amino-acid sequence of the protein.
- FTNOTE lines, which contain footnotes.
- HET and FORMUL lines, which list the cofactors, prosthetic groups, inhibitors or other nonprotein substances present in the structure.
- HELIX SHEET and TURN lines, which list the elements of secondary structure in the protein.
- CRYST1, ORIG and SCALE lines, which contain some general information about the protein crystals from which this structure was obtained by the technique of x-ray crystallography.
- ATOM and HETATM lines, which contain the atomic coordinate data needed to display the structure of the protein. Notice that no hydrogen atoms are listed in the file. Most protein crystals do not diffract well enough to allow hydrogen atoms to be resolved. Positions of hydrogen atoms must be inferred from the positions of other atoms.

  Note that residue numbers sometimes do not begin with #1 (and sometimes end before the known last residue number of the protein). In a crystallographic model, this usually means that some of the terminal residues were not dound in the electron-density map that was interpreted to produce the model, perhaps due to disorder in these parts of the molecule.
- CONNECT lines, which list bonds involving nonprotein atoms in the file.
- MASTER and END lines, which mark the end of the file.

Remember that this PDB file is the source of all the information that a modeling program uses to produce a display of the protein structure. Most modeling programs can display any protein for which you provide coordinate data in the format of ATOM lines in a PDB file.

Look again at the ATOM lines. Each line lists one atom. In order across the row after the line number and the word ATOM are:

(*a*) atom number: atoms are numbered in sequence through the fule;

(*b*) atom type: n = amide N, ca = alpha C, c = carbony1 C, o = carbony1 O, cb = beta carbon, and so forth;

(*c*) residue name: three-letter amino acid abbreviation;

(*d*) residue number;

(*e*) x-coordinate of the atom, in angstroms from the unit-cell origin;

(*f*) y-coordinate of the atom;

(*g*) z-coordinate of the atom;

(*h*) occupancy: the fraction of unit cells that contain the atom in this particular location, usually 1.00, or all of them (can be used to represent alternative conformations of side chains);

(*i*) temperature factor: an indication of uncertainty in this atom's position due to disorder or thermal vibrations (can be used by graphics programs to represent the relative mobility of different parts of a protein);

(*j*) every line ends with the PDB file identification code.

Columns *e*, *f*, and *g* locate the atom in three-dimensional space, giving each atom a set of Cartesian coordinates (*x*, *y*, *z*). Molecular modeling programs use this information to produce three dimensional displays of molecules on the computer screen, During rotations of the image, modeling programs continually recalculate all coordinates to establish new atom positions for all atoms as they move.

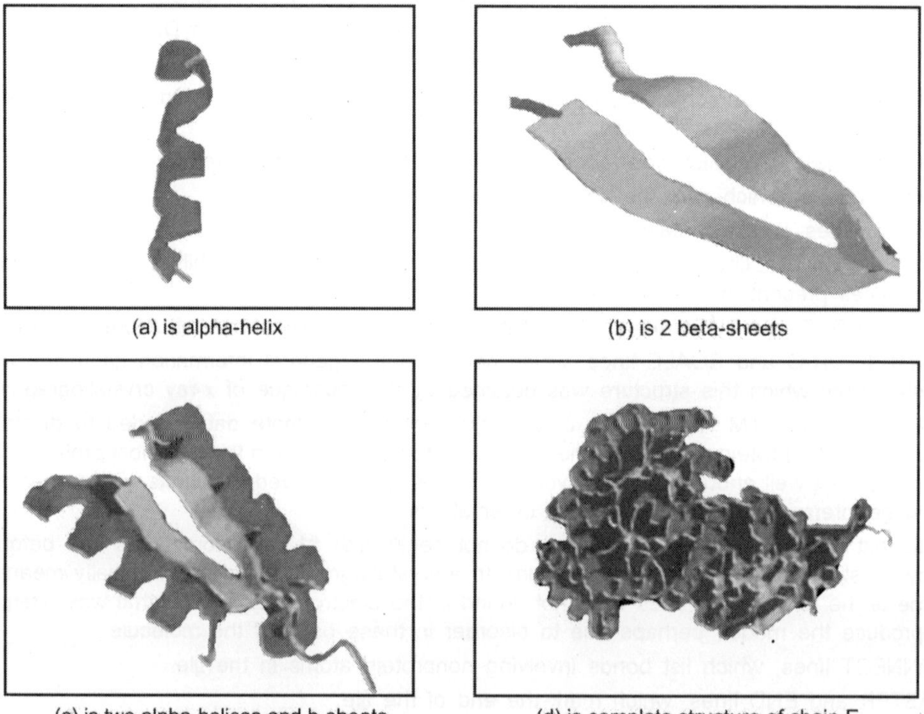

(a) is alpha-helix

(b) is 2 beta-sheets

(c) is two alpha-helices and b-sheets

(d) is complete structure of chain E

Fig. 15.7: Different levels of structure of RUBPcase

In RUBP case we see presence of TIM barrel: This fold is characterized by a central barrel formed by parallel β-strands surrounded by seven or eight α–helices which shield the barrel from solvent. The β-strands of the barrel form an intrinsic network of hydrogen bonds with the neighbouring strands and is oriented in the same direction. The overall twist associated with all the strands cause the first and the eighth strand to register in parallel held in place by hydrogen bonds causing the closure of the barrel. The overall sequence topology can therefore be either $(\beta/\alpha)_8$ where the protein begins with a strand or $(\alpha/\beta)_8$ where helix is the first secondary structure. In addition, many $(\alpha/\beta)_8$ enzymes have additional domains that are not a part of this fold. The TIM barrel formed by β-sheets refers to supersecondary structure given the names as fold. All these pictures are drawn with rasmol. The display selected here is cartoon and colour according to structure. When the end of an alpha-helix or beta-sheet is clicked the rasmol information window depicts the amino acid number and name. Once this information is collected these residues one selected from PDB file

Table 15.1: The scop classification of RUBPcase enzyme subunit

Domain Info	*Class*	*Fold*	*Superfamily*	*Family*	*Domain*
d8ruca1	Alpha and beta proteins (a/b)	TIM beta/alpha-barrel	RuBisCo, C-terminal domain	RuBisCo, large subunit, C-terminal domain	Ribulose 1,5- bisphosphate carboxylase-oxygenase
d8rucc1	Alpha and beta proteins (a/b)	TIM beta/alpha-barrel	RuBisCo, C-terminal domain	RuBisCo, large subunit, C-terminal domain	Ribulose 1,5-bisphosphate carboxylase-oxygenase
d8ruce1	Alpha and beta proteins (a/b)	TIM beta/alpha-barrel	RuBisCo, C-terminal domain	RuBisCo, large subunit, C-terminal domain	Ribulose 1, 5-bisphosphate carboxylase-oxygenase
d8rucg1	Alpha and beta proteins (a/b)	TIM beta/alpha-barrel	RuBisCo, C-terminal domain	RuBisCo, large subunit, C-terminal domain	Ribulose 1,5-bisphosphate carboxylase-oxygenase
d8ruca2	Alpha and beta proteins (a+b) domain	Ferredoxin-like	RuBisCO, large subunit, small (N-terminal)	Ribulose 1,5-bisphosphate carboxylase-oxygenase	Ribulose 1,5-bisphosphate carboxylase-oxygenase
d8rucc2	Alpha and beta proteins (a+b) domain	Ferredoxin-like	RuBisCO, large subunit, small (N-terminal)	Ribulose 1,5-bisphosphate carboxylase-oxygenase	Ribulose 1,5-bisphosphate carboxylase-oxygenase
d8ruce2	Alpha and beta proteins (a+b) domain	Ferredoxin-like	RuBisCO, large subunit, small (N-terminal)	Ribulose 1,5-bisphosphate carboxylase-oxygenase	Ribulose 1,5-bisphosphate carboxylase-oxygenase
d8rucg2	Alpha and beta proteins (a+b) domain	Ferredoxin-like	RuBisCO, large subunit, small (N-terminal)	Ribulose 1,5-bisphosphate carboxylase-oxygenase	Ribulose 1,5-bisphosphate carboxylase-oxygenase
d8ruci_	Alpha and beta proteins (a+b)	RuBisCO, small subunit	RuBisCO, small subunit	RuBisCO, small subunit	Ribulose 1,5-bisphosphate carboxylase-oxygenase
d8rucj_	Alpha and beta proteins (a+b)	RuBisCO, small subunit	RuBisCO, small subunit	RuBisCO, small subunit	Ribulose 1,5-bisphosphate carboxylase-oxygenase
d8ruck_	Alpha and beta proteins (a+b)	RuBisCO, small subunit	RuBisCO, small subunit	RuBisCO, small subunit	Ribulose 1,5-bisphosphate carboxylase-oxygenase
d8rucl_	Alpha and beta proteins (a+b)	RuBisCO, small subunit	RuBisCO, small subunit	RuBisCO, small subunit	Ribulose 1,5-bisphosphate carboxylase-oxygenase

with a different name. The new file when reopened depicts the structure of any alpha-helix. The structure of a beta-sheet, α-β motif and tim barrel are drawn in the same way is shown in Fig. 15.7

Table 15.3: The CATH classification of RubPcase subunit

Domain	*Class*	*Architecture*	*Topology*	*Homology*
8rucA01	Alpha Beta	2-Layer Sandwich	Alpha-Beta Plaits	
8rucA02	Alpha Beta	Alpha-Beta Barrel	TIM Barrel	Rubisco
8rucC01	Alpha Beta	2-Layer Sandwich	Alpha-Beta Plaits	
8rucC02	Alpha Beta	Alpha-Beta Barrel	TIM Barrel	Rubisco
8rucE01	Alpha Beta	2-Layer Sandwich	Alpha-Beta Plaits	
8rucE02	Alpha Beta	Alpha-Beta Barrel	TIM Barrel	Rubisco
8rucG01	Alpha Beta	2-Layer Sandwich	Alpha-Beta Plaits	
8rucG02	Alpha Beta	Alpha-Beta Barrel	TIM Barrel	Rubisco
8rucI00	Alpha Beta	2-Layer Sandwich	Ribulose 1,5-Bisphosphate Carboxylase/Oxygenase	
8rucJ00	Alpha Beta	2-Layer Sandwich	Ribulose 1,5-Bisphosphate Carboxylase/Oxygenase	
8rucK00	Alpha Beta	2-Layer Sandwich	Ribulose 1,5-Bisphosphate Carboxylase/Oxygenase	
8rucL00	Alpha Beta	2-Layer Sandwich	Ribulose 1,5-Bisphosphate Carboxylase/Oxygenase	

15.4 THE PROTEIN STRUCTURE CLASSIFICATION

Just as zoologists used categories like kingdom, genus and species to create an hierarchical classification of organisms, so molecular and structural biologists found it helpful to develop systematic methods for the categorisation of protein structures. There are a number of different schemes for classifying protein tertiary structures. These databases include SCOP (Structural classificalist of proteins), CATH (class, architecture, topology, homologus superfamily) and FSSP/DDP (fold classification based on structure-structure alignment of proteins (Dali-Domain Dictionary). Amongst the most successful is CATH which provides an hierarchical classification of the three dimensional structures of domains.

Nearly all proteins have structural similarities with other proteins and, in some of these cases, share a common evolutionary origin. Knowledge of these relationships is crucial to our understanding of the evolution and development of proteins. It will also play an important role in the analysis of the sequence data that is being produced by worldwide genome projects.

Both Scop and Cath databases attempt to classify completed protein structure. In Table 15.3 several other web site having other protein structure database are listed. FFSP refers to fold classification based on structure structure alignment of protein. It is based on a structural alignment of all pairwise combinations of proteins in PDB database by structural alignment program DALI.

A protein family comprises proteins with the same function in different organisms (orthologous sequences) but may also include proteins in the same organism (paralogous sequences) derived from gene duplication and rearrangements. Multiple sequence alignment of a protein family reveals a common level of similarity throughout the lengths of the proteins. The aligned region is referred to as a homeomorphic domain, and this region may comprise several smaller homology domains that are shared with other families may be further subdivided into subfamilies or grouped into superfamilies based on respective higher or lower levels of sequence similarity.

Table 15.3: List of protein structure databases

Sl.No.	*Site*	*Description*	*Host*
1.	3DALI	Protein Structural Sequence Information Library	EMBL Heidelberg (Germany)
2.	BioMagResBank	Database of NMR-derived Protein Structures	BIMAS-NIH (US)
3.	CATH	Protein Structure Classification	U College London (UK)
4.	DSMP	Database of Structural Motifs in Proteins	CDFD (India)
5.	ENTREZ Structure	Biomolecule 3D Structure Search	NCBI (US)
6.	Enzyme Structure Database		UCL (UK)
7.	FSSP	Families of Structurally Similar Proteins Database	EBI (UK)
8.	LOOP	Protein Loop Classification Database	BMM (UK)
9.	MMDB	Entrez Molecular Modeling Database	NCBI (US)
10.	MSD	Macromolecular Structures Database	BioMedNet (US)
11.	PDB	Protein Database of 3-D Structures of Biological Macromolecules	
12.	SCOP	Structural Classification of Proteins Database	

When the sequence of proteins with the same function are examined in greater details, some are found to share high sequence similarity. They are obviously members of the same family by the above criteria. However, others are found to have very little, or even insignificant, sequence similarity with other family members. In such cases, the family relationship between two distant family members A and C can often be demonstrated by finding an additional family member B that shares significant similarity with both A and C. Thus, B provides a connecting link between A and C. Another approach is to examine distant alignments for highly conserved matches.

Family in structural context as used in the FSSP database and the DALI/FSSP web site refers to two structures that have a significant level of structural similarity but not necessarily significant sequence similarity. At a level of identity of > 50%, proteins are likely to have the same three-dimensional structure, and the identical atoms in the sequence alignment will also superimpose within approximately 1 Å in the structural model. Thus, if the structure of one member of a family is known, a reliable prediction may be made for a second member of the family, and the higher the identity level, the more reliable the prediction. Protein structural modeling can be performed by examining how well the amino acid substitutions fit into the core of the three-dimensional structure.

Superfamily is a group of protein families of the same or different lengths that are related by distant yet detectable sequence similarity. Members of a given superfamily thus have a common evolutionary origin. Originally, Dayhoff defined the cut off for superfamily status as being the chance that the sequences are not related of $<10^{-6}$, on the basis of an alignment of the sequences but with a convincingly common number of tructural and functional features are placed in the same superfamily. At the level of three-dimensional structure, superfamily proteins will share common structural features such as a common fold, but there may also be differences in the number and arrangement of secondary structures. The PIR resource uses the term homeomorphic superfamilies to refer to superfamilies that are composed of sequences that can be aligned from end to end, representing a sharing of single sequence homology domain, a region of similarity that extends throughout the alignment. This domain may also comprise smaller homology domains that are shared with other protein families and superfamilies. Although a given protein sequence may contain domains found in several superfamilies, thus indicating a complex evolutionary history, sequences will be assigned to only one homeomorphic superfamily based on the presence of similarity throughout a multiple sequence alignment. The superfamily alignment may also include regions that do not align either within or at the ends of the alignment. In contrast, sequences in the same family align well throughout the alignment. The SCOP web site reports 820 superfamilies and the CATH web site (version 1.7 beta) reports 900 superfamilies.

15.4.1 SCOP database

The SCOP (The structural classification of proteins) database aims to provide a detailed and comprehensive description of the structural and evolutionary relationships between all proteins whose structure is known, including all entries in the Protein Data Bank. It is available as a set of tightly linked hypertext documents which make the large database comprehensible and accessible. In addition, the hypertext pages offer panoply of representations of proteins, including links to PDB entries, sequences, references, images and interactive display systems. In Fig. 15.8 a protion of entry for enzyme 1a6m in scop database and the structure a seen by rasmol and chime is shown in Fig. 15.9. Software rasmol is very popular for observing the protein structures. Rasmol is available free and can be easily downloaded from its website.

Structure shown in Fig. 15.9 is by ribbon display. There are several other display options like CPK, cytochrome etc. Existing automatic sequence and structure comparison tools does not make it possible identify all structural and evolutionary relationships between proteins. The SCOP classification of proteins has been constructed manually by visual inspection and comparison of structures, but with the assistance of tools to make the task manageable and help provide generality. The job is made more challenging and theoretically daunting by the fact that the entities being organized are not homogeneous: sometimes it makes more sense to organize by individual domains, and other times by whole multi-domain proteins.

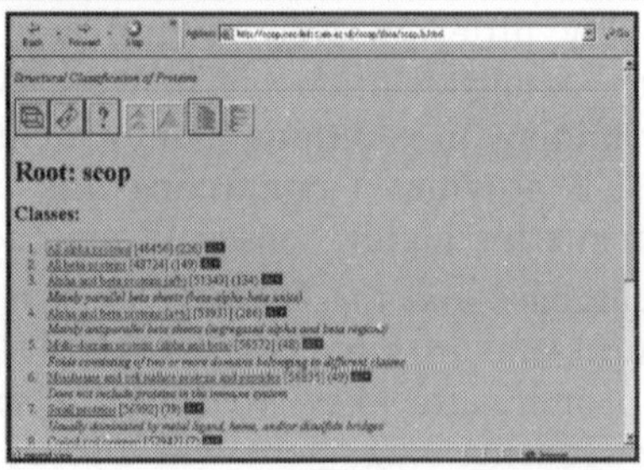

Lineage:
1. ROOT: scop
2. Class: All alpha proteins [46456]
3. Fold: Globin-like [46457] core:
4. helices; folded leaf, partly opened
5. Superfamily: Globin-like [46458]
6. Family: Globins [46463] Heme-binding protein
7. Protein: Myoglobin [46469] Species: Sperm whale (Physeter catodon) [46470]

Fig. 15.8: A portion of entry for enzyme 1a6m in scop database

Proteins are classified to reflect both structural and evolutionary relatedness. Many levels exist in the hierarchy, but the principal levels are family, superfamily and fold, described below. The exact position of boundaries between these levels are to some degree subjective. Our evolutionary classification is generally conservative: where any doubt about relatedness exists, new divisions at the family and superfamily levels is made. Thus, some researchers may prefer to focus on the higher levels of the classification free, where proteins with structural similarity are clustered.

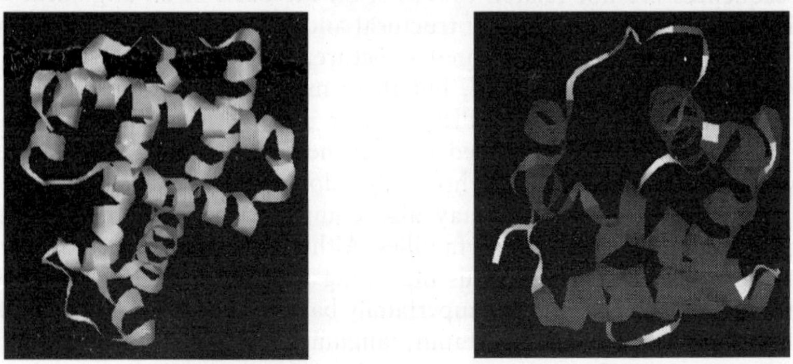

(a) (b)

Fig. 15.9: Structure of enzyme as seen in Rasmol (a) and Chime (b)

The different major levels in the hierarchy are:

15.4.1.1 Family: Clear evolutionarily relationship: Proteins clustered together into families are clearly evolutionarily related. Generally, this means that pairwise residue identities between the proteins are 30% and greater. However, in some cases similar functions and structures provide definitive evidence of common descent in the absense of high sequence identity; for example, many globins form a family though some members have sequence identities of only 15%.

15.4.1.2 Superfamily: Probable common evolutionary origin: Proteins that have low sequence identities, but whose structural and functional features suggest that a common evolutionary origin is probable are placed together in superfamilies. For example, actin, the ATPase domain of the heat shock protein, and hexakinase together form a superfamily.

15.4.1.3 Fold: Major structural similarity: Proteins are defined as having a common fold if they have the same major secondary structures in the same arrangement and with the same topological connections. Different proteins with the same fold often have peripheral elements of secondary structure and turn regions that differ in size and conformation. In some cases, these differing peripheral regions may comprise half the structure. Proteins placed together in the same fold category may not have a common evolutionary origin: the structural similarities could arise just from the physics and chemistry of proteins favoring certain packing arrangements and chain topologies.

SCOP classifies protein into 11 classes:

1. All alpha proteins includes proteins with only α-helix. The protein in Figs. 15.8 and 15.9 is an example of all α-protein
2. All beta proteins
3. Alpha and beta proteins (α/β)
   Mainly parallel beta-sheets (β-α-β units)
4. Alpha and beta proteins (α+ β)
   Mainly antiparallel beta-sheets (segregated α and β regions)
5. Multi-domain proteins (α and β)
   Folds consisting of two or more domains belonging to different classes
6. Membrane and cell surface proteins and peptides
   Does not include proteins in the immune system
7. Small proteins-Usually dominated by metal ligand, heme, and/or disulfide bridges
8. Coiled coil proteins
   Not a true class
9. Low resolution protein structures
   Not a true class
10. Peptides
    Peptides and fragments. Not a true class
11. Designed proteins
    Experimental structures of proteins with essentially non-natural sequences. Not a true class.

The protein structure classification databases have a broad utility that will attract a wide range of users. Experimental structural biologists who wish to explore the region of "structure space" near their proteins of current research, while theoreticians would find it most useful to browse the wide range of protein folds currently known. Molecular biologists would find the classification helpful because the categorization assist in locating proteins of interest and the links make exploration easy.

15.4.2 The CATH database

The CATH database is a hierarchical domain classification of protein structures in the Protein Data Bank (PDB, 2003). Only crystal structures solved to resolution better than 4.0 angstroms are considered, together with NMR structure. All non-proteins, models, and structures with greater than 30% "C-alpha only" are excluded from CATH. Protein structures are classified using a combination of automated and manual procedures. There are four major levels in this hierarchy: Class, Architecture, Topology (fold family) and Homologous superfamily. The four levels are discussed below:

15.4.2.1 Class, C-level: Class is determined according to the secondary structure composition and packing within the structure. Three major classes are recognised; mainly-alpha, mainly-beta and alpha-beta. This last class (alpha-beta) includes both alternating alpha/beta structures and alpha + beta structures, as originally defined by Levitt and Chothia (1976). A fourth class is also identified which contains protein domains which have low secondary structure content.

15.4.2.2 Architecture, A-level: This describes the overall shape of the domain structure as determined by the orientations of the secondary structures but ignores the connectivity between the secondary structures arrangement e.g. barrel or 3 layer sandwich.

15.4.2.3 Topology (Fold family), T-level: Structures are grouped into fold groups at this level depending on both the overall shape and connectivity of the secondary structures. This is done using the structure comparison algorithm SSAP and CATHEDRAL. Parameters for clustering domains into the same fold family have been determined by empirical trials throughout the databank. Structures which have a SSAP score of 70 and where at least 60% of the larger protein matches the smaller protein are assigned to the same T level or fold group. Some fold groups are very highly populated particularly within the mainly beta 2-layer sandwich architectures and the alpha-beta 3-layer sandwich architectures.

15.4.2.4 Homologous superfamily, H-level: This level groups together protein domains which are thought to share a common ancestor and can, therefore, be described as homologous. Similarities are identified either by high sequence identity or structure comparison using SSAP.

15.5 PROTEIN STRUCTURE PREDICTION

The function of macro-molecules such as DNa, RNA and proteins rely on their structure. The three-dimensional structure of these macromolecules depends or their linear sequences of their bulding blocks. Currently 3-D structures of macromolecules are determined primarily through the use of biophysical techniques such as X-ray crystallography and nuclear magnetic resonance (NMR). These methods are technically difficult and time consuming. Protein sequencing by comparison require much less effort. Therefore, since 3-D structures building blocks, it would be for easier if we could predict the structure and function proteins from building blocks.

The amino acid sequence of a protein dictates its three-dimensional structure. When placed in a medium of suitable solvent and temperature conditions, such as provided by a cell interior, protein fold spontaneously to their nature active sturcture. Some proteins require chaperones to fold, but these catalses the process rather than direct it. In the 1960's, C.B. Anfinsen and his coworkers at the NIH performed a series of seminal experiments in vitro that answered a key part of the problem. The original work led Anfinsen to propose the thermodynamic hypothesis, which states that the native conformation of a protein is adopted spontaneously. In other words, there is sufficient information contained in the protein sequence to guarantee correct folding from any of a large number of unfolded states.

If amino acids sequences contain sufficient information to specify 3-D structures of proteins, it should be possible to device an algorithm to protein structure from amino acid sequence. This method is specifically very useful in case of proteins whose structure is not yet described but amino acid sequence is known. Protein structure prediction methods here studied under following heads:

1. Secondary structure prediction
2. Homology modelling
3. Fold recognition
4. *Ab initio* prediction

15.5.1 Secondary Structure Predictions

Several computer based approaches attempt to predict secondary structure from the primary amino acid sequences. These programs base their predictions on different types of parameters. Some programs rely on the physical and energetic properties of amino acids and polypeptide backbone. More commonly, however, secondary structure predictions are based on statistical frequency of amino acids within secondary structures that have been crystallized for example Peter Chou and Gerald Fasman at Brandeis University have compiled X-ray crystallographic data to calculate the likelihood that an amino acid will be found in an α-helix and Aβ-sheet. Certain amino acids, such as glutamate and alanine are likely to be found in α-helix; others such as valine and isoleucine are more likely to be found in a β-sheet structure. Such information can be used to predict whether a sequence of amino acids is likely to be folded in an α-helix or β-sheet conformation.

Secondary structure preduction is correct for approximately 60 to 70% of all sequences. While this degree of accuracy is promising it is generally not sufficient to predict protein secondary structure reliably. Therefore, one must be cautious in interpreting the result of a secondary structure prediction program.

In recent years, the computational methodology known as neural nets has been applied to protein structure prediction. Neural net is already discussed in connection with models for gene finding in chapter 13. For secondary structure prediction amino acid sequences of protein with known crystal structures can be used to train a network (i.e., adjust to parameters) to predict secondary structure for new amino acid sequences. Thus neural network have yielded a small improvement in secondary structure predictions 907 predict protein.

15.5.2 Comparative Modelling

It is also called as homology modelling. Comparative modelling exploits the fact that evolutionarily related proteins with similar sequences, as measured by the percentage of identical residues at each position based on an optimal structural superposition, have similar structures. The similarity of structures is very high in the so-called "core regions", which typically are comprised of a framework of secondary structure elements such as alpha-helices and beta-sheets. Loop regions connect these secondary structures and generally vary even in pairs of homologous structures with a high degree of sequence similarity.

The process of building a comparative model is conceptually straight forward. First, an alignment is performed between the sequence for which the structure has been determined by experimental methods (the parent) with the sequence to be modelled (the target). The Completeness and quality of result depend cruelly on low; similar the sequences are. As a thumb rule of sequences of two related proteins have 50% or more identical residues in an optimal alignment, the proteins are likely to have similar conformations over more than 90% of the structures.

A common example correct protein structure prediction based on comparative modeling is preduction of structure of HIV protease, which was homologous to other proteases with structures that have been solved by X-ray crystallography both the models out to be fairly accurate.

15.5.3 Fold Recognition or "Threading"

Threading uses a database of known three-dimensional structures to match sequences without known structure with protein folds. This is accomplished by the aid of a scoring function that assesses the fit of a sequence to a given fold. These functions are usually derived from a database of known structures and generally include a pairwise atom contact and solvation terms. Threading methods compare a target sequence against a library of structural templates, producing a list of scores. The scores are then ranked and the fold with the best score is assumed to be the one adopted by the sequence. The methods to fit a sequence against a library of folds can be extremely elaborate computationally. In many situations, even a crude or approximate model can help an experimentalist significantly in guiding his/her experiments. Thus, even though the current methods fold recognition are still in their infancy, prediction of structures for all protein sequences of complete genomes in conjunction with experimental work is a realistic goal. These methods method can use structural data and methods for structure prediction to probe protein and organismal function and evolution.

15.5.4 *Ab initio* Prediction

The *ab initio* approach is a mixture of science and engineering. The science is in understanding how the three-dimensional structure of proteins is attained. The engineering portion is in deducing the three-dimensional structure of the given sequence. Some methods for *ab initio* prediction include: (i) Molecular Dynamics (MD) simulations of proteins-substrate complexes provide a detailed and dynamic picture of the nature of inter-atomic interactions with regards to protein structure and function; (ii) Monte Carlo (MC) simulations that do not use forces but rather compare energies, via the use of Boltzmann probabilities; (iii) Genetic Algorithms which tries to improve on the sampling and the convergence of MC approaches, and (iv) exhaustive and semi-exhaustive lattice-based studies which are based on using a crude/approximate fold representation (such as two residues per lattice point) and then exploring all or large amounts of conformational space given the crude representation.

EXPERIMENT 15B

Comparing Proteins: Hemoglobins using Deep View Software of ExPASY

About Deep View: Deep view can find hydrogen bonds within proteins and between proteins and ligands. It allow you to examine electron-density maps from crystallographic structure determination, to judge the quality of maps and models, and to identify many common types of problems in protein models. It allows you to view several proteins simultaneously and superimpose them to compare their structures and sequences. It computes electrostatic potentials and molecular surfaces, and carries out energy minimization. For proteins of known sequence but unknown structure, Deep View submits amino acid sequences to ExPASy to find homologous proteins, onto which you can subsequently align your sequence to build a preliminary three-dimensional model. Then Deep View subits your alignment to ExPASy, where the SWISS-MODEL server builds a final model, called a homology model, and returns it to you by e-mail.

About the hemoglobins: The present study is two homologous proteins: alpha and beta chains of hemoglobin are compared.

TESTING THE HYPOTHESIS

1. Obtain the files 3HHB.pdb from the PDB in the same way as done in Experiment 15A. Load the file 3HHB.pdb, deoxyhemoglobin. Select chain A from the control panel as shown in Fig. 15.10. Save selected residues. Name the file 3HHBAlpha.pdb, direct it to a convenient folder, and click SAVE. Next, select chain B only. Save it as before, naming it 3HHBeta.pda. Close 3HHB and open your new files, 3HHBAlpha.pdb first.

2. Go to menu window: Sequence Alignment. The Align window appears at the bottom left with the sequences of the two chains in one letter abbreviations as seen in Fig. 15.11. The name of the currently active layer is shown in red, with the reference layer at the top.
3. Select menu Colour: Layer Fit: Magic Fit. The two chains overlap as shown in Fig. 15.11.
4. Go to menu Fit: Calculate RMS Deviation ... Accept the default settings in the dialog box by clicking OK. Deep calculates the root-mean-square distance between all corresponding CA atoms in the alpha and beta chains, and reports the number of atoms used (138) and the RMS distance (1.57 angstroms). You can make this calculation only with the non-reference layer active.
5. Fit: Best Fit (with Struct. Align.): Again accept the default settings in the dialog and click OK. This time, Deep View first carries out an alignment of the sequences that is guided by the magic Fit superposition. Then it carries out the superposition using only the residues that are structurally aligned. Calculate the RMS deviation, based on a smaller number of better corresponding residues.

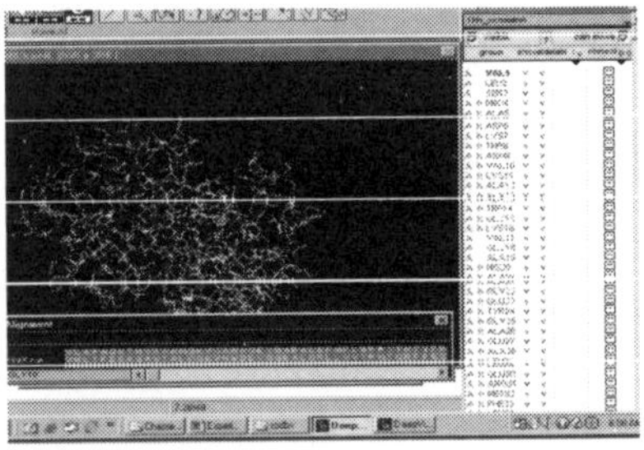

Fig. 15.10: Web view of deep view

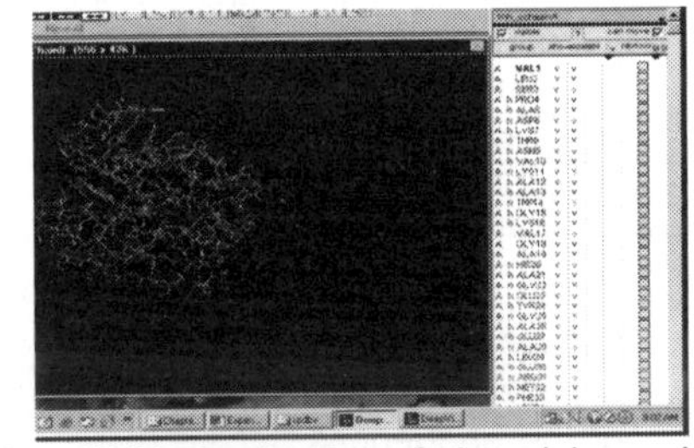

Fig. 15.11: Overlapping chains alpha and beta

6. Scroll across the sequences in the Alignment window. Notice that certain pairs of aligned residues are highlighted (gray). these are the residues that were aligned with each other by the fitting operations you conducted. Place the pointer on (but do not click) a residue of the reference layer. Its identity appears at the left end of the window, and the residue blinks in the graphic window (you may have to look carefully to find a blinking residue). Move the pointer to a residue in the lower sequence. Now you learn its identity and its distance from the corresponding residue in the reference layer.
7. Select: aa Identical to ref. In both layers, Deep View selects and highlights the aligned pairs that are identical. Scroll through the sequence alignment. Approximately what percentage of residues is identical in the beta chain to those in the alpha chain ? The alpha and beta chains probably evolved by divergence from a common ancestral globin gene. The highlighted residues are those that have been conserved during this evolution. Conserved residues are likely to be essential to protein function.
8. Select: aa Similar to ref. Now a larger number of pairs are selected. These pairs are either identical (the residues selected before) or they similar in chemical properties. Find similar pairs in which both are positively charged, negatively charged, small, polar, or nonpolar. mutations that replace one residue with one of similar properties, such as lysine replaced by arginine, are called conservative mutations.
9. Select: Inverse selection in the beta-chain sequence, the highlighted residues are those that are neither identical nor similar to those in the alpha chain. These residues may serve different functions in the two subunits, or their specific side chains may not be essential to the function of the chain. Color all these residues green, and hide side chains. Are most of the differences that distinguish alpha from beta buried or on the surface of this subunit?
10. Display the Layer Infos window and notice the number of residues selected in the 3HHBBeta layer (Sel column). This number, 47, is the number of amino-acid differences between the alpha and beta chains of human hemoglobin. Use the Select menu and this Layer Infos window to determine the numbers of identical and similar residues in the two chains.

11. At the left end of the Align window, click on the small text icon. The text page that appears summarizes the current structural alignment. Below aligned pairs are asterisks (*) if the pairs are identical, dots if the pairs are similar, and no mark if they are neither. You can save this text file with File: Save: Sequence Alignment, and then open and print it with any word processor.
12. Other Alignment Options. Examine the commands under the Fit menu. Notice that you can also align models using only selected residues. This allows you to choose specific regions or chains of models to align, with unselected residues following the selected ones, but not affecting the alignment. For example, you might want to see how the beta chains of hemoglobin move with respect to the alpha chains in an alpha-beta dimer.

    You could accomplish this by selecting the alphas only and carrying out Fit: Fit Selected Residues. Deep View would superimpose on the basis of the alphas, and comparison of the models would highlight differences between the positions and orientations of the betas. The only restriction on fitting selected residues is that you must select exactly the same number of residues in each layer.

INTERPRETATION OF DATA

These two chains are very similar in overall conformation, but differ substantially in amino-acid sequence. With Deep View, you can superimpose the chains and quickly locate amino acids that are identical, similar, and different in the two chains. In addition, you can print out an alignment of the sequences.

The deep view software can also be used for analysing binding sites, using Ramachandran Plots, building protein models to specifications, comparing proteins, viewing and interpreting electron-density maps and computing surfaces and electrostatic potentials, and building homology models by threading the sequences of unknown proteins onto the known structure of homologous proteins.

SUMMARY

A protein is a polymer comprised of amino acid residues. The primary structure of a protein chain is the ordered sequence of its amino acid residues; the secondary structure describes the three- dimensional shapes taken up by group of residues close to one another. The common secondary structures are alpha-helices, beta-strands, beta sheets, turns and random coils. One or more of secondary structural elements very often follows a regular pattern like α-β-α or β-α-β, etc. These strucures are called as supersecondary structures. Tertiary structure describes the ways in which various elements of secondary structures are arranged to form overall three dimensional structure. Quaternary structure is formation of larger ligomeric protein molecules by interaction of multiple protein chains. Revolutionary techniques for protein structure study are X-ray crystallography and NMR. Perutz and Kendrew (Historical Profile 17) gave the X-ray sructure of Hemoglobin. In Experiment 15A primary, secondary, tertiary and quaterary structure of enzyme Rubpcase is studied. SCOP and CATH are important protein structure databases.

EXERCISE

1. Which of the following pairs are orthologous, which are paralaogs and which are neither:
   (*a*) Human hemoglobin α and human hemoglobin β.
   (*b*) Human hemoglobin β and horse hemoglobin β.
   (*c*) Human hemoglobin α and horse hemoglobin β.

2. Many protein from pathogens have homologous sequences in human beings. Suppose you had a method for comparing the determinants of specificity in the binding sites of two homologues proteins, how could you use this method to select the target for drug design?
3. The bacterium *Pseudomonas fluoreoscens* and the fungus *Curvularia enequalis* each possess a chloroperoxidase, an enzyme that catalyses halogenation reactions. Do these enzymes have same folding patterns?
4. Compare and contrast SCOP and CATH classification.
5. Supersecondary structures and structural domains in a protein are classified neither in secondary or tertiary level of classification of structure of proteins. Comment.

16

NUCLEIC ACID STRUCTURE PREDICTION

Nucleic acids are polymers of nucleotides and may be described as either polydeoxyribonucleotides (DNA) or polyribonucleotides (RNA). Nucleotides are composed of nitrogenous base, pentose sugar and phosphate group. Different types of nitrogenous bases, their structure and role in different nucleotide composition is given in chapter 4. In both DNA and RNA the nucleotides are linked by phosphodiester bonds joining the 5′ and 3′ carbons of adjacent nucleotide sugars. In case of DNA the sugar is deoxyribose and in RNA it is ribose sugar. The structure of a nucleotide is shown in Fig. 16.1 and a sequence of nucleotides in 5′–3′ direction is called as primary structure of nucleic acid is shown in Fig. 16.2. The sugar-phosphate linkages in nucleic acids have a directional sense. The convention when writing them is to place the C-5′ to the left and the C-3′ to the right as indicated in Fig. 16.2.

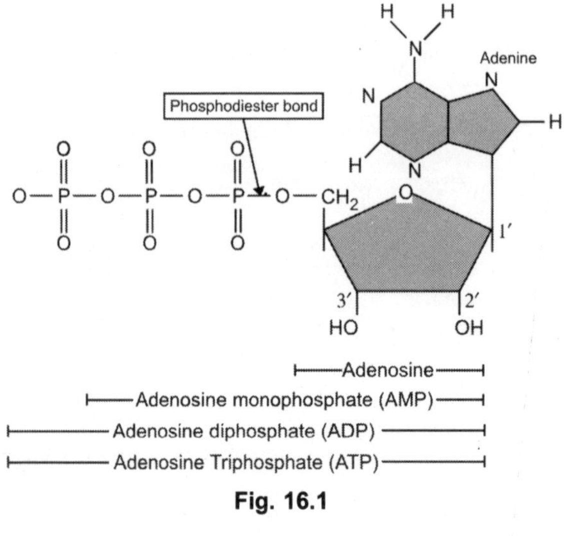

Fig. 16.1

16.1 STRUCTURE OF DNA

Like protein structure (chapter 3), DNA structure is also classified into primary, secondary and tertiary structure. The primary structure as stated above refers to base sequence in polynucleotides

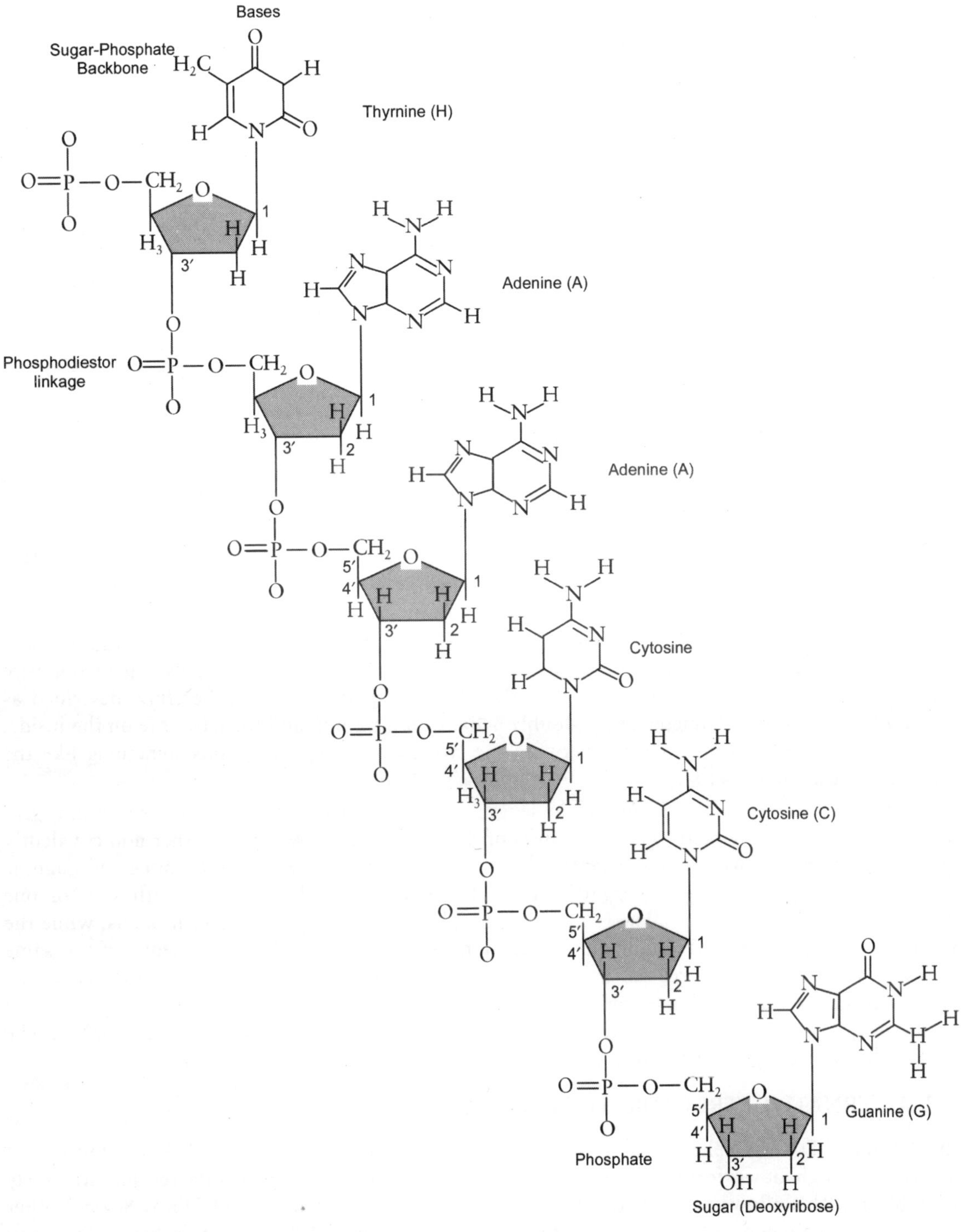

Fig. 16.2: Structure of one DNA strand. Note sugar here is deoxyribose

(Fig. 16.2). The secondary structure refers to the spatial arrangement of nucleolides that are near one another in the linear sequence. The tertiary structure refers to coiling and supercoiling of the DNA.

The accepted structure of typical chromosomal DNA was first suggested in 1953 by Watson and Crick (see Historical Profile 18). Watson and Crick built an accurate molecular model representing the structure of DNA using experimental data of other workers, mainly Ervin Charagoff (1950) and Wilkins (1940). Charagoff showed that ratio of purines (A + G) and pyrimidiyes (T + C) is equal to one. X-ray diffraction patterns obtained from fibres of DNA by Wilkins and coworkers at Kings College, London, showed DNA to have at least two possible secondary structures, dependent upon the relative humidity at which the data were collected. These were called B-DNA and A-DNA. The other forms of DNA have different secondary structures are Z-DNA, triplex DNA, and quadruplex DNA.

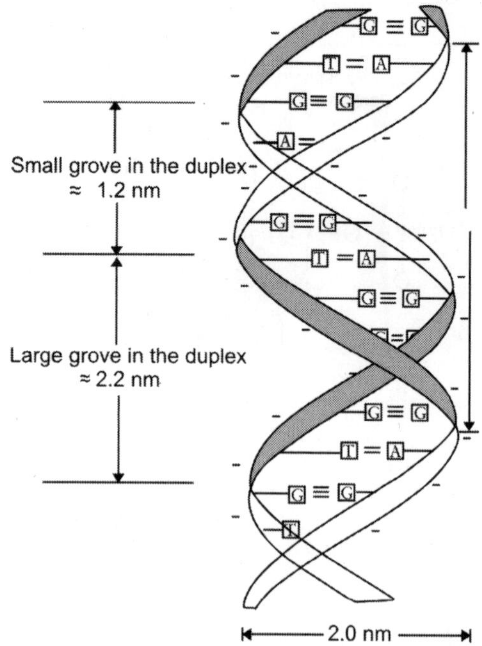

Fig. 16.3: Double helical structure of DNA

Fig. 16.3 shows the structure of the B-form of DNA proposed by Watson and Crick. The structure consists of two polydeoxyribonucleotide chains, with the sugar-phosphate groups on the outside. Both chains form right-handed helices, each chain twisting around an imaginary central axis. Right-handed helices are distinguished by the strands rising towards the right as they twist around an imaginary central axis. These chains run in opposite directions, one in the $5' \rightarrow 3'$ direction, the other $3' \rightarrow 5'$. The chains are, therefore, described as antiparallel. The overall diameter of the double helix is about 2 nm, and the bases are on the inside, in a stacked configuration, extending between the two sugar-phosphate chains something like the steps of a spiral staircase.

Each base pair occupies 0.34 nm of the length of the molecule and 10 base pairs (3.4 nm) give one complete turn of the helix. The two halves of the double helix are held together non-covalently by hydrogen bonds between complementary bases. Adenine always pairs with thymine and guanine with cytosine, thus generating Chargaff's ratios. The space in the helix is only sufficient for one purine pairing with pyrimidine. The G : C pair is held together by three hydrogen bonds, while the A : T pair has only two, although the overall dimensions of both pairs are essentially the same allowing the formation of a regular double helix. The geometry of the B-DNA is characterized by having two grooves alternating along its length: the major groove is wider, while the other is called the minor groove. Hydrophobic interactions play a very important role in structure of DNA. The hydrophillic side chain is exposed and hydrophobric nitrogenous bases are sheilded.

16.2 ALTERNATIVE STRUCTURES OF DNA

The B form of DNA is believed to be the major secondary structure type of DNA in solution and *in vivo,* although other forms also occur. Six bonds between each nucleotide residue are freely rotatable and this allows for considerable variation in the conformations of DNA. Several other forms of DNA known are: A-DNA, Z-DNA, D-DNA, E-DNA, P-DNA tripple helix DNA and quadruplex DNA. Structure of DNA vary according to hydration and salt conditions in the cell. A-DNA is less hydrated than B-DNA. This allows the polydeoxy nucleotides to tilt extensively and

HISTORICAL PROFILE 18

Sir James Watson and Sir John Francis Crick for Giving DNA Structure

James Watson

Harry Compton Crick

James Watson was only 25 years old when he and his older colleague, Francis Crick in 1953, discovered the structure of deoxyribonucleic acid (DNA) the building block of all life on Earth. Modern biology, and the biotechnology industry it has spawned, would be unthinkable if these two had not determined the structure of the DNA molecule. Their model of this structure—the double helix-has become a universal symbol of the scientific profession, and the title of Watson's 1968 best-seller.

Watson and Crick won the Nobel Prize for Physiology and medicine in 1962, but this was not the end of Watson's career in the public eye. As Director of the Cold Spring Harbor Laboratory he has continued to lead the way in genetic research. From 1988 to 1992 James Watson served as the first Director of the Human Genome Project at the National Institutes of Health. He is now President of the Cold Spring Harbor Laboratory.

Francis Harry Compton Crick was born on June 8th, 1916, at Northampton, England, being the elder child of Harry Crick and Annie Elizabeth Wilkins. He has one brother, A.F. Crick, who is a doctor in New Zealand.

Crick was basically a physicist and shifted to study biology in 1947. Crick at that time knew no biology and practically no organic chemistry or crystallography, so that much of the next few years was spent in learning the elements of these subjects. During this period, together with W. Cochran and V. Vand he worked out the general theory of X-ray diffraction by a helix. In 1953 along with J.D. Watson he gave structure of the double-helical structure of DNA and the replication scheme. Crick and Watson subsequently suggested a general theory for the structure of small viruses. Crick in collaboration with A. Rich has proposed structures for polyglycine II and collagen and a structure for polyadenylic acid. In recent years Crick, in collaboration with S. Brenner, has concentrated more on biochemistry and genetics leading to ideas about protein synthesis (the <<adaptor hypothesis>>), and the genetic code, and in particular to work on acridine-type mutants. Francis Crick died on July 28, 2004.

form shorter and fatter duplex, with about 11 base pair per complete helical turn. The bases are tilted to the extent of about 20° with respect to the extended helical axis, rather than the 6° or so in B-DNA. This occurs under the conditions of decreased hydration, may be due to interaction with hydrophobic molecule or changing cellular conditions. Major and minor grooves are modified. DNA structure also varies depending upon base composition. In 1980, X-ray diffraction analysis of the self-complementary dodecamer dCdGdCdGdAdAdTdTdCdGdCdG showed it to have a typical B-DNA structure. In contrast, the hexamer dCdGdCdGdCdG and the tetramer dCdGdCdG have been shown to have completely different secondary structures.

Deoxyribonucleic acids containing alternating G-C pairs form left handed double helical structures called Z-DNA. This structures considerably slimmer than B-DNA, with 12 base pairs per turn and is 18 Å in diameter. The sugar-phosphate backbones form a zig-zag pattern along the length of the molecule, rather than showing the smooth rotation of B-DNA that's why given the name Z-DNA. Major and minor grooves are also eliminated. Observations suggest that Z-DNA exists in specific regions of naturally occurring DNA. These regions apparently consist of about 12 to 24 sets of 12 alternating purine and pyrimidine residues in sequence: a feature which favours Z-conformation. Z-DNA was discovered by Andrew Wang and Alexander Rich (1979). It is left handed DNA, formed under high salt conditions. Negative supercoiling also favours formation Z-DNA. The biological functions of Z-DNA are not known with certainty, although it is thought that it might have a role in regulating expression of the genetic information. Z-DNA is thought to be important in regulation of gene activity for example transcription enhancer in simian virus 40. It is very common in plasmids, tumor cells and virus. Some of the structural features of the A-, B- and Z-forms of DNA are summarized in Table 16.1. In Experiment 16A different form of DNA are studied using nucleic acid database.

Table 16.1: Some Helical Parameters Associated with DNA

Parameter	*A-DNA*	*B-DNA*	*Z-DNA*
Form	Fibre	Fibre	Crystal
Handedness	right-handed	right-handed	left-handed
Base pairs/turn	11	10	12
Helix length/base pair (nm)	0.29	0.34	
Rotation/base pair (range)	33° (16-44°)	36° (27-42°)	

Alexander Rich and Gary Felsenfeld gave the structure of triplex DNA. It is formed in synthetic DNA. If double helical DNA and complementary single stranded DNA mixed, this forms triplex DNA, by binding in the major groove region. The pairing rule is that a thymine in synthetic DNA will bind near an A-T pair of biological DNA and a cytosine will bind to G-C pair. Thus technique is implied in antisense therapy. There is very little replication or transcription of this DNA. Structure of triplex DNA is shown in Fig. 16.4.

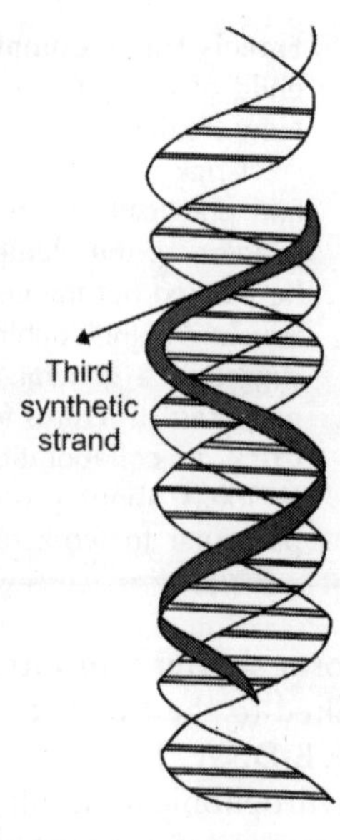

Fig. 16.4: Triplex DNA

G-Quadruplex or tetraplex DNA is abundant in telomer region of chromosome and is widely spread in eukaryotic genome. It makes DNA inactive gel like structure. Guanine tetrads i.e. tetrameric arrangement of guanine bases in GMP gel, is called G-teramers. Chains of there G tetramers form quadraplex DNA.

Cruciforms are another 'unconventional' group of structures, adopted by DNA, which are dependent upon specific base sequences. Cruciform structures are formed at inverted repeats (Fig. 16.5). An inverted repeat occurs where a sequence of bases on one strands is followed, on the same strand, by the complementary sequence in reverse order. If the strands of DNA separate (or 'denature'), then each strand will form a hairpin loop because of intrastrand base pairing. Together, both loops give the cruciform structure. Since cruciform formation in only possible when unwinding of the DNA occurs, any stress on DNA that causes unwinding will favour the formation of these unusual structures. It has, indeed, been suggested that cruciforms could serve as recognition sites for the binding of proteins to specific regions of DNA.

Supercoiling is a tertiary structural form of DNA that is common in the circular DNA molecules of bacteria. Supercoiling, superhelical and supertwisting are all synonyms for the same phenomenon.

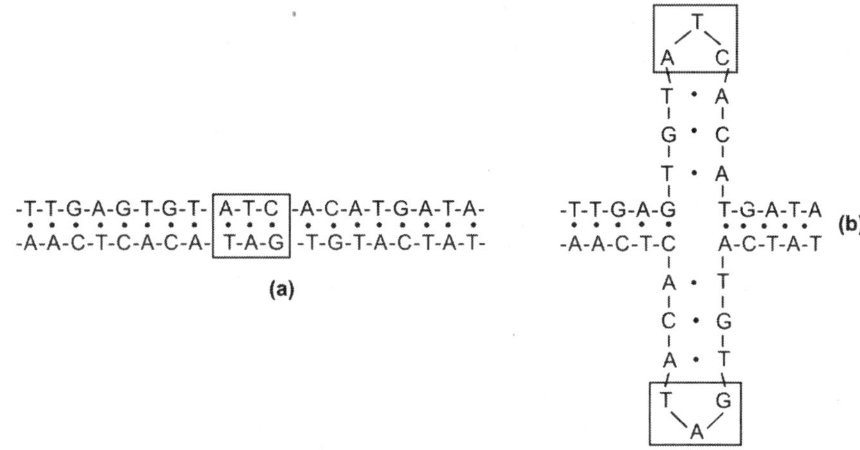

Fig. 16.5: (a) An inverted repeat sequence of bases in DNA that can give rise to (b) a cruciform structure

Circular DNA. lacking supercoiled regions is called relaxed DNA, and will lie in a plane. Supercoiling is perhaps best envisaged by imagining the closed circle of DNA to be broken, and the two strands of the duplex slightly unwound. The ends of the sugar-phosphate chains are then resealed. The molecule will now try to return to its original twist, but this can only be achieved if it becomes distorted such that is no longer lies in its original plane (Fig. 16.6).

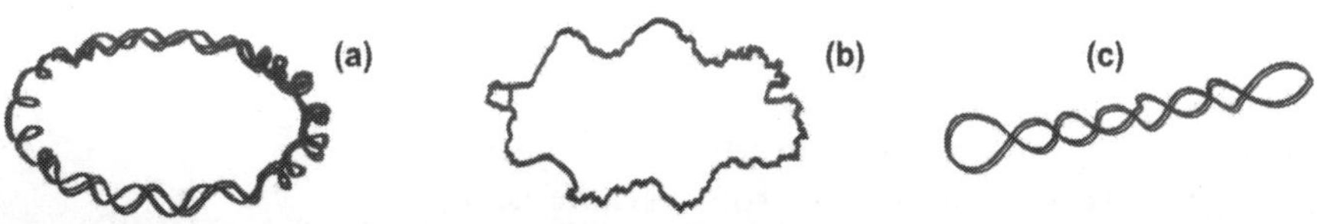

Fig. 16.6: Transition from (a) relaxed, to (c) negatively supercoiled DNA, via structure (c) where the main backbone has been hydralysed, slightly unwound and the phosphodiester bonds rejoined.

16.3 RNA (RIBONUCLEIC ACID) STRUCTURE

RNA broadly is classified into (1) genetic, (2) non-genetic (refer to chapter 4). Genetic RNA is used for transfer of information from one generation to other. Non-genetic RNA are not involved in information transfer. Three main types of genetic RNA are mRNA, rRNA and tRNA. Basic structure of a RNA strand is same to that of a DNA. Difference lies in sugar type, in RNA it is ribose sugar instead of deoxyribose. Nitrogenous base here is uracil instead of Thymine. Main characters of the three types of RNAs in a prokaryote are summarised in Table 16.2. Structure of a typical RNA is shown in Fig. 16.7. RNA is not necessarily single stranded. Sometimes it can form loops and also be double stranded.

Table. 16.2: Three main types of RNA (Data from a typical prokaryote)

Type	*Main character*	*Relative amount*	*Sedimentation coefficients*	*Mass (kD)*	*No. of NTP*
rRNA	is most abundant	80	23	1.2×10^3	3700
			16	0.5×10^3	1700
			5	3.6×10^1	120
tRNA	is smallest	15	4	2.5×10^3	73
mRNA	is most hetrogenous	5	Heterogenous	Heterogenous	Average length 1.2 kb.

Several other types of RNA present in eukaryotes are given below:

1. sn-RNA refers to Small nuclear RNA. This participate in splicing of RNA (introns region). They are usually 100-200 NTP, are often complexed with protein to form snRNP on Snrups. They are found only in nucleus.
2. Small RNA is present in cytosol. This plays a role in targeting of newly synthesized proteins.
3. Pre mRNA is the initial transcript of RNA is usually bigger than mRNA.
4. Heteronuclear (Hn) RNA is a group of RNA molecule found only in nucleus.
5. Sense RNA/Anti-Sense RNA. Duplex DNA is double helical. Only one strand serves as template for synthesizing mRNA or is transcribed. This strand of DNA is called sense strand and other strand is called anti-sense strand. If by chance the antisense DNA is also transcribed this leads to formation of RNA called as antisense RNA. Antisense is also called MIC RNA or messanger inhibitory complementary RNA. RNA synthesized from sense DNA is called sense RNA. In nature this sense and antisense RNA forms double helical structure. This double stranded RNA or duplex RNA has 2 implications: (a) There is no binding of ribosomes to this double helical RNA, so no translation and thus no protein synthesis. (b) This also triggers the degradations of sense RNA because double stranded RNA is attacked by intracellular ribonucleases. This antisense approach is very specific and has a therapeutic value. A protein necessary for virus reproduction or cancer cell growth but not needed for normal cells can be stopped from being synthesized. For example Vitranene™ is an antisense drug from ISIS a pharmaceutical and CIBA vision corporation is used to treat cytomegalovirus induced retinitis (CMV) in AIDS patients. About 40 per cent of AIDS patients develop this retinitis (blindness). Introduction of this drug into eyes, trigger the destruction of CMV mRNA and interfere with the adsorption of virus to the surface of host cell.
6. Autocatalytic RNA's were discovered by Thomas Cech and his colleagues in 1982. Autocatalytic RNAs are those RNAs which can splice themselves. They are also called as Ribozymes. They are mainly non-genetic.

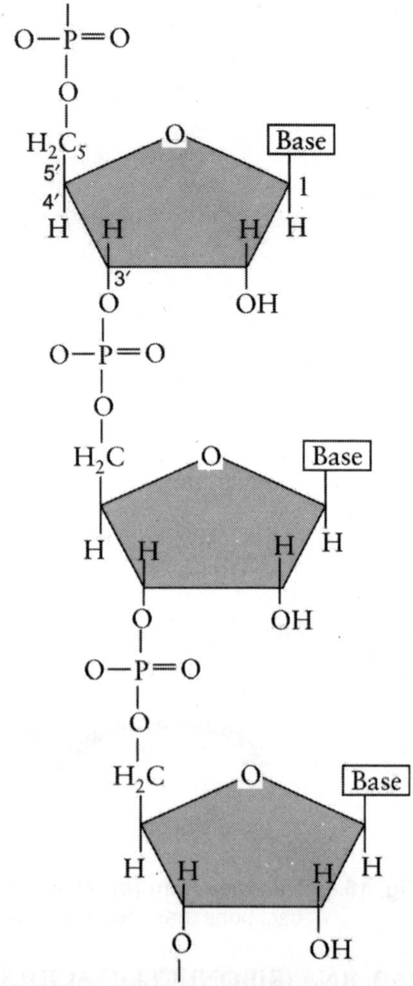

Fig. 16.7: Structure of part of an RNA chain

16.4 PREDICTION OF DNA CONFORMATION

As discussed above the predominant form of DNA double helix is B-DNA. Since DNA is a dynamic structure, short segements of DNA may change from one conformation to another depending upon external condition. Though many determinants influence DNA conformation, one important factor is sequence of nucleotides. B-DNA is favoured form of the double helix and repeats of AA or TT tend to hold DNA in a B-DNA conformation. However, repeat of GC sequence under certain

EXPERIMENT 16A

To Study Different Forms of DNA Structures using Nucleic Acid Database (NDB)

About NDB: The goal of the Nucleic Acid Database Project (NDB) is to assemble and distribute structural information about nucleic acids. The NDB was founded in 1992 by Helen M. Berman, Rutgers University, Wilma K. Olson, Rutgers University, and David Beveridge, Wesleyan University. The NDB Project is funded by the National Science Foundation and the Department of Energy.

Hypothesis: It would be possible to extract all possible information including DNA sequence coordinate and structure data and examples of A, B, Z and quadruplex DNA.

TESTING THE HYPOTHESIS

Computer set up: A normal PC with internet connection in windows environment:

1. Log on to NDB web site. http://ndb server. rutgersedu. This takes into the web site shown in Fig. 16.8.
2. Enter NDB 001 in search text box. This opens the Data Box 16.1. This is an example of B-DNA study the NDB entry thoroughly.
3. Find other examples of A, Z and quadruplex DNA.

Fig. 16.8: Web view of NDB

DATA

Data Box 16.1: NDB entry

NDB ID: BD0001

Title:	5'-D(*AP*CP*CP*GP*AP*CP*GP*TP*TP*CP*GP*GP*T)-3'
Structural	B DOUBLE HELIX
Nucleic Acid sequence:	Chains A, B: A C C G A C G T C G G T
Primary Citation:	Rozenberg, H., Rabinovich, D, Frolow, F., Hegde, R.S., Shakked, Z. Structural code for DNA recognition revealed in crystal Structures of papillomavirus E2-DNA targets. Proc.Natl.Acad.Sci.USA, 95, pp. 15194 - 15199, 1998.
Experimental Information:	X-RAY DIFFRACTION
Space Group:	H 3
Cell Constants:	a = 64.067 b = 64.067 c = 44.679 (Ångstroms) a = 90.00 b = 90.00 y = 120.00 (degrees)
Crystallization	Method: VAPOR DIFFUSION, HANGING DROP
Conditions:	Drop: mgCl$_2$, SPERMINE, Na Cacodylate, MPD Reservoir: Na Cacodylate, MPD
Refinement:	The structure was refined using the SHELXL-97 program. The R value is 17.3 for 8888 reflections in the resolution range 23.60 tp 1.60 Ångstroms with I > 0.00 sigma (I).

Coordinate + Structure Data **Derivative Data**

Asymmetric Unit coordinates (pdb format, Unix compressed (.Z) Nucleic Acid Backbone Torsions
Asymmetric Unit coordinates (cif format, Unix compressed (.Z) Base Pair Parameters
Biological Unit coordinates (pdb format) Base Pair Step Parameters
Structure Factors (cif format) Hydrogen Bonding .Classification
XML | Complete with coordinates (xml format, GNU compressed (.gz)
XML | Coordinates only (xml format) GNU compressed (.gz)

XML | Header only (xml format) GNU compressed (.gz)

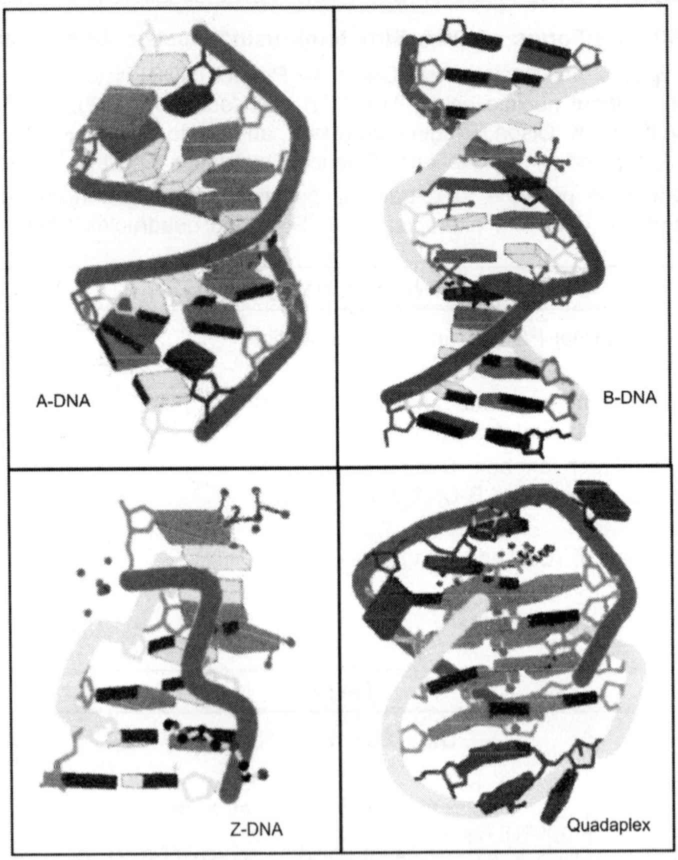

Fig. 16.9: Different forms of DNA

INTERPRETATION OF DATA

Entry of NDB database for NDB ID 0001 is shown in Data Box 16.1. This is a B-double helix. The structure is derived by crystallization method of vapor diffusion, hanging drop method and refined by using SELXL-Py program. Atomic coordinates are available in both .pdb and cif format.The coordinates obtained can be saved in different file and used for further structural study. In Fig. 16.9 different types of DNA structure as extracted from NDB site is shown. Different aspects of DNA structures could easily be studied using NDB web site.

condition couse B-DNA to change its conformation to Z-DNA. Alternatively GC or CC tend to change B-DNA to A-DNA. Many studies suggest that B to Z and B to A transitions are important biologically since they affect the ability of protein to bind within major or miner groove.

Based on primary sequence, computer programs can predict how likely such transitions are to occur. However, these predictions are not completely reliable. For this reason structural predictions must be combind with their experimental evidence. For example, a computer program may predict that segment of DNA is likely to be found in a Z-DNA conformation. To confirm this prediction, antibodies that specifically recognise Z-DNA and not B or A-DNA, can be used. If the DNA is really in Z-DNA conformation, the antibodies will bind to this molecule.

16.5 PREDICTION OF RNA SECONDARY STRUCTURE

RNA is transcribed or synthesized in cells as single strands of (ribose) nucleic acids. However, these sequences are not simply long strands of nucleotides. Rather, intra-strand base pairing will produce structures such as the one shown in Fig. 16.10. There are several motifs commonly found in RNA secondary structures. Those are: helix, hairpin loop, interior loop, buldge loop, multi loop and pseudo knot. The stability of a particular secondary structure is a function of several constraints:

1. The number of GC versus AU and GU base pairs. (Higher energy bonds form more stable structures.)
2. The number of base pairs in a stem region. (Longer stems result in more bonds.)
3. The number of base pairs in a hairpin loop region. (Formation of loops with more than 10 or less than 5 bases requires more energy.)
4. The number of unpaired bases, whether interior loops or bulges. (Unpaired bases decrease the stability of the structure.)

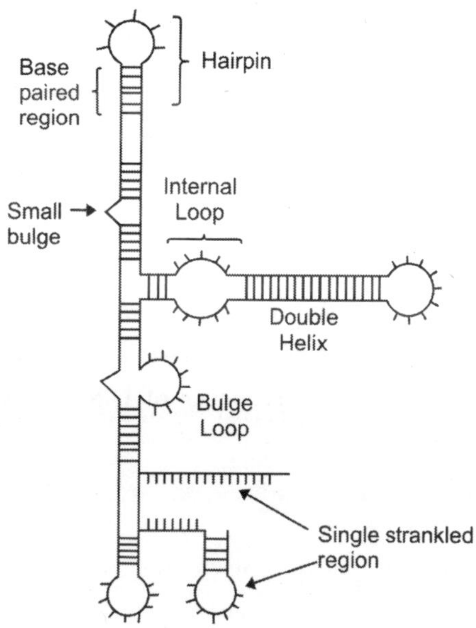

Fig. 16.10: Different secondary structure of RNA

Several methods are used to predict the RNA structure. Once the primary structure is known, the next step would be to identify the function of it. The first step in achieving this would be to predict its secondary structure. The secondary structure of RNA can expose biologically relevant features. There are two main approaches used to predict the secondary structure.

1. The stability of a secondary structure is quantified as the amount of free energy released or used by forming base pairs. Positive free energy requires work to form a configuration: negative free energies release stored work. Free energies are additive, so one can determine the total free energy of a secondary structure by adding all the component free energies (units are kilocalories per mole). The more negative the free energy of a structure, the more likely is formation of that structure, because more stored energy is released. This fact is used to predict the secondary structure of a particular sequence. Discovering a base pair configuration with the minimum possible free energy is the goal of most secondary structure prediction algorithms. To compute the minimum free energy of a sequence, empirical energy parameters are used. These parameters summarize free energy change (positive or negative) associated with all possible pairing configurations, including base pair stacks and internal base pairs, internal, bulge and hairpin loops, and various motifs which are know to occur with great frequency.
2. Comparative Sequence analysis: This uses multiple sequence alignments of homologous sequences to predict the structure. This needs many aligned sequences. Most of the RNA folding algorithms predict the secondary structure. These are not so that successful in predicting tertiary structure.

Tertiary structure of RNA is investigated by X-ray crystallography and NMR. Secondary structure predictions can be used along with NMR and X-ray crystallography in determining 3-D structure.

SUMMARY

Study of nucleic acid structure is central to molcular biology and bioinformatics. The molecular structure of nucleic acid underlies their function. Nucleotides which are composed of sugar, phosphate and nitrogenous base, form the repeating structural unit of nucleic acids. The primary structure is a strand that contains a linear sequence of nucleotides. The formation of secondary structure occurs because strands of DNA (and RNA) can form complementary regions due to hydrogen bonding between adenine and thymine (uracil) and between guanine and cytosine. Base stacking also stabilizes a double-stranded structure. The most common form of DNA is a right-handed double helix called as B-DNA. In DNA other different helical conformations have been identified including: A-DNA, Z-DNA, P-DNA, C-DNA, triplex helix and quadruplex helix. In Experiment 16A different conformations of DNA are studied using nucleic acid database. RNAs usually are single stranded, however it is common for RNA to form double helicale structures such as hairpins, bulges, looks and functions.

EXERCISE

1. Use the NDB and find out example structures of different types of RNAs as done in Experiment 16A.
2. Find out a list of software used for displaying nucleic acid structures.
3. A naturally occurring DNA has following sequence

   5′ –A – A – G– G A–A–A–A– G–G–G–A–G–G–A–G–A–G–3’

   3′ T–T –C–C–T–T–T–T–C–C–C–T–C–T–V–5′

   What is the sequence of DNA molecule could form triplex helix from double helix?

SECTION - 3

ORGANISM SPECIFIC BIOINFORMATICS

17

GENOME PROJECTS: AN INTRODUCTION

A genome is the entire DNA in an organism, including its genes. Genes carry information for making all the proteins required by all organisms. These proteins determine, among other things, how the organism looks, how well its body metabolizes food or fights infection, and sometimes even how it behaves. DNA is made up of four similar chemicals (called bases and abbreviated A, T, C and G) that are repeated millions or billions of times throughout a genome. The human genome, for example, has 3 billion pairs of bases. The particular order of As, Ts, Cs, and Gs is extremely important. The order underlies all of life's diversity, even dictating whether an organism is human or another species such as yeast, rice, or fruit fly, all of which have their own genomes and are themselves the focus of genome projects. Because all organisms are related through similarities in DNA sequences, insights gained from non-human genomes often lead to new knowledge about human biology.

The term genome refers to the total genetic composition of an organism. For example, the nuclear genome of humans is composed of 22 different autosomes and two sex chromosomes. In addition, humans have a mitochondrial genome that is composed of a single circular chromosome. Genomics refers the molecular analysis of the entire genome of a species. Genome analysis is a molecular dissection process applied to a complete set of chromosomes. During genome analysis, segments of chromosomes are cloned and analyzed in progressively smaller pieces, the locations of which are known on the intact chromosomes. This is the mapping phase of genome analysis. The mapping of the genome ultimately progresses to the determination of the complete DNA sequence. A complete DNA sequence provides the most detailed description of an organism's genome at the molecular level. Genome bioinformatics refer to analysis of the information content of entire genome.

Science of genomics is further divided into following branches:

1. **Comparative genomics:** A comparison of gene numbers, gene functions and biological functions of genes in the genome of diverse organisms, the objective being to identify group of genes that play a unique biological role in particular organism.
2. **Functional genomics:** Functional genomics deals with genome wide gene expression and regulation. Function of gene is identified by genome comparison. The function of a newly identified gene is tested by introducing mutations into the gene and then examining the resultant mutant organism for an altered phenotype.
3. **Structural genomics:** Comparison of partial amino acid sequence data for a protein, to the genome sequence, permits the immediate recognition of the corresponding gene and therefore

the complete amino acid sequence. With this knowledge it will be possible to predict protein folding and ultimately function. This is studied under structural genomics.

4. **Evolutionary genomics:** When different genomes are compared to study evolution of different species. It is evolutionary genomics. In chapter 20 evolution of grass species is studied as an example.

For many years before anyone tackled a whole genome, most of the sequences that were deposited in internet databases such as the GenBank were full length genes that had a known function. Genome sequences posses several difficult challenges as it tries to isolate complete gene sequences of an organism. Genome sequences are of two types depending on how they are obtained.

1. **Genome Clone:** These sequences are fragment of a genome containing a gene of interest (GOI) and are called a genomic clone. Coding sequences in these clones contain introns and the clones may also contain sequences adjacent to the GOI. For genomes that contain long, seemingly random stretches of DNA between the genes and vector contamination makes finding a particular gene among the many fragments of a genome quite a challenge.
2. **cDNA Clone:** Instead of starting with genomic DNA one could isolate mRNAs could then be converted to DNA (called cDNA because they are complementary to mRNA) using the enzyme reverse transciptase. Reverse transciptase is a key enzyme in genome sequencing. Hamilton Smith in 1978 purified this enzyme from *Hemophilus influenzae* and ultimately used this in sequencing first ever complete genome. (Historical Profile 19) A cDNA from the GOI could then be isolated and sequenced. These cDNA clones do not contain introns because the introns had been spliced out prior to maturation of the mRNA.

The basic techniques needed for sequencing entire genomes, cloning vectors for making genomic libraries, PCR for amplifying genes and the DNA sequencing machines were already available in 1980s. In the early 1990s the development of automated sequencers and computer programs capable of analyzing lots of DNA made possible two new approaches to obtaining sequence information. Automation of sequencing techniques is discussed in details in chapter 2 (Box 2.1. Instead of going specifically after a gene of interest, people created rich cDNA libraries (containing many of the expressed genes of an organism), picked cDNA clones randomly, and rapidly determined some of the sequence of nucleotides from the end of each clone. These expressed sequence tags or ESTs could then be compared to all known sequences using similarity search program such as BLAST. An exact match to a sequenced gene meant that the gene encoding that EST was already known. If the match was close but not exact one could conclude that the EST was derived from a gene with a function similar to that of the known gene.

In this section while dealing bioinformatics at organism level, chapters are classified and studied under the following heads:

1. Bacterial Bioinformatics
2. Fungal bioinformatics
3. Plant bioinformatics
4. Animal bioinformatics including human genome

17.1 GENOME SEQUENCING PROJECTS

Genome sequencing was first started in the model organisms used in genetic studies that have relatively smaller genomes. These include several bacterial species, *Caenorhabditis elegans* (a nematode), *Drosophila melanogaster, Arabidopsis thaliana* (s small plant), mice and humans.

In Fig. 17.1 major milestones in genome sequencing are given. 1st milestone was sequencing of chromosome number 3 of *Saccharomyces* in May 1992 by Andrzu Goffeau and co-workers. In 1995, a team of researchers headed by Craig Venter at the Institute for Genomic Research in Gaithersburg, Maryland, and Hamilton Smith at Johns Hopkins obtained the first complete DNA sequence of the bacterial genome from *Hemophilus influenzae* (Historical Profiles 19 and 20). Its genome is composed of a single circular chromosome 1.83 million base pairs (bp) in length and contains approximately 1743 genes (Fig. 17.1). In 1996, the first entire DNA sequence of a eukaryote, *Saccharomyces* (Baker's yeast), was completed. This work was carried out by a European led consortium of more than 100 laboratories, including some in the United States, Japan and Canada. The overall coordinator was Andreu Goffeau of the University of Louvain in Belgium. The yeast genome contains 16 linear chromosomes; these 16 chromosomes have a combined length of about 12 million bp, which contain approximately 6000 genes.

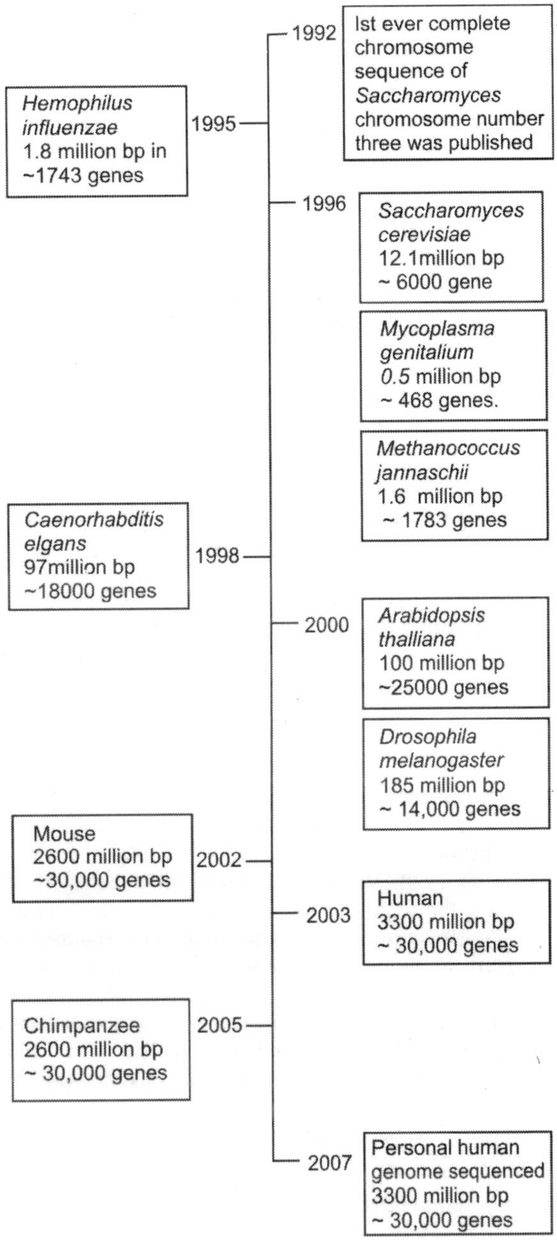

Fig. 17.1: Milestones in genome bioinformatics

The 97 million base genome of a tiny animal the round worm *Caenorhabditis elegans* was deciphered in 1998 by an international team led by Robert Waterson and John Sulston. Genome of other model animal species the fruitfly *Drossophila* was fully described in 2000 by collaboration in five countries. This collaboration was led by Gerald Rubin and J. Craig Venter. *Arabidopsis* genome initiative (AGI) is the name of multinational effort to sequence the *Arabidopsis* genome, which sequenced the complete genome of this plant in 2000. This was the first ever plant whose genome was completely sequenced.

The human genome is considerably larger. The human genome consists of 3 billion base pairs organized into 23 distinct chromosomes. The next landmark in genome projects was in 2002 when mouse genome was sequenced completely and was shown to contain 30,000 genes. After this in 2003 human genome project started earlier, was completed and shown to contain similar number of genes. A typical chromosome can contain several hundred to several thousand genes, arranged in linear order. The sequences that make up the protein-coding part of these genes actually account for only about 4 per cent of the entire genome. The other 96 per cent of the sequences do not code

HISTORICAL PROFILE 19

Sir J. Craig Venter for Visionary Contributions in Genome Research

Dr. Craig Venter

Dr. J. Craig Venter is one of leading scientists known for his visionary contributions in genomic research. He is founder and president of the J. Craig Venter Institute and the J. Craig Venter Science Foundation. The Venter Institute conducts basic research that advances the science of genomics; specializes in high volume genome sequencing, and explores the ethical and policy implications of genomic discoveries and advances. The J. Craig Venter Science Foundation supports both the Venter Institute and the Institute for Genomic Research (TIGR), an affiliated research organization founded by Venter in 1992. While on faculty at the National Institute of Health, Venter developed expressed sequence tags or EST's, a revolutionary new strategy for discovering genes.

In 1991, working with Nobel laureate Hamilton Smith, Dr. Craig Venter created a bold new sequencing process called DNA shotgunning method. The 'shotgunning' method created a controversy among Venter's colleagues who called it crude and inaccurate, like so many other technologies that change traditional standards, therefore creating disgust among peers. However, Venter crosschecked his results by sequencing the genes in both directions, achieving a level of accuracy that greatly impressed his initial skeptical rivals. Within a year, TIGR published the entire genome of *Hemophilus influenzae*, a bacterium with nearly 2 million nucleotides that causes meningitis and ear and respiratory infections.

In 1998, Venter became the first president of Celera Genomics to sequence the human genome using the whole genome shotgun technique, new mathematical algorithms, and new automated DNA sequencing machines. The completed sequence of the human genome was published in February 2001 in the journal, *Science*. In addition to the human genome, Venter and his team at Celera sequenced the fruit fly, mouse, and rat genomes. In 2003, Venter launched a global expedition to obtain and study microbes from environments ranging from the world's oceans to urban centers. In September 2007 he created swap genome from one species of bacteria to another. Genome of Dr. Craig Venter is already sequenced. This mission, now in progress, is yielding insights into genes that make up the vast realm of microbial life.

for proteins. Other non-coding sequences are relatively short sequences that are found in hundreds of thousands of copies scattered throughout the genome. Still other noncoding sequences are remnants of genes called pseudogenes. Characterising whole genome is important to a fundamental understanding of the design principles of organisms and for the discovery of new gene such as these having role in human genetic disease. Human and chimpanzee genomes are very similar the primary difference between the two is that humans have lesser number of chromosomes. In human evolution lineage, two ancestral ape chromosomes fused at their telomeres to produce human

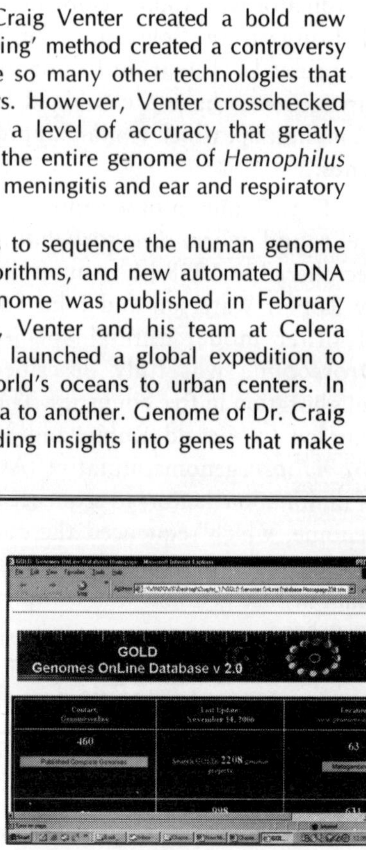

Fig. 17.2: Web view of gold online database

chromosome number 2. There are only nine other major chromosomal differences between chimpanzee and human chromosome segment inversions on human chromosome 1, 4, 5, 9, 12, 15, 16, 17 and 18. Chimpanzee genome sequence analysis was published in 2005, by the Chimpanzee Sequencing and Analysis Consortium, a group of scientist supported in part by National Human Genome Research Institute. A database now exists which marks the genetic differences between human and chimpanzee genome.

Aug. 2007 marks 150th anniversary of the discovery of the first Neanderthal, a fossil human who lived in Europe and parts of Asia until they disappeared about 30,000 years ago. Pååbo, a pioneer in the field of ancient DNA research, brought the world closer to understanding of evolution modern human to Neanderthal when he sequenced Neanderthal mitochondrial DNA in 1997.

The 99.9 per cent of nucleotide sequence of human genome in any two individuals are the same. In 2003 human genome complete sequencing costed $3 billion. Genome of Sir James D. Watson was completely sequenced in 47 days in 2007 and costed only 1 million. Today the cost of a personal genome sequencing is comparable to that of Mercedes Maryback ($350, 000). This is our evolutionary legacy; it contains the genetic information that makes us human beings. In reality, however, there are many different human genomes. Geneticists have great expectations that are responsible for genetic diseases such as phenylketonuria and other inborn errors of metabolism, as well as the mutations that increase the risk of more complex diseases such as heart disease, breast cancer, and diabetes.

Table 17.1: Some examples of genomic sequencing

Domain	*Organism*	*Size (Mbp)*	*Comments/disease association*
Eubacteria			
Gamma (Enterobacteriaceae)	*Escherichia coli*	4.64	Normal intestinal flora
Gamma (Pasteurellaceae)	*Hemophilus influenzae*	1.83	Meningitis
Gamma (Pseudomonas group)	*Pseudomonas aeruginosa*	6.26	Respiratory infections
Gamma (Vibrionaceae)	*Vibrio cholerae*	4.0	Cholera
Epsilon subdivision	*Helicobacter pylori*	1.67	Duodenal ulcers
Beta subdivision	*Neisseria meningitidis*	2.18	Meningitis
Bacillus Clostridium group	*Bacillus subtills*	4.20	Normal soil bacterium
	Mycoplasma genitalium	0.58	Smallest cellular genome
Actinobacteria	*Mycobacterium tuberculosos*	4.41	Tuberculosis
Spirothaetales	*Treponema pallidum*	1.14	Syphilis
	Borrelia burgdorferi	0.91	Lyme disease
Chlamydia group	*Chlamydia pneumoniae*	1.23	Mild form of pneumonia
Aquificales	*Aquifex aeolicus*	1.55	Heat-tolerant, oxidizes hydrogen
Cyanobacteria	*Synechocystis sp.*	3.57	Blue-green algae
Deinococcus Thermus	*Deinococcus radiodurans*	3.61	Radiation-resistant
Thermatogales	*Thermatoga maritima*	1.86	Anaerobic, hot marine mud
Archaea			
Euryarchaeota	*Methanococcus jannaschii*	1.66	Marine hydrothermal vents
Crenarchaeota	*Aeropyrum pernix*	1.67	Marine sulfur vents
	Sulfolobus solfataricus	2.25	Yellowstone, USA, sulfur hot springs
Eukaryota			
Fungi	*Saccharomyces cerevisiae*	12.07	Budding yeast
Protozoa	*Plasmodium falciparum (2 chsms)*	2.01	Malaria
Nematode	*Caenorhabditis elegans*	97	Soil organism
Insect	*Drosophila melanogaster*	185	Fermenting fruits
Primate	*Homo sapiens (chsm 22)*	32	Ubiquitous; first chromosome sequenced

HISTORICAL PROFILE 20

Hamilton O. Smith for Sequencing 1st ever Genome

Hamilton O. Smith

Hamilton Othanel Smith shared the 1978 Nobel Prize in physiology or medicine with fellow biologists Werner Arber and Daniel Nathans for the set of linked discoveries that started of the boom in biotechnology and ultimately leads to sequencing of Ist ever genome of bacterium *Hemophilus influenzae*. Because of these discoveries, researchers can more easily elucidate the structure and coding of DNA molecules and they hope to correct many genetic illnesses in the future. His research also made it possible to design new organisms, a controversial but potentially beneficial technology. Smith purified and explained the activity of the first restriction enzyme, which became the principal tool used by genetic engineers to selectively cut up DNA.

Smith was born on August 23, 1931, in New York. He obtained a medical degree from the Johns Hopkins School of Medicine in 1956. During the years 1956 to 1962, he held various posts, including an internship at Washington University in St. Louis, Missouri, a two-year Navy stint in San Diego, California, and a residency at Henry Ford Hospital in Detroit, Michigan. He gradually taught himself genetics and molecular biology in his spare time. In 1962 he began a research career at the University of Michigan on postdoctoral fellowship from the National Institute of Health, before finally returning to Johns Hopkins in 1965 as a research associate in the microbiology department.

After his return to the United States, Smith purified the first Type II restriction endonuclease, which he obtained from the bacterium *Hemophilus influenzae*, and identified the nucleotide sequence which the enzyme would cut. He gave a supply of the enzyme to Daniel Nathans, who used it in his own work. The three men eventually won the 1978 Nobel Prize.

Smith is a co-founder of Celera Genomics, and in 1995 he led the team that accomplished the first sequencing of a bacterial genome, at the Institute of Genomic Research, where he serves as a trustee. Since 2002 he has been at the forefront of a project, funded by the U.S. Department of Energy, that seeks to create a genetically-engineered single-cell organism with the fewest genes necessary to sustain life, capable of feeding and reproducing itself.

Genomes Online database (GOLD) holds the databases which provide information about all published and unpublished genome projects. Till May 2007 there are 566 published genome projects of which 41 archaeal, 476 bacterial and 49 eukaryotic genomes. Such a huge number of projects have yielded enormous data which needs to be managed properly. In Table 17.1 a brief summary of genomic sequencing project is given. They are classified in bacterial, fungal, plant and animal genomes.

The first genome database called, ACEDB is a *C. elegans* database. The database and methods to access this database were developed by Mike Cherry and colleague in 1993. ACEDB is a public domain software package, developed by Richard Durbin and Jean Thierry Mieg, with additional input from a number of others. It runs on UNIX (including Linux), Windows 95/NT and macintosh systems. You can download the ACEDB software from several different sites.

17.2 INTRODUCTION TO PROTEOMICS

The power of proteins has been appreciated for nearly 200 years. In the early 1800s, chemists chose the name "protein" for these substances after the Greek word "proteios", which means, "holding first place". The term "Proteome" was coined in 1994 by Marc R. Wilkins, Vice president and Head of Bioinformatics at Proteome System in Sydney, Australia, to mean the total protein complement encoded by a genome. In other words, the Proteome of an organism includes all the proteins contained in the organism. In Fig. 17.3 a link between genomics and proteomic studies is depicted. The study of these proteomes. is known as proteomics. Proteomics a field, is growing in importance in this new millennium. A large number of initiatives have been launched worldwide to conduct research in this field. The ultimate utility of proteomics would be realized when a new generation of safe and effective drugs are developed based on the study of various proteomes. This would help treat and cure the various diseases that afflict humans, animals, and even plants.

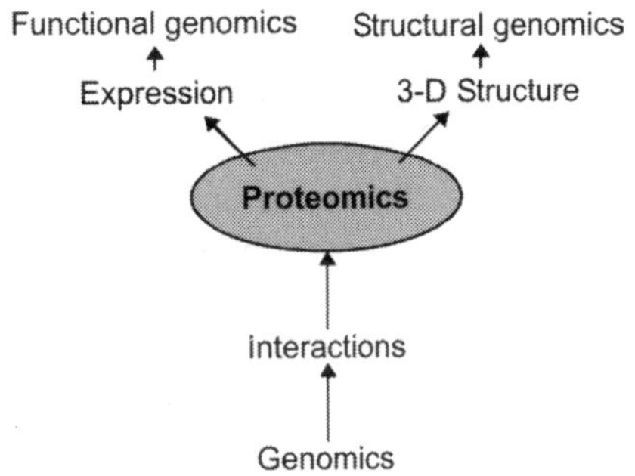

Fig. 17.3: Diagrammatic representation of link between genome and proteome

Proteomics usually pertains to four broad categories, identification and quantification of all the proteins; study of protein-protein interactions that affect the various complex pathways and networks and structural characterization. Identifying the proteins helps in getting a complete picture of the proteome under study. Protein interaction study helps to define the function of a specific protein in relation to the other proteins that together form the pieces in the network of pathways the control cullular processes. Finally, structural characterization gives a clue to the active sites in the protein thus elucidating its functional attributes, in addition to providing valuable information in terms of drug targets. Each of these endeavours serves different purposes, though ultimately when the different units are pieced together, it serves the common goal of a better understanding of cellular processes; in turn leading to development of better drugs.

Human Proteome Organization (HUPO, http://www.hupo.org/) classified proteomics in a more precise manner as identification of all proteins coded by human genome. Expression proteomics deals with protein expression in different cell types of the organism protein distribution in subcellular compartments of the organelles, post-translational modifications of the proteins and protein-protein interactions. Functional proteomics study relation between protein structure and function. Recent advances in automated sample preparation, mass spectrometry and bioinformatics, has made proteomic analysis very easily.

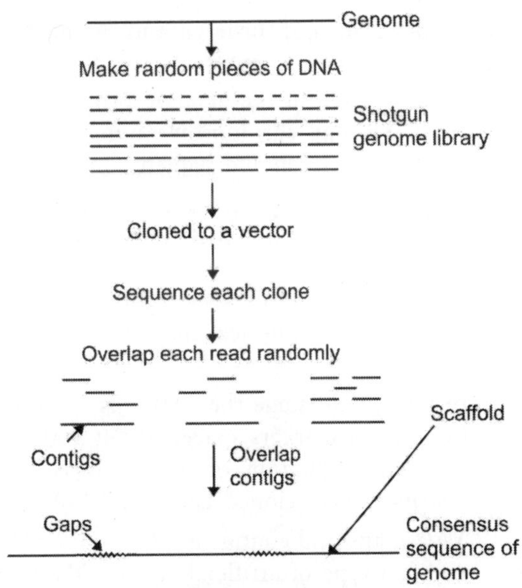

Fig. 17.4: Whole genome shot gun sequencing approach

17.3 METHODS IN GENOMICS

Like geographers oceanographers and astronomers who whenever encounter a new territory, their first activity is to create a map that they can use as a common reference when discussing that territory. Genetists use many kinds of maps to explore the terrain of genome.

All current genome-sequencing strategies are clone based. The general meaning of the term clone is to have many identical copies of something. In genetics this term has several meanings: a single cell that has been divided to produce a colony of genetically identical cells; or an individual who has been to produce from a somatic cell of another individual such as the sheep Dolly. Here in genomics it refers to many copies of a DNA fragment that are propagated within a vector or produced by PCR. The approach in genome sequencing is to break the genome in thousands or millions of more random pieces, and then read the sequence of each little piece. After this the sequence of each little piece is overlapped, to know where their sequences are identical. This process is continued until a consensus sequence is reached i.e. where it is agreed that the present representation of sequences is what is present in genomes (Fig. 17.4). When there are lots of gaps and sequences are badly overlapped it is called as draft level sequencing. However, when there are no typos and every base pair is absolutely in consensus from typos to telomere sequencing is said to be complete.

However, there are two ways to assemble a consensus sequence of a genome. One is called whole genome shotgun sequencing (WGS) and the other is called ordered clone sequencing. The logic of whole genome shotgun sequencing is sequence first, map later. First, sequence reads are obtained from clones randomly selected from a whole genome library without any information on where these clones map in the genome. Such a library is called a shotgun library. Then, these sequence reads are assembled into a consensus sequence covering the whole genome by matching homologous sequences shared by reads from overlapping clones. Occasional gaps are encountered whenever a region of the genome is by chance not found in the shotgun library—some DNA fragments do not grow well in particular cloning vectors. Such gaps are filled in by techniques such as Primer walking—that is, by using the end of a cloned sequence as a primer to sequence into adjacent uncloned fragments. With the use of the WGS approach, 112 prokaryotic species have been fully sequenced, and more than 200 additional prokaryotic sequencing projects are in progress.

The logic of ordered clone sequencing is the opposite of that for the whole genome shotgun approach: map first, sequence later. Individual cloned inserts from a genomic library are screened to look for similarities in restriction enzyme recognition sites, indicating that two inserts are contiguous in the genome. This results in a set of ordered and oriented clones that together span the genome. Such an ordered and oriented set of clones covering the whole genome is called the physical map of the genome.

Strategy used in the genome projects is explained with the help of human genome project in Fig. 17.5. Human chromosome number 21 is used in the present example. There are 4 steps in completion of its genome sequence:

1. Ist goal genome project was to have a genetic map of each chromosome with makers spaced at distances about 1 MB. This work was accomplished by finding markers such as RFLPs and STSs and assigning them to chromosomes. Genetic maps for each chromosome incorporating about 15,000 markers were completed in 1995. This map is derived from recombination studies. In Fig. 17.6 a physical map or three hypothetical clones A, B and C is drawn with the help of RFLP.
2. In the second stage the goal was to prepare physical map of each chromosome containing the location of markers spaced about 100,000 bp apart. Physical maps are maps of the order, over lap and orientation of physically isolated pieces of genomes—in other words maps of the distribution of cloned genomic DNA from genomic clone libraries.
3. YAC maps and contigs has 0.5 – 1.0 MB resolution YAC refers to yeast artificial chromosome and is a type of artificial vector. DNA segment of more than 1 megabase can be inserted into this vector and are used in human genome project contigs, on the other hand are continueous segments are of DNA recovered by overlapping clones.

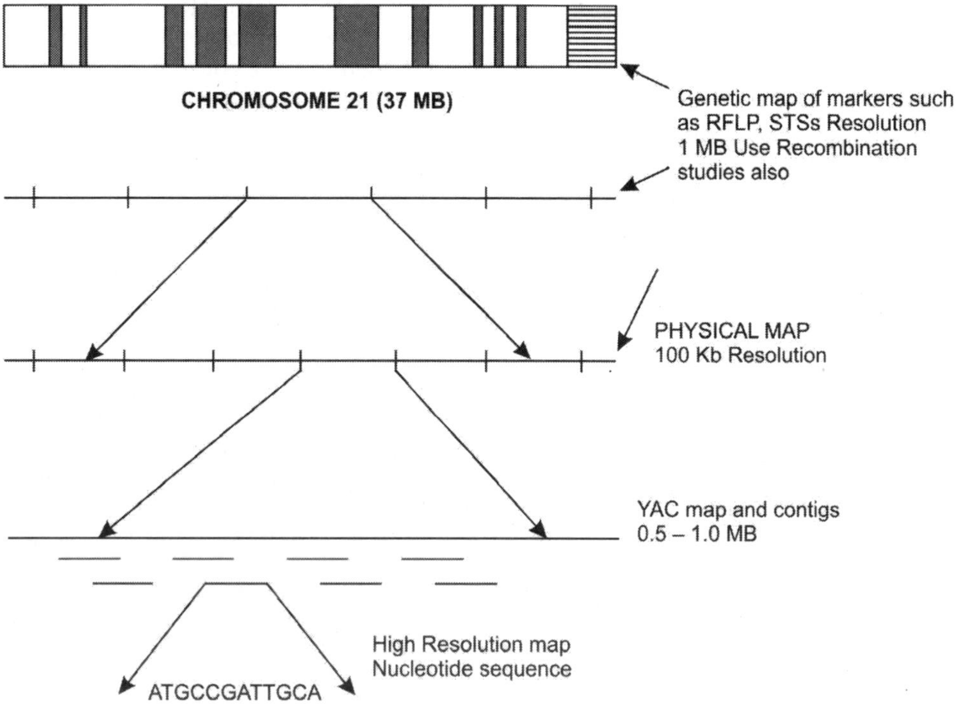

Fig. 17.5: Overview of human genome project

4. Nucleotide sequence: Sequences determined are assembled into genomic sequence of 3.2×10^9 nucleotide. The highest resolution map is the complete DNA sequence of the genome that is the complete sequence f nucleotides A, T, C and G of each double helix in a genome. The approach in constructing a high resolution map is to (1) break a genome up into thousands to million of more or less random little pieces and make random clones, (2) read the sequence of each little piece, (3) overlap the little pieces where their sequence are identical. These represents the contigs which are a series of clones that contain over lapping pieces of chromosomal DNA, and, (4) continue overlapping ever-larger pieces called scaffold until you've accounted for all the little pieces i.e. we have reached to a consensus sequence, a sequence for which there is a consensus (agreement) that it is an authentic representation of the sequence for each DNA molecule in that genome.

In Fig. 17.6 a physical map is created by clone fingerprint mapping. Three clones A, B, C are digested with multiple restriction enzymes, and the resulting complex mixture of restriction fragments is separated on the basis of size, by gel electrophoresis. The bands containing the fragments are stained to show their location. The number of identically sized bands for each pair of digests is determined. Clone A shows 11 bands of which seven resembles to clone B. Clone B shows 13 bands of which ten resembles to clone C. A and B digests share more than 50 percent of the bands, as do the B and C digests, indicating that they come from overlapping regions of the genome. Several bands (four bands) are found in all three of A, B and C, suggesting that some part of the three clones overlap. The physical map derived from this data is shown in part (b) Fig. 17.6. The map shows some bands are overlapping along A, B, C. Some bands are exclusively overlapping among A and B, are B and C.

There are several types molecular markers. These may be single nucleotide polymorphism called SNP or small sequence length polymorphism or RFLPs. The molecular landmarks on linkage may be the neutral, polymorphic variants such as RFLPs, STS and VNTRs. In cytogenetic maps, a molecular landmark is any piece of DNA, whose position we can locate on the microscopically visible chromosomes through DNA-DNA hybridization techniques. The potential density of molecular markers on a linkage map is much higher than that of classical phenotypic markers.

1. Restriction fragment length polymorphisms: Restriction fragment length polymorphisms (RFLPs) represent the presence or absence of restriction-enzyme recognition sites in different individuals. Restriction enzymes recognise a single unique restriction site which is recognised wihtout regard to the source of the DNA No. of cuts in the DNA from a particular organism is determined by number of restriction sites. Any two individuals will differ by a great many RFLPs. An RFLP is assayed by using a cloned DNA fragment as a probe. The probe must bind to one site alone: In a Southern blot, it reveals different-sized restriction fragments thus defining an RFLP locus for each individual.

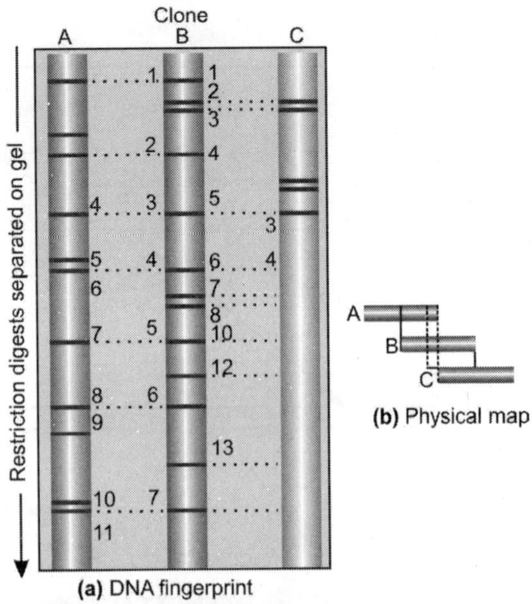

Fig. 17.6: Restriction map

2. Single nucleotide polymorphisms (SNP): describes variation in sequence between individuals caused by a change in a single nucleotide. This is responsible for most of the genetic variations among individuals. SNPs are useful in mapping because they are numerous throughout the genome. In humans, SNPs are spaced at intervals between 11 and 300 bases. They can be detected in several ways, some of which are adaptable to high throughput. One simple way is simply to sequence one region from several persons and compare the regions for differences. Most SNPs are not within genes and are simply used for marking chromosome regions. A few are within genes and act as specific tags for disease alleles.
3. Simple sequence length polymorphism (SSLP): The presence of different number of short repetitive elements at a particular locus in different homologus chromosome. They serve as very useful markers in genome mapping. SSLPs have one basic advantage over RFLPs. A RFLP usually has only 1 or 2 "alleles", or morphs, in a pedigree or population under study. In contrast, SSLPs commonly have multiple alleles, and as many as 15 alleles have been found for an SSLP locus.

   There are two types of SSLPs routinely used in genomics—minisatellite and microsatellite markers:

   (*a*) Minisatellite markers are based on variation in number of VNTRs (variable number of tandem repeats). Minisatellite DNA consists of ~10 copies of short repeating sequences from 15 to 100 NTP long. The length of the repeating unit is measured in 10s of base pairs. The number of repeats varies between individual genomes.

   (*b*) Microsatellite DNAs consists of repetitions of extremely short typically < 10 bp units.

   There are several techniques for placing these molecular markers on cytogenetic maps:

   (*i*) *In situ* hybridisation mapping: Insitu hybridisation is also called as cytological hybridisation. It is performed by denaturing the DNA of cells squashed on a microscope slide so that reaction is possible with an added single stranded RNA or DNA; the added preparation is radioactively labeled and its hybridisation is followed by autoradiography.

(*ii*) FISH: In the process called fluorescent *in situ* hybridisation, or FISH, the probe is labeled with a fluorescent dye and the location of the homologous fragment is revealed by a bright fluorescent spot on the chromosome. The probe sequence can be mapped t o the approximate position on the chromosome to which it hybridizes by noting its position in relation to banding patterns, centromere, or other cytological features. The location is approximate because this technique does not have the resolving power of recombinational mapping.

(*iii*) Chromosome painting: An extension of FISH is chromosome painting. Rather than using visible chromosome features as landmarks, this technique uses a standard control set of probes homologous to known locations to establish the cytogenetic map. The probes are labeled with different fluorescent dyes, which thus "paint" specific regions and make them identifiable under the microscope.

To connect the different map types molecular markers from linkage and cytogenetic maps are anchored to physical map and the genome sequence maps.

17.4 THE PROTEOMIC METHODS

17.4.1 2-D Protein Electrophoresis

Now a day the two-dimensional (2-D) protein electrophoresis is a method of choice for protein separation. This method enables to distinguish up to 10000 proteins. The method is based on two distinct physical and chemical features of proteins: first, the proteins are separated according to their isoelectric point (pI). Isoelectric point is such a pH value, where the overall protein charge equals to zero. This can be obtained by creating a pH gradient in the gel where protein is loaded and electric current is applied. Proteins then migrate towards cathode or anode according to their total charge up to the point where the gel pH equals pI of a given protein. After the separation in the first direction, common electrophoresis on a polyacrylamide gel (PAGE) is applied, but the electric current is applied perpendicular to the original orientation of electrodes. Proteins then migrate in the second direction through gel only according to their size. After both phases of 2-D electrophoresis, it is necessary to visualize proteins by one of the staining or labeling methods (chemical or radioactive). The resulting maps of proteins can be compared in between the experimental and control sample or among the samples from patients with specific disease and their healthy controls and thus identify differentially expressed proteins that can be linked with the pathogenesis of the studied disease. The identity of differentially expressed proteins is verified after cutting out the area of the gel with given difference and subsequent analysis using mass spectrometry.

17.4.2 Mass Spectrometry (MS)

Mass spectrometry is a method that enables precise measurement of molecular weight of a broad spectrum of substances. As the studied substance has to be intact in gas phase, the use of mass spectrometry for protein analysis (also of polysaccharides and oligonucleotides) was only enabled by development of soft ionization techniques of mass spectrometry, like matrix assisted laser detection of desorption/ionization (MALDI) and electrospray ionization (ESP) Protein identification is generally performed in two ways: Protein is digested by trypsin or by other proteolytic enzyme to smaller peptides and their precise molecular weights are measured using MS. The spectrum of those molecular weights in then compared with theoretical spectra that are calculated from protein sequences from available databases (using bioinformatics tools). Tandem MS enables to choose the peptide which is then fragmented by the collision with inert gas. The fragmentation pattern gives either full of partial information about protein sequences that is subjected to the search in databases.

Apart of protein identification, MS is a priceless tool for protein postranslation modification analysis, because it enables to localize given modifications within the protein and also helps to find out the nature of such modification.

17.5 DERIVING MEANING OF GENOMIC SEQUENCES THROUGH BIOINFORMATICS TOOLS

17.5.1 Mapping of Docking Sites on to Genomes

Genome encodes a series of docking sites for different proteins and RNAs. Docking is synonymous to binding site. Many proteins dock at sites located in the DNA itself, whereas other proteins and RNAs dock at sites located in mRNA. The sequence and relative positions of those docking sites permit genes to be transcribed, spliced, and translated properly at the appropriate time in the appropriate tissue. The association of a translating ribosome with the translocation channel in the membrane of the endoplasmic reticulum is also included in docking. Determination of docking site for transcription, gene regulation, translation and gene splicing in a genome sequence has been possible through bioinformatics tools. Several docking sites located in genome sequence are depicted in Fig. 17.7.

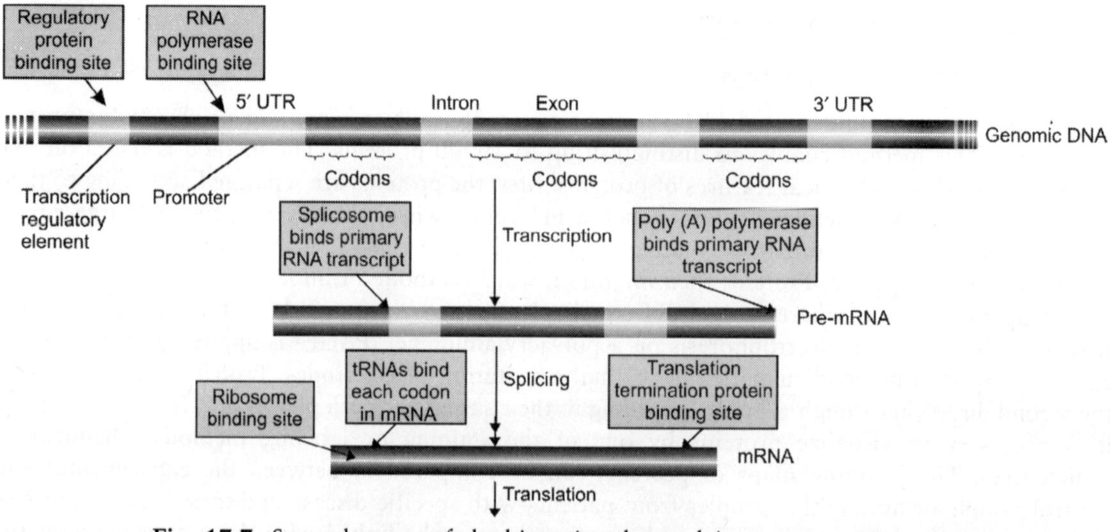

Fig. 17.7: Several types of docking sites located in genome sequences

There are now statistical gene finding computer programs that search for the predicted sequence of the various docking sites used for transcription start sites, for 3' and 5' splice sites, and for translation initiation codons within genomic DNA. These predictions are based on consensus motifs for such known sequences, but they are by no means perfect.

17.5.2 ORF Detection

An ORF or open reading frame is a gene-sized sequence of sense codons with the appropriate 5′ and 3′ end sequences such as start and stop codons. Possible introns are processed out computationally. Such an ORF seems like a good candidate for being a gene. To find candidate ORFs, the computer scans the DNA sequence on both strands in each reading frame. Because there are three possible reading frames on each strand, there are six possible reading frames in all. This aspect is discussed in details in chapter 12, on gene finding there is no specific difference in ORF finding in genomes.

17.5.3 Codon Bias Prediction

In a given species, not all synonymous codons for an amino acid are used with equal frequency. Rather, certain codons are present much more frequently in the ORFs. For example, in *Drosophila melanogaster*, of the two codons for cysteine, UGC is used 73 per cent of the time, whereas UGU is used 27 per cent. This usage is a diagnostic for *Drosophila* because, in other organisms, this codon bias pattern is quite different. Codon biases are thought to be due to the relative abundance of the rRNAs complementary to these various codons in a given species. In the data for codon usage matches with the predicted ORF then this is a supporting evidence that the proposed ORF is genuine.

17.5.4 Finding Exon and Introns in a Genome Sequence

cDNA sequences are extremely valuable in identifying the exons of a gene because cDNAs are DNA copies of mRNAs. The alignment of cDNAs with their corresponding genomic sequence clearly delineates the exons, and hence introns are revealed as the regions falling between the exons. Thus, cDNA sequences can greatly assist in identifying the correct reading frame, including the initiation and stop codons. Full-length cDNA evidence is taken as the gold-standard proof for identification of the sequence of a transcription unit, determination of how the transcript is processed, and localization of the ORF that it encodes.

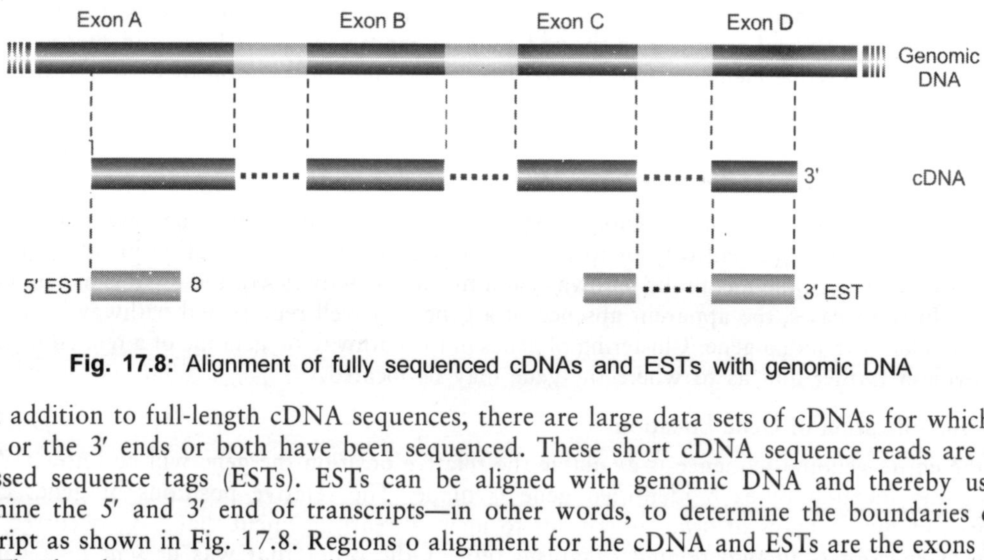

Fig. 17.8: Alignment of fully sequenced cDNAs and ESTs with genomic DNA

In addition to full-length cDNA sequences, there are large data sets of cDNAs for which only the 5′ or the 3′ ends or both have been sequenced. These short cDNA sequence reads are called expressed sequence tags (ESTs). ESTs can be aligned with genomic DNA and thereby used to determine the 5′ and 3′ end of transcripts—in other words, to determine the boundaries of the transcript as shown in Fig. 17.8. Regions o alignment for the cDNA and ESTs are the exons of the gene. The dots between segments of cDNA or expressed sequence tags (ESTs) indicate regions in the genomic DNA that do not align with cDNA or EST sequences; these regions are the locations of the introns. For the ESTs, only a short sequence reads establish the boundaries of the transcription unit, but they are not informative about the internal structure of the transcript unless the EST sequences cross an intron.

17.5.5 Gene Identification

Gene identification in prokaryotic organisms in simplified by their lacking introns. Once the sequence patterns that are characteristic of genes in a particular prokaryotic organism (e.g. codon usage, codon neighbour preference) have been found gene locations in the genome sequence can be predicted quite accurately. The presence of introns in eukaryotic genomes make gene prediction more involved because in addition above features locations of intron-exon and exon-intron splice junction must also be predicted. Methods of gene prediction in eukaryotes and prokaryoties are discussed in details in chapter 13. Since gene prediction methods are only partially accurate, gene identification is facilated by high throughput sequencing of part cDNA copies (called EST.). Presence of ESTs confirms that predicted gene is transcribed. A more thorough sequencing of full length cDNA clones may be necessary to confirm the structure of genes chosen for a more detailed analysis.

17.5.6 Genome Annotation

The amino acid sequence of proteins encoded by predicted genes is used as query of protein sequence databases in a database similarity search. A match of a predicted protein sequence to one or more database sequences not only serves to identify the gene function but also validates the genes prediction. Pseudogenes, gene copies that have lost function may also be found in this analysis. The genome sequence is annotated with information of gene content and predicted structure, gene location and functional prediction. The procedure for searches starting with genome, EST and CDNA sequence is already described in section 17.5.4. Usually not all query proteins will match a database hence it is important to extend the analysis by searching the predicted protein sequence for characteristics domains (conserved amino acid pattern) that serve as signature for a protein family or domains that represent a structural fold using computational tools described in chapters 3 and 16.

17.5.7 Repeat and Transposable Elements Identification

Eukaryotic genomes comprise classes of repeat elements including tandem repeats (tandem repeats present in centromers and telomers, dispersed tandem repeats (minisatellites and macrosatellites) interspersed transposable elements (see chapter 20 for detailed description of these). Transposable elements can comprise one half or more the genome sequence. Identification of classes of repeat elements is aided by searchable databases is discussed in chapters 20 and 21.

17.5.8 Metabolic Pathway and Regulation Analysis of a Gene

As genes are identified in a new genome sequence, some will be found that are known to act sequentially in a metabolic pathway or to have a known role in gene regulation in other organism. From this information, the metabolic pathways and metabolic activities of the organisms will become apparent. In some cases, the apparent absence of a gene in a well represented pathway may lead to a more detailed search of a gene. Clustering of genes in the pathway on genome of a related organisms can provide a further hint as to where the gene may be located.

17.5.9 Identification of Gene Location

Once the entire genome sequence is available the relative position the gene will be known. A map showing the location of each identified gene is made. The relative positions of genes can be compared in similar maps of other organisms to identify rearrangement that have occurred in the genome. Gene orders in two related organism reflects the order that was present in a common ancesstor genome. Chromosomal breaks followed by resassembly of fragments in a different order can produce new gene gaps.

Populations of an organism show sequence variation that are readily detected by DNA sequencing and other analysis method. The inheritence of genetic diseases in humans and animals (cancer and heart disease) and of desirable traits in plants, can be traced genetically predigree analysis. Sequence polymorphism that are very close to a trait may be used to trace the trait by virtue of the fact that the polymorphism and trait are seldom separated from one generation to the next. These linked polymorphisms may then be used for mapping and identifying important genes.

17.5.10 Using Polypeptide and DNA Similarity is Mainstary of Evolutionary Genomics

Candidate ORFs predicted by the above techniques often can be anchored in reality by comparing them with all the other genes that have ever been found. This is done by submitting candidate sequences as "query sequences" to the public databases, such as BLAST. Because organisms share common ancestors, gene sequences are generally very similar between them. Hence a query sequence that represents a real gene will likely have relatives in the genes isolated and sequenced in other organisms, especially in the closely related ones. The computer scans the database and returns a list of full or partial hits, starting with the closest matches. If the candidate sequence closely resembles that of a gene previously identified from another organism, then provides a strong indication that the ORF is a real gene. Evolutionary trees are drawn, and phylogeny of the organisms can be studied.

Genomic sequencing and computational analysis could solve several evolution related questions. For example comparisons of the genomes of archaeans, bacteria, and eukaryotes indicate that the machinery for DNA replication and transcription in archaeans resembles that of eukaryotes, whereas the metabolism and the genes for metabolism strongly to resemble those of bacteria. About half of the genes are unique to the Kingdom Archaea. In *Hemophilus*, or example, there are 1743 protein-coding sequences. These sequences encode 1247 unique proteins. Each family consisting of proteins that are significantly similar in structure to the other members. The individual members of each protein family are known as paralogs; their coding sequences all derive from a common ancestor through gene duplication and subsequent divergence. In the case of *Hemophilus*, the proteome consists of 1247 unique sequence and 178 families of protein that makes the proteome consisting of 1425 proteins. The proteins of the several organisms have been analyzed to identify proteins that are orthologs; this term refers to highly similar proteins from different organisms that derive from a common ancestral coding sequence through speciation rather than gene duplication. The similarities in sequence of proteins from different organisms are known to reflect conservation of function.

17.5.11 Vector Contamination Identifition

Genome sequences are assembled from DNA sequence fragcaments of approximate length 500 bp. Chromosomes of a target organism are purified, fragmented and subcloned in fragments of size of kbp in bacterial artificial chromosomes (BACs). The BAC fragments are then further subcloned into plasmid vector for DNA sequencing. Full chromosomal sequences one then assembled from overlaps. Vector contamination identification is done by similarity searches specially contextual searches which went nearly identical matches.

17.5.12 Deciphering the Encoded Information Through Comparative Genomics

Area of genomics that compare and contrasts genomes from different organisms is called comparative genomics. It provides a powerful method of identifying essential sequences in different organisms. Knowledge of the genomes of model organisms and that of human being may reveal important new information about disease, development, neurobiology, aging, and many other biological processes. In terms of practical usefulness, a knowledge of the genomes of pathogens may help us devise better ways to make vaccines and hence to limit the death and destruction that these organisms cause.

Studies in model organisms have contributed to our understanding of genes and pathways that are implicated in human disease. Among a set of 289 genes that are known to be mutated or amplified in human diseases, 230 of the encoded proteins are similar to proteins found in *D. melanogaster*, 212 to ones in *C. elegans*, and 121 to ones in *S. cerevisiae*. These include proteins involved in DNA synthesis, DNA repair, general metabolism, and the cytoskeleton. Among the human genes associated with cancer, 68 per cent have an ortholog in *Drosophila*; somewhat fewer have an ortholog in *Caenorhabditis*, and the similarity with genes is not usually as great as that found in fly genes.

17.5.13 Functional Genomics

Functional Genomics as defined earlier deals with genome-wide patterns of gene expression and the mechanisms by which gene expression is coordinated. Thus rather than concentrating on single gene we here focus on expression of a set of genes. As changes take place in the cellular environment—for example, through changes in the external conditions, development, or aging—the patterns of gene expression also change. But genes are usually deployed in sets not individual. As the level of expression of one coordinated set may be decreased, the level of expression of a different coordinated set may be increased.

The study of genome-wide patterns of gene expression became feasible with the development of the DNA chip (or microarray), a flat surface about the size of a postage stamp on which 10,000–100,000 distinct spots are present, each containing a different immobilized DNA sequence suitable for hybridization with DNA or RNA isolated from cells growing under different conditions, from cells not exposed or cells exposed to a drug or toxic chemical, from different stages of development or from diferent types or stages of a disease such as cancer. Two types of DNA chips are presently in use:

1. A chip arrayed with oligonucleotides synthesized directly on the chip, one nucleotide at a time, by automated procedures; these chips typically have hundreds of thousands of spots per array.
2. A chip arrayed with denatured, double-stranded DNA sequences of 500–5000 bp, in which the spots, each about a millionth of a drop in volume, are deposited by capillary action from miniaturized fountain-pen like devices mounted on the movable head of a flatbed robotic workstation: these chips typically have tens of thousands of spots per array. Fig. 17.9 shows and explain the method by which DNA chips are used to assay levels of gene expression in one experimental sample relative to a control. Only six spots one shown here for the sake of convenience. Number of spots in practice varies from 10,000 to 100,000.

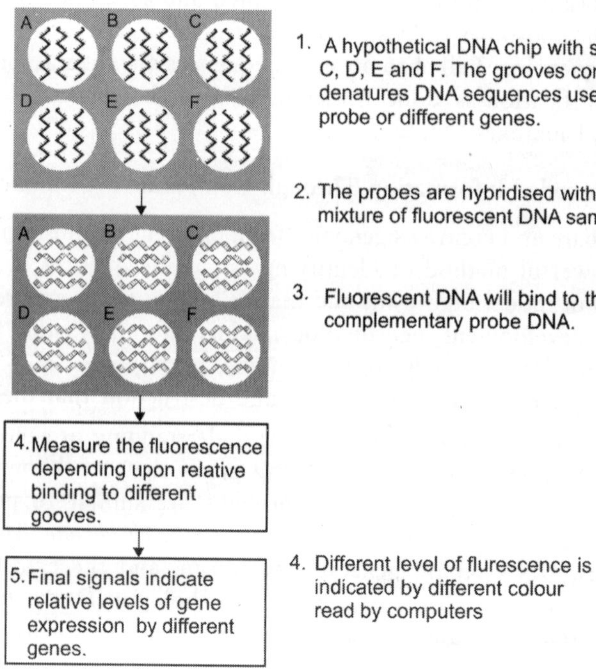

Fig. 17.9: DNA microarrary

Application of DNA microarrays includes:

1. Investigating cellular states and processes: Patterns of expression that change with cellular state can give clues to the mechanisms of processes such as sporulation, or the change from aerobic to anaerobic metabolism.
2. Diagnosis of disease. Testing for the presence of mutations can confirm the diagnosis of a suspected genetic disease, including detection of a late-onset condition such as Huntington disease, to determine whether prospective parents are carriers of a gene that could threaten their children.
3. Genetic warning signs: Some diseases are not determined entirely and irrevocably by genotype, but the probability of their development is correlated with genes or their expression patterns. A person aware of an enhanced risk of developing a condition can in some cases improve his or her prospects by adjustments in lifestyle.
4. Drug selection: Detection of genetic factors that govern responses to drugs, that in some patients render treatment ineffective and in others even cause serious adverse reactions.
5. Classification of disease: Different types of leukaemia can be identified by different patterns of gene expression. Knowing the exact type of the disease is important in selecting optimal treatment.
6. Target selection for drug design: Proteins showing enchanced transcription in particular disease states might be candidates for attempts at pharmacological intervention.

17.5.14 Clinical Proteomics

Clinical proteomics is trying to identify via the analysis of the protein spectrum in the biological samples (urine, saliva, blood) new markers that can be used in future early diagnosis and treatment of a specific disease. The major role in all spheres in early diagnosis and treatment of a given disease is played by bioinformatics (digitalization, 2-D picture analysis of the gels, modelling 3-D protein structures and the networks of protein interaction, creation, care and development publicly available databases). One of the interesting proteomic projects is Swedish human protein Atlas (http://www.proteinatlas.org/), that groups the information about expression of individual proteins in different tissues but its uniqueness lies in the accompanying archive of immuno histochemically stained samples—at the time of writing of this text, there were more than 400,000 available.

SUMMARY

Genome is all the DNA in an organism including its genes. These sequenced genes are of two types depending on how they are obtained. First a genomic clone and the other is cDNA clone. It is the advancement in sequencing techniques that has made genomic sequencing of several model organisms possible. These model organisms include bacterial species *Hemophilus influenzae*, *Caenorhabditis elegans*, *Drosophila melanogaster*, *Arabidopsis thalliana*, nice and human. Craig Venter and Hamilton Smith in 1995 first gave the complete sequence of *Hemophilus influenzae*. Gold holds the database which provides information about all published and unpublished genome projects. At present there are 566 published genome projects. ACEDB is the first genome database of *C. elegans*.

Proteome includes all the proteins complement encoded by a genome. All current genome sequencing strategies are clone based. There are 2 ways to assemble to consequence of a genome. One is called whole genome shotgun sequencing and the other is called ordered clone sequencing. The molecular landmarks on linkage maps are several molecular variants such as RFLPs, STS and VNTRs. The proteomic methods discussed in this chapter are protein electrophoresis and mass spectrophotometry.

Application of bioinformatic tools means deriving information from genomic sequences. These include determination of docking sites, ORF delection, direct evidence from cDNA sequences, using polypeptide and DNA similarities, codon bias prediction, comparative and functional genomics.

EXERCISE

1. In sequencing 1 million genome with fivefold coverage by using BACs, how many clones would you be handling? Assume that an average BAC contain 200 kb.
2. What is docking element? Is a bacterial operator a docking element? In what sense a codon with an ORF is a docking element?
3. A certain cDNA of size 2 kb hybridized to eight genomic fragments of total size 30 kb and contained 2 ESTs. The ESTs were also found in two of genomic fragments each of size 2 kb. Sketch a possible explanation for this result?
4. What is comparative genomics? A sequenced fragment of DNA in *Drosophila* was used in BLAST search. The best match was to a kinase gene from *Neurospora*. Does this mean that *Drosophila*. Sequence contain a kinease gene? If yes what further information compartive genomics would provide about this data?
5. You have the following sequence reads from a genomic clone of *Drosophila melanogaster* genome. Use these six sequences reads to create a sequence contig of this part of the *D. melanogaster* genome.

   Read 1: T G G C C G T G A T G G G C A G T T C C G G T G
   Read 2: T T C C G G T G C C G G A A A G A
   Read 3: C T A T G C G GG C G A A C T T T T G G C C G
   Read 4: C G T G A T G G G C A G T T C C A A T G
   Read 5: T T G G C C G T G A G G G C A G T T
   Read 6: C G A A C T T T G G C C G T G A T G G G C A G T T C C

   Use these six sequences reads to create a sequence contig of this part of *D. melanogaster* genome.
6. Sometimes cDNAs turn out to be monsters i.e. fusion of DNA copies of two different mRNA accidentally inserted adjacent to each other in the same clone. You suspect that a DNA clone from the nematode *Caenohabditis elegans* to such a monster because the cDNA insert predicts a protein with two structurally domains not normally observed in same protein. How would you use the availability of the entire genomic sequence to assess if this cDNA clone is a monster or not?

18

BIOINFORMATICS OF PROKARYOTES

Prokaryotes includes both bacteria and archaea. During evolution bacteria and archaea have evolved at different times. Prokaryotic bioinformatics is application of informatics tools for extracting useful information both from bacterial and archaeal genome and proteome data. It mainly depends on three branches of study:

1. Genomics is a research field that studies the genome of bacteria and archaea or studies prokaryotes using approaches or technology derived from understanding the bacterial genome and its DNA sequence.

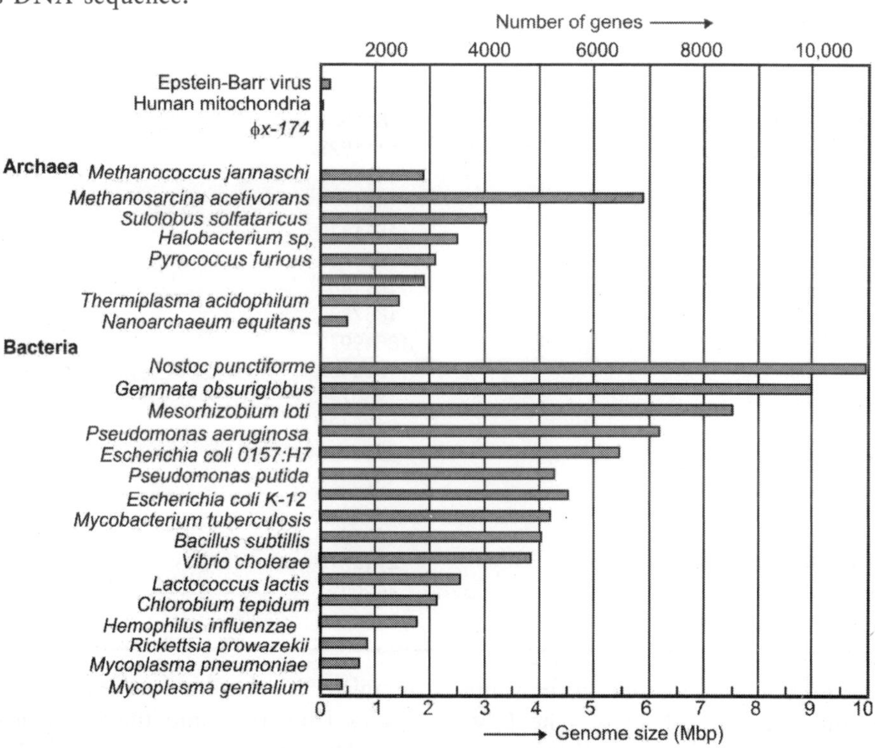

Fig. 18.1: Genome size (Mbp)

2. Prokaryotic proteomics is a study area that mainly deals with proteome of prokaryotes.
3. Prokaryotic nomenclature, chemical pathway, taxonomic and bibliographic databases are also formulated using bioinformatic tools.

Not long ago it was thought that all prokaryotic genomes (both bacteria and archaea) were much smaller than eukaryotic genomes. However, the application of new techniques for constructing physical maps and whole genome sequencing has demonstrated that there is tremendous diversity in the size and organization of prokaryotic genomes. Fig. 18.1 shows some examples of genome size of bacteria and archaea. The size of bacterial chromosomes ranges from 0.6 Mbp to over 10 Mbp, and the size of archaea chromosomes range from 0.5 Mbp to 5.8 Mbp.

The smallest archaea genome identified far is from *Nanoarchaeum equatns*, a tiny obligate symbiont with a genome size of 0.491 Mbp. This organism lacks genes required for synthesis of liquids, amino acids, nucleotides, and vitamins, and hence must grow in close association with another organism which provides these nutrients. The smallest bacterial genome identified thus far is from *Mycoplasma genitalium*, an obligate intracellular pathogen with a genome size of 0.58 Mbp. *M. genitalium* is restricted to the intracellular niche because it lacks genes encoding enzymes required for amino acid biosynthesis and the peptidoglycan cell wall, genes encoding TCA cycle enzymes, and many other biosynthetic genes. In contrast to such obligate intracellular bacteria, free-living bacteria must dedicate many genes toward the biosynthesis and transport of nutrients and building blocks. The smallest free-living organisms have a genome size over 1 Mbp.

Table 18.1: Genome sizes of different bacteria

	Organism	*Number of Base pairs*	*Number of Genes*
1.	ϕX–174	5386	10
2.	Human mitochondrion	16569	37
3.	Epstein-Barr virus (EBV)	172282	80
4.	*Mycoplasma genitalium*	5800t0	468
5.	*Mycoplasma pneumoniae*	816394	630
6.	*Rickettsia prowazekii*	1111523	878
7.	*Treponema pallidum*	1138011	1039
8.	*Borrelia burgdorferi*	1471725	1738
9.	*Aquifex aeolicus*	1551335	1749
10.	Thermoplasma acidophilum	1564905	1509
11.	*Campylobacter jejuni*	1641481	1708
12.	*Helicobacter pylori*	1667867	1589
13.	*Methanococcus jannaschii*	16649970	1783
14.	*Hemophilus influenzae*	1830138	1738
15.	*Thermotoga maritima*	1860725	1879
16.	*Archaeoglobus fulgidus*	2178400	2437
17.	*Deinococcus radiodurans*	3284156	3187
18.	*Synechocystis*	3573470	4003
19.	*Vibrio cholerae*	4033460	3890
20.	*Mycobacterium tuberculosis*	4411529	4275
21.	*Bacillus subtill*	4214814	4779
22.	*Escherichia coli*	4639221	4406
23.	*Pseudomonas aeruginosa*	6264403	5570

Mycoplasma genitalium is one of the simplest organism. Its genome was sequenced in 1995 by a collaboration of groups at TIGC The Johns Hopkins University and the University of North Carolina. The genome is dense in coding regions and 468 genes have been identified. Introns are absent here.

M. genitaliun so far is the closest we have to a minimal organism, the smallest capable of minimal life. Viruses in contrast require the cellular machinery of their hosts. Virus genome is much smaller and is in the range of organelle genome (Fig. 18.1). They are of a particular interest because of their application in pathogen-host interaction. Epstein-Bar virus which causes mononucleosis has genome in the range of 172282 and contains 80 genes. Details of fx-174 a virus infecting *E. coli* and human mitochondrial DNA are given in Table 18.1 for comparison.

The micro-organism *Methanococcus jannaschii* belongs to the archaea, one of the three major divisions of the life along with the bacteria and eukarya (chapter 14). The archaea comprise group of prokaryotes adapted to extreme environmental conditions such as high temperature and pressure, or high salt concentration.

Genome of *M. jannaschii* was sequenced in 1996 by TIGR. It was the first archael genome sequenced. It contains a large chromosome containing a circular double stranded DNA molecule 1.6 MB and two extra chromosomal elements of 58407 and 16550 bp respectively. There are 1743 predicted coding regions of which 1682 on the chromsome and 44 and 12 on large and small extra chromosomal elements respectively. Some RNA genes contain introns.

Genomics of prokaryotes is different in a number of ways from genomic studies of eukaryotes. Genome size of a bacterium is much smaller to that of unicellular eukaryotic genome. A yeast may have a genome size of 1.3×10^7 bp which still above the size of largest bacterial genome. For comparison, eukaryotic chromosomes range from 2.9 Mbp (*Microsporidia*) to well over 4,000 Mbp, although the largest genomes are littered with a tremendous amount of repetitive junk.

Gene order in prokaryotes is conserved to a much lesser extent than protein sequences. Only several operons. primarily that code for physically interacting proteins, are conserved in all or most of the bacterial and archaeal genomes. Nevertheless, even the limited conservation of operon organization that is observed can provide valuable evolutionary and functional clues through multiple genome comparisons. In most pairwise genome comparisons, < 10% of the gene string ranged from < 5% for the cyanobacterium *Synechocystis* sp to 24% for the minimal genome of *Mycoplasma genitalium*, and 23% in *Thermotoga maritima.* The coverage of the archaeal genomes was only slightly lower than that of bacterial genomes. The majority of the conserved gene strings are known operons, with the ribosomal superoperon being the top-scoring string in most genome comparisons.

A feature often used to distinguish prokaryotic and eukaryotic genomes is the amount of junk DNA because large amounts of introns and repetitive sequences are present (see chapter 20). Noncoding DNA is probably largely remnants of genes that have been lost over the course of evolution. As a general rule, prokaryotes tend to have very little junk DNA (typically less than 15% of the genome) and eukaryotes have substantial amounts of junk DNA. However, genome sequences of certain bacteria with very limited ecological niches (e.g., the human pathogen *Mycobacterium leprae*) indicates that, like in eukaryotes, a substantial amount of bacterial genomes can consist of junk DNA. In Table 18.1 genome size and gene number for few prokaryotic and eukaryotic members is given. Genome sequencing show that there are 500 to 1200 genes in parasitic bacteria, 1500 to 7500 genes in free living bacteria and 1500 to 2700 genes in archaea.

In contrast to the linear chromosomes found in eukaryotic cells, the strains of bacteria initially studied were found to have single, covalently closed, circular chromosomes. The circularity of the bacterial chromosome was elegantly demonstrated by electron microscopy in both Gram-negative bacteria (such as *Escherichia coli*) and Gram-positive bacteria (such as *Bacillus subtilis*).

Bacteria plasmids were also shown to be circular. In fact, the experiments were so beautiful and the evidence was so convincing that the idea that bacterial chromosomes are circular and eukaryotic chromosomes are linear was quickly accepted as a definitive distinction between prokaryotic and eukaryotic cells. However, like most other distinctions between proaryotic and eukaryotic cells, it is now clear that this dichotomy is incorrect. Not all bacteria have a single circular chromosome: some bacteria have multiple circular chromosomes, and many bacteria have linear chromosomes and linear plasmids.

Some bacteria have multiple chromosomes others might have linear chromosomes. *Borrelia* have linear chromosomes and most strains contain both linear and circular plasmids: most of the bacteria in the genus *Streptomyces* have linear chromosomes and plasmids and some have circular plasmids as well. In addition, in some cases there may be a dynamic equilibrium between linear and circular forms of a DNA molecule. Linear chromosomes and plasmids were not discovered in bacteria until relatively recently. The first published evidence for linear chromosomes was in 1979, but because the techniques used at that time were limited and because the dogma that all bacterial chromosomes are circular was so entrenched, few people believed that linear chromosomes and plasmids occurred in bacteria until 1989. By that time pulsed field gel electrophoresis had been developed, and this new technique provided convincing evidence that the chromosome of *Borrelia burgdoferi* was linear.

The important take-home point is that we are just beginning to appreciate the similarity of many processes once thought to be completely different between bacteria and eukaryotes, partly because we now have better tools for studying these processes and partly because most of the earlier studies focused on relatively few types of bacteria. The more we study a wider diversity of bacteria, phages, and plasmids, the more obvious it becomes that *E. coli* is an excellent model for dissecting broad features of molecular and cell biology, but not all bacteria do everything the same way. Furthermore, we have only recently begun to attack the molecular genetics of the archaea, and what we have learned so far suggests that this diverse group of prokaryotes share even more common features with the eukaryotes.

Since a bacterial genome contains all the information for building and running a cell, there has been considerable interest in determining the DNA sequence for the genomes of bacteria, particularly for those bacteria that cause disease. We hope that by knowing the code for what makes bacteria tick, that we may be better able to come up with means to control them (for example, if they cause disease), maintain them (for example, if they are important for our environment), use them for beneficial purposes (for example, industrial spill cleanup or making products).

For bacteria, genome sequences had been obtained for approximately 35 different bacteria by the turn of the millennium, with the first genome sequence for a bacteria determinated in 1995. In 1995, a team of researches helped by Craig venter at Institute for Genomic Research in Geithersberg, Maryland and Hamilton Smith at Hoptium obtained first complete DNA sequence of the bacterial genome *Hemophilus influenzae*. (see Historical Profiles - 19 and 20). Many more bacterial genomes are being sequenced and so databases have been formed to contain all the DNA sequence information (as long strings of A's, T's, C's and G's). In addition, early research determined what triplets of the DNA bases encode what amino acids, and so that we can predict from a bacterial genome sequence what proteins it may encode. Both the DNA sequence of bases, and the protein sequences of amino acids, contains additional "strings" of sequence within the DNA or protein that specify particular functions and/or act as signals for the bacteria.

Bacterial genomic approaches to studying life include determining the DNA sequence of a given bacterial genome, comparing the DNA sequence with other genome sequences, and deciphering what the genomes encode. The whole gene complement of different organisms can be compared with one another, detecting, for example, genes specific to certain disease-causing bacteria that may be targets for drug therapy or for diagnostics. This field of research also attempts to study bacteria on a whole-organism scale. It takes a more holistic view regarding how the genes and proteins function and work together, or what genes or proteins are necessary for a certain function (such as causing disease). This whole-organism approach is quite different from the early approaches in bioinformatic used to study genes and proteins. Classically, single genes or proteins were studied on a (relatively) individual basis, with little understanding of how all the genes and proteins in a cell worked together. Genomics attempts to gather information regarding how the full complement of genes in an organism work together and it is thought that such an approach to studying bacteria, and in fact all life, will complement the classical, but still very necessary, studies that are focused on particular genes or proteins.

Bacterial bioinformatics can give us a broader understanding of how a bacteria functions, a bacteria's origins, and what bacteria live in our world that we can't study by other means (i.e., through obtaining their DNA from the environment and studying it). Of medical interest, bacterial genomics is also anticipated to play a significant role in speeding up the development of better therapies and vaccines for controlling disease-causing bacteria. It will also be the cornerstone of anticipated DNA-based diagnostic tools that will hopefully enable doctors to make quicker, more accurate diagnoses of infectious disease.

18.1 GENOME DATABASES

18.1.1 *E. coli* Databases

The *E. coli* Genome Center is a laboratory of the Genetics Department, College of Agricultural and Life Sciences, at the University of Wisconsin. This is the complete sequence of *E. coli K-12* genome and several of the *E. coli* phages. The data is being analysed by several bioinformatic tools. GenProtEC is another database dedicated to *E. coli* genome and proteome. It aims to provide biochemists with the most updated and consolidated information about *E. coli* genes and proteins resulted from both the traditional experimental research and our computational analysis. A database on transcriptional regulation in *E. coli*. Regulon DB is a database that integrates biological knowledge of the mechanisms that regulate the transcription initiation in *E. coli*, as well as knowledge on the organization of the genes and regulatory signals into operons in the chromosome. The operon is the basic structure used in RegulonDB to describe the elements and properties of transcriptional regulation. The current version contains information around 500 regulatory mechanisms, mainly for sigma 70 promoters. Vital statistics of *E. coli* genome is genes in Table 18.2. This was published in 1997 by the group of F. Blattner at University of Lalisconsis.

Table 18.2: Vital statistics of genome of *E. coli* strain MGIGSS

Genome size	4.6 Mb
Chromosomes	1, circular
Number of genes	4000
Percentage with human homologs	8%
Average gene size	1 kb, no introns
Transposons	Strain specific, ~ 60 copies per genome
Genome sequenced in	1997

18.1.2 *Mycobacterium* Databases

Purpose of *Mycobacterium* spp. genome data is to collate and integrate various aspects of the genomic information from *M. africanum, M. bovis, M. Bovis* BCG, *M. canetti, M. microti,* and above all, *M. tuberculosis*. TubercuList provides a complete dataset of DNA and protein sequences derived from the paradigm strain *M. tuberculosis* H37Rv. linked to the relevant annotations and functional assignments. It allows one to easily browse through these data and retrieve information, using various criteria (gene names, location, keywords, etc.).

18.1.3 *Bacillus* Database

Bacillus subtilis genome database is a database dedicated to the analysis of the *Bacillus subtilis* genome. The purpose of this database is to integrate various aspects on the genomic information of *B. subtilis* and, it provides a clean dataset of non-redundant DNA sequences of *B. subtilis* (strain 168), associated to relevant annotations and protein sequences. It allows one to easily browse through these data and retrieve information, using various criteria (gene names, keywords, location, etc.). The data contained in SubtiList originates mainly from the *B. subtilis* genome sequencing project, but this dataset also benefits from the *B. subtilis* entries present in the EMBL/GenBank/DDBJ databanks. MICrobial Advanced Database Organization is another *Bacillus* databases. The database is primarily devoted to the *Bacillus subtilis* genome sequencing program. It links the genetic map of the microbe with its sequences, together with other bacteria. DNA comes from primary databank entries, plus data from the SubtiList database.

18.1.4 *Hemophilus influenzae* Genome Database

The *Hemophilus influenzae* Rd genome is the first genome of a free living organism to be completely sequenced. This page offers access to the latest versions of the sequence data and related annotation.

18.1.5 *Mycoplasma genitalium* Genome Database

This page offers access to the latest versions of the sequence data and related annotation.

18.1.6 Photosynthetic Bacteria Database

Cyanobase provides an easy way of accessing the sequence and all-inclusive annotation data through image maps, keyword searches and the gene category list. The cyanobacterium carries a complete set of genes for oxygenic photosynthesis, which is the most fundamental life process on the earth. This organism is also interesting from an evolutional viewpoint, for it was born in a very ancient age and has survived in various environment. Chloroplast is believed to have evolved from cyanobacterial ancestors which developed in endosymbiontic relationship with a eukaryotic host cell.

18.1.7 Archaeal Database

Methanococcus Jannaschii Functions database provides an updating of the functional content of the first completely sequenced Archaeal genome of *Methanococcus jannaschii.*

18.2.8 General Prokaryotic Databases

Genome Information Broker for Microbial Genomes: The GIB holds information on the following genomes: *Saccharomyces cerevisiae, Aquifex aeolicus, Bacillus subtilis, Borrelia burgdorferi, Chlamydia trachomatis, Escherichia coli, Hemophilus influenzae, Helicobacter pylori, Mycobacterium tuberculosis, Mycoplasma genitalium, Mycoplasma pneumoniae, Synechocystis PCC6803, Treponema pallidum, Archaeoglobus fulgidus, Methanobacterium thermoautotrophicum, Methanococcus jannaschii. and Pyrococcus horikoshii.* The following services and views of the data are available: Genomic View: displays genome information in diagram, Retrieve Clone: retrieves clone information, Retrieve ORF: retrieves ORF information, Retrieve Gene: retrieves gene information. Thumbnail sketch of this server: illustrates a brief overview of the server system. Genome sequence FTP service: Allows you to obtain nucleotide sequences. TIGR Microbial Database: TIGR Microbial Database provides a listing of microbial genomes completed and in progress. HOBACGEN Homologous Bacterial Genes Database: HOBACGEN is a database system that contains all the protein sequences of bacteria organized into families. It allow one to select sets of homologous genes from bacterial species and to visualize multiple alignments and phylogenetic trees. Thus HOBACGEN is particularly useful for comparative genomics, phylogeny and molecular evolution studies on bacteria.

EXPERIMENT 18A

To Study the Occurrence and Structure of Gene named DNA across the Genome of *E. coli* using Genome Study Tools at CMR Microbial Website

About CMR: The Comprehensive Microbial Resource (CMR) is a tool that allows the researcher to access all of the bacterial genome sequences completed to date. For each genome not sequenced at TIGR two kinds of annotation are displayed: the Primary annotation taken from the genome sequencing center and the TIGR annotation generated by an automated annotation process at TIGR. To get to the tools available on the CMR click on the menu options in the lower right-hand corner of the header at the top of the page. Each menu option takes you to a listing of the tools available under that main category.

It is currently partially funded by a Microbial Sequence Center (MSC) grant from the National Institute of Allergy and Infectious Diseases (NIAID). CMR is a free website used to display information on all of the publicly available, complete prokaryotic genomes. In addition to the convenience of having all of the organisms on a single website, common data types across all genomes in the CMR make searches more meaningful, and cross genome analysis highlight differences and similarities between the genomes. Below is an explanation and link for each of these menu options.

1. Genome Tools: Find organism lists as well as summary information and analyses for selected genomes.
2. Searches Search: CMR for genes, genomes, sequence regions, and evidence.
3. Comparative Tools: Compare multiple genomes based on a variety of criteria, including sequence homology and gene attributes. SNP data is also found under this menu.
4. Lists: Select and download gene, evidence and genomic element lists.
5. Downloads: Download gene sequences or attributes for CMR organisms, or go to our FTP site.
6. Carts-Select genome preferences from our Genome Cart or download your Gene Cart genes.

Several genome-comparing tools are available. These are:

1. Protein Homology Tools:
   (*i*) Multi-Genome Homology Comparison: Shows the number of proteins in a reference genome that have hits to up to 15 selected comparison genomes. Similarity between proteins is based on All vs All Data. This tool produces a circular display to allow the user to see the protein the genomes have in common. (*Note*: This tool was formerly called Genome v/s Genome Protein Hits).
   (*ii*) Genome Homology Graph: Shows the number of protein hits a reference genome has in common with all of the genomes in the CMR (not including itself), based on blast searches.
   (*iii*) Protein Scatter Plot: Compares the number of proteins hits between a reference and comparison organism in a scatter plot.
   (*iv*) Protein Scatter Plot: Compares the %GC content between a reference and comparison organism.
   (*v*) Region Comparison: This tool takes the protein encoded for by a locus and finds similar proteins in other organisms on the CMR using the v/s All search.
2. CMR Pathway Tool:
   (*i*) Genome Properties: Provides information on the characteristics of organisms derived from genomic data and literature sources.
   (*ii*) Genome Properties Detailed Comparison: Displays detailed step information for the selected set of organisms and property or set of properties.
   (*iii*) KEGG Pathway Display: Highlights pathway steps based on the presence or absence of supporting EC evidence in the CMR.

3. Alignment Comparison Tools:
   (*i*) Role Category Bar Chart: Shows the number and percentage of genes in the CMR assigned to each of the role categories.
   (*ii*) Terminator Bar Chart: Shows the number and percentage of genes in each genome predicted to have Rho-independent terminators.
   (*iii*) Multi-genime Query: Queries across all genomes in the database based on various type of search criteria.

Hypothesis: It is assumed that DNA A is query common word for genes in prokaryotes and there would be several genes with this name in *E. coli* genome.

TESTING THE HYPOTHESIS

Computer set up: A normal PC with internet connection in windows environment:

1. Log on to CMR website at http"// cmr. jcvi. org.tigr-scripts/ CMR/CMRhomepage.cgi. This takes you to the welcome page as shown in Fig. 18.2.
2. Enter DNA in search text box.

Fig. 18.2: Web view of CMR

3. This gives result of 763 entries as shown in Fig. 18.3. A portion of this is shown in Data Box 18.1.

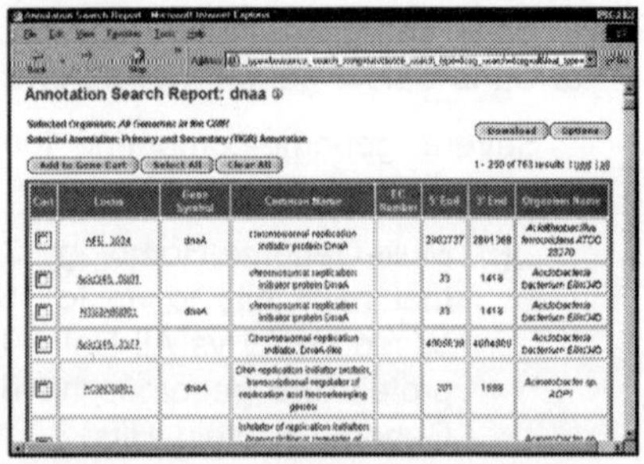

Fig. 18.3: Obtained by CMR

DATA

Data Box 18.1

Selection	Locus	*Gene* Symbol	*Common* Name	*EC* Number	5' End	3' End	*DNA* Molecule
	NT01EL2163	DNAA	chromosomal replication initiator protein DNAA		2142449	2143858	chromosomal Erythrobacter litoralis HTCC2594
	NT12EC2575	DNAA	chromosomal replication initiator protein DNAA		2616903	2616157	chromosomal Escherichia coli 536
	ECP_3903		chromosomal replication initiator protein DNAA		4068157	4066823	chromosomal Escherichia coli 536

INTERPRETATION OF DATA

Searching CMR database gave 763 entries which contained the word DNAA. The first entry was of *Erythrobacter litoralis* where its common name is replication initiator protein dna A gene which is located on the chromosomal DNA. It's also reported in several of *E. coli* entries. it is doing the same function in other entries.

SUMMARY

Bacterial bioinformatics is application of information resources for entreatign useful information and proteome data of bacteria. The size of bacterial chromsomes range for to 10 Mbp. Bacterial genome is different from eukaryotic genome in a number of ways.

EXERCISE

1. If you were given a sample of chromosomal DNA and were asked to determine if it is prokaryotic or eukaryotic, what experiment would you perform and what would be your expected results.
2. The chromosome of *E. coli* contains 4.7 million base pairs. How long will it take to replicate its DNA; Assuming that pol III is the primary enzyme involved, this enzyme can actively proofread during DNA synthesis, how many base pair mistakes will be made in one round of DNA replication in bacterial population containing 1000 bacteria?

   Hint: Rate of Syn by pol II is 850 nucleotide per second and is bidirectional error rate is 1 in 100,000,000 NTP synthesis.

19

FUNGAL BIOINFORMATICS

Fungi are traditionally defined as eukaryotic, spore producing, achlorophyllous organism with absorptive nutrition that generally reproduce both sexually and asexually and whose usually filamentous branches structure is called hyphae. These lower eukaryotic particularly unicellular fungi spend greatest part of their life cycle in haploid state and have been extensively employed in genetics and genomic studies. The fungal kingdom includes some of the earliest cultivated organisms for food production and some of the most damaging pathogens to impact society. The production beer, wine bread and many cheeses requires fungal metabolism. Fungi have also critically influenced the developed of modern medicine. The antibiotic penicillin and the cholesterol lowering agent lovastatin are from fungi *Penicillium notatum* and *Aspergillus terreus* respectively, are examples of medically significant fungal metabolites. Fungal metabolites can also be potent toxins *e.g.*, the food and feed molds *Aspergillus flavus* and *A. parasiticus* produce the carcinogenic agent aflatoxin and corn pathogen *Fusarium moniliforme* produces a toxin that causes serious and often fatal health problem in humans and animals. Many fungi are phytopathogens and causes severe loss of major agricultural crops. Fungi are also the agents an increasing number of opportunist human infections.

Originally fungi were classified with plants because of presence of cell wall in their cells. Taxonomic classifications of fungi have always been controversial because some of their characters resemble to plants and others to animals. In an attempt to recognize monophyletic groups (groups that contain an ancestor and all its descendants) the organism once classified as fungi now are considered in three different groups, the monophyletic kingdoms, fungi and stramenophila and four protist phyla. A phylogenetic tree drawn primarily on small subunit of ribosomal DNA sequence analysis data and depicts fungi and animals are monophyletic. Chytidiomycota, Zygomycota, Ascomycota and Basidiomycota form an inclusive group with ancestor represented by the node that joins them are included in one monophylogenetic group called Kingdom fungi. Inclusion of Oomycota and slime molds in fungi make it polyphyletic. Because of this reason phylas Oomycota, Hyphochytiomycota, and Labrinthulomycota are included in a separate kingdom called Stramenophila. The slime molds do not form a monophyletic group and are problematic in the classification. Major groups of organisms covered in the general term fungi are given in Fig. 19.1. In Fig. 19.1a the phylogenetic tree that emerged as rDNA sequence analysis is drawn. In Fig. 19.1b the major groups that emerged as a result of new phylogenetic classification of fungi as given by Alexopaulas, Mims and Blackwell in 2002 is summarized. This classification recognizes that organisms once classified, as fungi are all not closely related organisms.

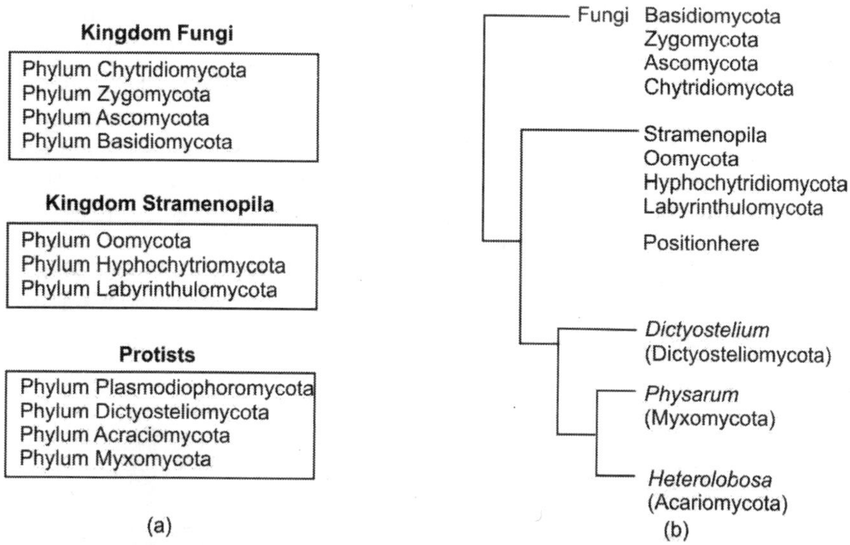

Fig 19.1: (a) Major groups of organisms that have once been classified as fungi enclosure in one box means monophyletic (b) Phylogenetic tree based on rDNA sequence analysis showing polyphyly of organisms once classified as fungi

19.1 RATIONALE FOR FUNGAL GENOMICS AND BIOINFORMATICS STUDIES

The term 'fungi' include a group of very diverse organisms. Fungal genomics and bioinformatics when approached in a kingdom-wide manner that is, by selecting a balanced collection of fungi (rather than choosing individual fungi in isolation) that maximizes the overall value for comparative genomics, evolutionary studies, eukaryotic biology, and medical studies. The sequence of each organism should not only enable research in a specific research community, but should enhance the value of each additional sequence for comparative studies of the evolution of genes, chromosomes, and regulatory and biochemical pathways.

A fungal sequencing program represents an ideal system for establishing the power and limits of comparative studies because:

1. Fungi usually have small genome size with an average of 30 MB (2 times that of yeast, 10 times that of *E. coli*, 1/100 that of mammals) and chromosome number range between 7 to 20. The small genome sizes allow for the comparison of many complete eukaryotic genomes for small cost and effort. In Fig. 19.2 genome size of various fungal organism is depicted. For comparison on left-hand size genome size of *Mycoplasm* and *E. coli* is shown and on right side genomes of *Arabidopsis* and Drosophila are depicted.
2. The fungal kingdom displays extraordinary diversity, with more than 1 million different representative species.
3. Genomes can be selected representing a wide variety of evolutionary distances, ranging from as little as 5 million years to approximately 1 billion years.
4. Genomes can be selected representing specific branch points in the phylogenetic tree to illuminate the molecular basis for key biological innovations.
5. Fungi offer outstanding opportunities to study natural populations and evolution. For example, there are over 4000 well-characterized natural isolates of *Neurospora* deposited at the Fungal

Genetics Stock Center, taken from widespread and ecologically diverse regions. These represent five heterothallic species, four that must out-cross and one that can effectively self-cross.

6. Fungal genome is an ideal tool for comparative genomics. Analyses of representative genomes distributed across the fungal tree are expected to provide the molecular basis for understanding the extraordinary diversity that has arisen over the estimated 1 billion years since divergence from a common ancestor. The ability to compare genomes that represent a variety of evolutionary distances and relationships is exactly what is needed to build effective tools for comparative analysis and to define what are the appropriate evolutionary distances for answering different questions. Such comprehensive studies are difficult to accomplish with large mammalian genomes. Our ability to perform them rapidly in fungi will establish the foundation for comparative genomics and inform optimal choices when sequencing capacity exists for broad, deep sampling of these larger genomes.

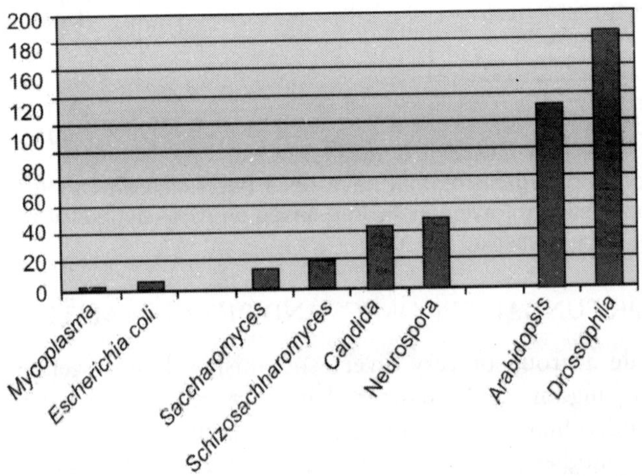

Fig. 19.2: Genome size range of several fungi in comparison to other prokaryotes and eukaryotes

Although *Sachharomyces cerevisiae* provides key insights into the function of many human proteins, the small size of its genome limits its value as a tool for this purpose. A deeper sampling of fungal genes will rapidly increase the number of human proteins for which we can access homologues in model organisms. A recent review indicates that homologues exist for approximately 30 per cent of all human proteins within the limited available fungal sequence data. This is more than twice the number of similar proteins found by examining the genome of *S. cerevisiae* alone. Genome sequence from many diverse fungi, coupled with comparative and functional genomics, will advance our understanding of the eukaryotic proteome, which in turn will help define what it means to be a eukaryote. In doing so, we will learn not only how to manipulate fungi for our own purposes, but how to therapeutically intervene to manipulate human physiology for the treatment of metabolic and infectious diseases.

19.2 FUNGAL GENOME DATABASES

The first eukaryotes chromosome ever sequenced was a fungal chromosome. It was chromosome number 3 of yeast in 1992 when first chromosome was sequenced and gave birth to this new promising field of genomics and bioinformatics. Between 1994 and 1996, all 15 other yeast chromosomes were completed and made public by 16 different teams of the international community of 'yeast scientists'. This series was only interrupted in mid-1995 by the publication of the bacterial chromosomes from *Hemophilus infuenzae* and *Mycoplasma genitalia*. In the Table 19.1 the breakthrough research in field of fungal Bioinformatics is given.

Table 19.1: Breakthrough research in fungal bioinformatics

1. In 1992 first fungal chromosome i.e. chromosome number 3 of *Saccharomyces* was sequenced. This was a historical ever as this was also the first chromosome ever sequenced.
2. In 1996 first complete fungal genome was sequenced. This was genome of the yeast, *Saccharomyces cerevisiae* that was sequenced through a worldwide collaboration. It was also the first example of knowing complete genome sequence of a eukaryote.
3. On 24 April 1996 the total yeast genome was put on the web by MIPS (Munich information center for protein sequences) in a fully organized and annotated form.
4. In early 2002 the Welcome Trust Sanger Institute, sequenced and annotated the genome of fission yeast, *Schizosaccharomyces pombe*, which contained the smallest number of protein-coding genes yet recorded for a eukaryote: 4,824.
5. In 2004 the genome of first filamentous ascomycetes was sequenced and annotated. It provided compelling evidences that the evolution of *S. cerevisae* includes a whole genome duplication or fusion of two related species.

Sequence information for fungi is accumulating. List of fungi whose genome is already sequenced or is in the process of being sequences is increasing every day. Fungi genome database is available for human pathogens (*Aspergillus fumigatus*, *Candida albicans*, *Cryptococcus neoformans*, and *Pneumocystis carinii*), crop pathogens (*Botrytis cinerea, Fusarium sporotrichioides*, *Magnoporthe gresea*, *Mycosphaerella graminocola Phytophthora infestans* and *Phytophthora sojae*), toxigenic fungi (*A. flavus, A. parasiticus*), a biodegrading fungi (*Phanerochaete chrysosporium*), and model fungi (*S. cerevisiae, Aspergillus nidulans, Neurospora crass, Schizosaccharomycse pombe*) (Fig. 19.3). This accumulate at such a fast speed that there is an urgent need to analyze and make sense out of this data. These sequencing projects represent species from Ascomycetes and Basidiomycetes mainly. Common fungal databases are given Table 19.2. Description of genomes and databases from few representative members is dealt in details here.

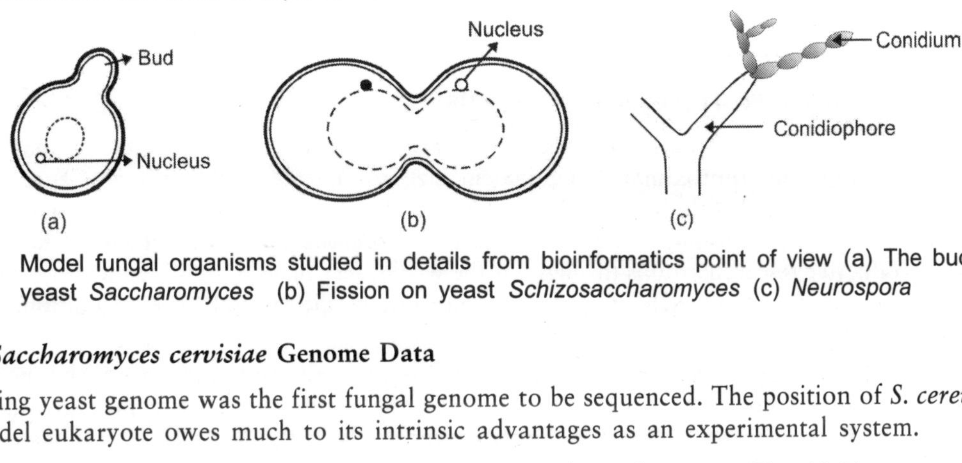

Fig.19.3: Model fungal organisms studied in details from bioinformatics point of view (a) The budding yeast *Saccharomyces* (b) Fission on yeast *Schizosaccharomyces* (c) *Neurospora*

19.2.1 *Saccharomyces cervisiae* Genome Data

The baking yeast genome was the first fungal genome to be sequenced. The position of *S. cerevisiae* as a model eukaryote owes much to its intrinsic advantages as an experimental system.

1. It is a unicellular organism, unlike many more complex eukaryotes (Fig. 19.3).
2. It can be grown on defined media, which gives the experimenter complete control over its chemical and physical environment.
3. *S. cerevisiae* has a life cycle that is ideally suited to classical genetic analysis.
4. This has permitted construction of a detailed genetic map that defines the haploid set of 16 chromosomes.
5. Moreover, very efficient techniques have been developed that permit any of the 6000 genes to be replaced with a mutant allele, or completely deleted from the genome, with absolute accuracy.

6. The combination of a large number of chromosomes and a small genome size meant that it was possible to divide sequencing responsibilities conveniently among the different international groups involved in the project. Fungi in general have a much smaller genome. In the Fig. 19.2 shows comparative genome size of fungi with eukaryotes.

Table 19.2: Fungal genome database available

Species	*Genome database*	*Name and URL*
1. *Saccharomyces cerevisiae*	*Saccharomyces*	http:/www.yeastgenome.org
2. *Schizosaccharomyces pombe*	*S. pombe*: Gene DB	www.genedb.org/genedb/pombe
3. *Ashbya gossypii*	Ashbya genome database	http://agd.unibas.ch/index.html
4. *Neurospora crassa*	*Neurospora crassa*	http:/www.braod.nit.edu/annotaton/genome/ neurospora/home.html

The genome of the yeast *Saccharomyces cerevisiae* has been completely sequenced through a worldwide collaboration of 600 scientists in Europe, North America and Japan. Goffaû and coworkers published the results (Historical Profile 21) in 1996 and in the same year the database was made public by MIPS (Munich information center for protein sequences). The sequence of 12,068 kilobases defines 5885 potential protein-encoding genes, approximately 140 genes specifying ribosomal RNA, 40 genes for small nuclear RNA molecules, and 275 transfer RNA genes. In addition, the complete sequence provides information about the higher order organization of yeast's 16 chromosomes and allows some insight into their evolutionary history. The genome shows a considerable amount of apparent genetic redundancy, and one of the major problems to be tackled during the next stage of the yeast genome project is to elucidate the biological functions of all of these genes.

Besides the nuclear genome, mitochondrial DNA from one strain of yeast is also sequenced. The yeast mt DNA is more than 10 kb.

19.2.2 *Schizosaccharomyces pombe* Genome Data

The fission yeast *Schizosaccharomyces pombe* is also a unicellular ascomycete. It has a simple eukaryotic genome of approximately equivalent size to that of budding yeast, *Saccharomyces cerevisiae*, at around 14 Mb (Fig. 19.3). Unlike *S. cerevisiae*, the *S. pombe* genome is spread between only three chromosomes. The fission yeast *Schizosaccharomyces pombe* is the favoured tool of many productive research groups throughout the world, serving as a useful model for fundamental principles and mechanisms, such as genome organization, differential gene regulation, cell-cycle control, signal transduction, or cellular morphogenesis.

The Wellcome Trust Sanger Institute have sequenced and annotated the genome of *Schizosaccharomyces pombe* in 2002, which contains the smallest number of protein-coding genes yet recorded for a eukaryote: 4,824. The centromeres are between 35 and 110 kilobases (kb) and contain related repeats including a highly conserved 1.8-kb element. Regions upstream of genes are longer than in budding yeast (*Saccharomyces cerevisiae*), possibly reflecting more-extended control regions. Some 43% of the genes contain introns, of which there are 4,730. Fifty genes have significant similarity with human disease genes; half of these are cancer related. They identified highly conserved genes important for eukaryotic cell organization including those required for the cytoskeleton, compartmentation, cell-cycle control, proteolysis, protein phosphorylation and RNA splicing. These genes may have originated with the appearance of eukaryotic life. Few similarly conserved genes that are important for multicellular organization were identified; suggesting that the transition from prokaryotes to eukaryotes required more new genes than did the transition from unicellular to multicellular organization.

19.2.3 *Neurospora crassa* Genome Data

The German sequencing project of *Neurospora crassa* chromosome II and V started in 1998. Chromosome II and Chromosome V are estimated to be 4.6 MB and 9.2 MB in length respectively. The total genome has a length of about 43 MB, which is more than 3 times larger than the first completely sequenced eukaryotic genome of *Saccharomyces cerevisiae*. The entire data set has been automatically analysed and annotated by the PEDANT software.

Biologists face an overwhelming richness of nucleotide and protein sequence data. As of the end of 2003, there were over 100 complete or almost complete nonviral genomes in publicly available databases. Most of these were bacterial, since prokaryotic genomes are generally much smaller in size than eukaryotic genomes. Among eukaryotes, fungi have some of the smallest genome sizes and, hence, represent the highest number of complete or almost complete genomes sequenced, with most of these released within the last 2 years. What are the genes that fungi have in common? Among these genes, which ones have homologs in plants, animals, or bacteria, and which ones are only found in fungi? Researchers are just beginning to be able to address these types of questions with data from high-throughput genomic sequencing. Fungal genomic data in comparative genome analyses, particularly is related to the study of fungal plant pathogens. Comparative genomics can facilitate research into phylogenetics (via whole genome comparisons), targeted drugs (via unique target sites in pests), gene discovery (via conserved sequences), and gene function (via guilt by association).

19.2.4 *Ashbya gossypii* Genome Database

The filamentous fungus *Ashbya gossypii* is a currently used in industry for the production of vitamin B2. It is also an attractive model to study filamentous growth, because of its small genome, haploid nuclei, efficient gene targeting, propagation of plasmids and growth on different media. The genome of *A. gossypii* is only 9.2 megabases located on seven chromosomes. This constitutes 4718 proteins, 199 tRNA and at least 49 small nuclear (snRNA) genes. The genome lacks transposons and subtelomeric gene repeates and gene duplications are rare. The number of protein coding sequences.

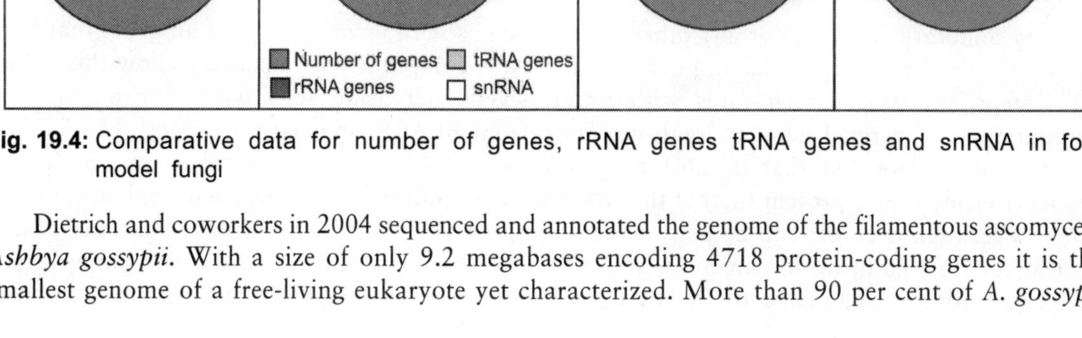

Fig. 19.4: Comparative data for number of genes, rRNA genes tRNA genes and snRNA in four model fungi

Dietrich and coworkers in 2004 sequenced and annotated the genome of the filamentous ascomycete *Ashbya gossypii*. With a size of only 9.2 megabases encoding 4718 protein-coding genes it is the smallest genome of a free-living eukaryote yet characterized. More than 90 per cent of *A. gossypii*

HISTORICAL PROFILE 21

Andrzu Goffaû, the Group Leader to have Completely Sequence Yeast Genome

Andrzu Goffaû has been one of the founders of the Biotechnology Programmes funded by Commission of the European Communities. His prominent achievement was to initiate and coordinate the European Project for Sequencing the Yeast Genome. Thirty-five European laboratories were involved in the sequencing of the entire chromosome III of *Saccharomyces cerevisiae*. This work published in May 1992 was the first chromosomal sequence completed ever. The sequencing of all other yeast chromosomes by a total of 96 European and other laboratories was subsequently completed and made public in April 1996. This was the first Eukaryotic genome sequence to be completed.

Andrzu Gofffaû has recently retired as "Proessor Extraordinaire" from the University de Loubvain in Belgium. His early training was in plant physiology. Further predoctoral and postdoctoral trainings in enzymology, bioenergetics, yeast genetics and molecular biology were provided by long stays in VErsailles (J. Bove), Paris (H. Heslot), Baltimore (A. Lehninger) and in Brussels (J. Brachet). In 1970, he created the Unit of Physiological Biochemistry at the Faculty of Agronomical Sciences of the University of Louvain from which he retired in 2000.

Andrzu Goffaû is a member of EMBO and the National Academies of Belgium, France and Brazil. He has been awarded several honorary degrees and medals such as: Prix Quinquenal Joseph Maisin du FNRS (Belgium), the Hansen Gold Medal for Microbiology (Copenhagen), the Gold Medals from the Universitites of Wroclaw (Poland) and Charles V (Praha), the Honorary Causa Degrees from the University of Manchester Institute of Technology, and from the University Pierre et Marie Curie (Paris 6). He recently was awarded the George Beadle Medal (2002) from the Genetics Society of America for distinguished services to the field of genetics and the community of genetictists.

---

genes show both homology and a particular pattern of synteny with *Saccharomyces cerevisiae*. Analysis of this pattern revealed 300 inversions and translocations that have occurred since divergence of these two species. It also provided compelling evidence that the evolution of *S. cerevisiae* included a whole genome duplication or fusion of two related species and showed, through inferred ancient gene orders, which of the duplicated genes lost one copy and which retained both copies.

Genome of *Ashbya gossypii* is available at Genbank (accession number AE016814 through AE016821) and annotation is available at supporting online material.

19.3 GENOME ANNOTATION FOR FUNGAL GENE DISCOVERY

Fungal gene discovery is divided into similarity searching, comparative genomics and gene prediction. This is an extension of information already given in section 2.

19.3.1 Similarity Searching

Genome annotation of fungi or any other genome begins with identification of all genes that have significant similarity to proteins in public databases. If computational resources allow this is best done stepwise, using search tools with progressively increasing sensitivity. The majority of computationally derived gene annotation comes from BLAST similarity matching. BLAST is a rapid search algorithm that is most frequently used tool in bioinformatics. Any contiguous sequence element or a protein thereof that have BLAST similarity alignment with public database above a reassigned cutoff can be considered annotated and is removed from further analysis. A typical cut for significant alignment is 35% identity over a stretch of at least 100 amino acids. Sequence with an alignment score below this cut off are again searched against public databases with a Smith Waterman (SW) dynamic programming implementation and sequence alignment above a predefined cutoff are selected as annotated. This step requires substantial cut of time unless a special hardware is available. The SW will increase annotation by about 10%. The sequence remaining are searches against a profile Hidden Markov Model (HMM) database such as Pfam. Hidden Markov Model searching method is more sensitive because it incorporates all the information in a multiple sequence alignment in to the search. A limitation with pfam database is that it only represents a fraction of all the proteins.

19.3.2 Comparative Genomics

Typically only about 50% of the predicted genes in a new genome can be typically annotated by performing similarity searches of public databases. The value is still lower for fungi. For example when *N. crassa* genome was searched against public databases less than 40 per cent of the genome could be annotated. For further annotation comparative genomics is can be taken. Comparative genomics can be especially useful for identifying conserved novel genes and for predicting similarity and differences in global biological processes. The database at the genome level consists of complete or extensive genome sequence or non-redundant assembly of EST sequence datasets. Comparison of genome with itself is very useful for identifying genes related by duplication within the genome that have diverged to different functions (paralogs). Comparison with another species genome genes that are accessorily related and have same function. Genome comparison can be used to identify conserved regulatory region.

BLASTN is often used for DNA comparisons of a genome of a related species with a unigene assembly or genome of a related species. Studies of genomics with blasts are discussed in details in chapter 12. The utility of BLASTN is limited however because it is unable to handle large insertions that will arise at intronic regions when genome and EST sequence are compared. BLASTN presents individual exon matches as separate pieces of overall hit and some form of computational post processing of aligned sequences is required for assembly of the complete alignment. Another limitation of BLASTN is its requirement for a contiguous word match. The default state of NCBI BLASTN program uses a minimum word size of 11. This can result in significant matches may be missed if there is a substantial divergence resulting from wobble at third place position of many amino acids codons. WABA overcomes these drawbacks of BLASTN because it handles introns and wobbles mismatches without compromising performance. Wobble aware bulk alignment was used to compare 8 million bp of nematodes *Caenorhabditis briggsae* genome with entire 97-million bp genome of related species *C. elegans*. A large number of uncharacterized open reading frames (ORFs) and regulatory regions are identified. Two species share about 80% of sequence identity with coding region and 30% similarity with non-coding region. This method would be valuable for comparison of fungal species that are related at a similar evolutionary distance to *C. elegans* and *C. brigasse*.

This algorithm compares DNA sequences at protein level by translating both the query to compare genome of divergent organisms TBLASNX could be used. This algorithm compares the DNA sequences at the protein level by translating both the query and the subject sequence in all six reading frames. Each DNA–DNA comparisons requires 36 independent proteins—protein comparisons in order for all pairwise protein comparisons be performed. BLASTN is often informative but it is computationally demanding because of the number of alignment required. Alignment genome sequence and protein sequence and protein from six divergent species can reveal highly conserved gene function and metabolic pathways. When yeast genome was compared with the genome of *C. elegans* the number of proteins responsible for fundamental biological processes are similar in both organisms. Analogous comparative genomic studies of divergent fungal species may identify conserved fungal-specific processes that can be used for development of broad spectrum antifungal compounds with little or no cross reactivity to humans. Toxicity to humans is main shortcoming of current generation of pharmaceutical anti-fungal agents.

19.3.3 Gene Prediction

Sequence similarity searching is a very discovery tool but it may not identify novel genes because they often lack significant similarity to known genes. Coding regions of the genome have features that can facilitate gene identification independent of similarity alignments. Coding regions usually a different alignment because of their protein translation information content. Often associated with coding regions are highly conserved nucleotide are nucleotide element that are critical for gene function. This includes transcription and translation start sites, introns, donors, acceptors and poly adenylation sites, gene prediction programmed has been developed to identify gene-coding regions based on these features. For example the pomes program for *Schizosaccharomycestes* has been used to identify genes based on multiple coding and coding associated features including exon base composition, GC ratio, intron donor, and acceptor site. Gene prediction program that are based on coding region features can be used to identify novel genes because they do not depend on the similarity to a known gene.

EXPERIMENT 19A

Study of Yeast Genome Database and Fungal Gene Annotation Methods

Hypothesis: It is assumed that the given DNA sequence is a sequence derived from a fungal genome. Using the gene annotation tools existing for fungi it would be possible to know that this gene is coding for which function and belongs to which fungi.

Computer set up: A normal PC with windows environment.

1. Have manual look at the sequence study to look for start and stop codons.
2. Log into the web site of Swiss-Prot and select esphet proteomic tool as shown Fig 13.11.
3. Translate the sequence in all six ORFs using translate tool software available at this site. A view of the web site is shown at Fig. 19.5.
4. We obtain a sequence result as shown in Data Box 19.2.
5. Analysis of amino acid sequence in all reading frame indicates that frame are rejected and are not expected to code for a protein. Most probable/ most likely reading frame selected is frame 1 5′ to 3′.

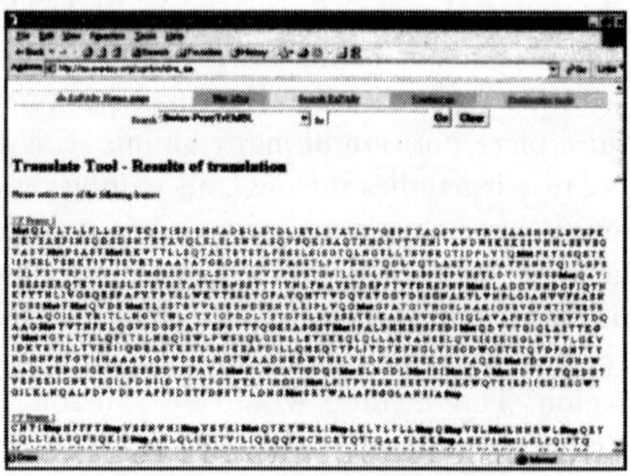

Fig. 19.5: Concept of ORF

6. Log on to web site of yeast genome database at http:/www.yeastgenome.org. BLAST available here multiple sequence alignment with other yeast sequences.
7. Blast is done for multiple alignment of the present amino acid sequence and identify the present sequence (Fig. 19.6). The result is given data Table 19.2. Only alignment to fungal sequences is selected for further consideration.

DATA

Data Box 19.1: The given sequence

ATGCAATTATATCTGACACTTCTTTTTCTATTAAGTTTCGTCGAATGTTCATATATAAGTTTCATATCGAATAATGCAGACGAAA
TATTGGAAACTGATTTAATTGAAACTTTATCTTACGCTACTTTGACAGTAGGTGAGCCTTATGTTGCACAATCGGTGGTTGTAA
CAAGAGTATCTGCAGCTTCTCATAGCCCTCTCTCAGTTTCGCCAAAAAATAGAGTGAGCGCATCTCCAATTAATTCACAAGAC
AGTGATTCTAATACAAGAACAGCAGTTCAACTGTCATTGTCGTTATCAAACTACGCAAGCCAAGTATCTCAAAAAATAAGCGC
ACAAACCAATAATGATCCTGTCACTGTTTCAAATATTTACGCAAATGATAATTCGAAAAGCAAAAGTTCAGTACACAATTTGA
GTTCGGTTTCCGGCGTCGCCTCAGTCATGCCAAGTGCATCAACAATGCGTAAAGTCACCACATTATTATCTCAAACTGCTTCGA
CTTCGACTTCGACTTTATTTTCTTCATCTTTATCTATTTCGGGGACACAATTAAATGGAACTTTATTGACTTCTGTTTCAAAAGG
TACAATCGATCCACTGGTAACTCAAATGCCTTCCTATTCTTCCCAAGAAACAAAAATTATTCCTTCTTCATTGACCTCGAATAA
AACGATATATACTATTTCAGTAAGGACAAATGCAGCTACAGCTACCGGCGAAGATTCCTTCATTGCTTCAACTCCTGCTTCCTC
GACTTTGTTCTATCCATCCAACTCAACCCAGGATTTAGTCCAGACGCTCGCATCTACAACAGCAAGCCCTGCTTACCCATCTAA
TAGGACACAAATCACTCTTTCTCCATCTGTATCACTATATTCGACAACTAGTCCAATTTACCCTTCAAACATTACAGAAAACGG
TTCTTCACCATCACCATCACTATCATCGACTGTTAGTCCAGTTTATCCATCGAGTTCAACAGGAAACATTCTTTTGTCATCACTA
TTTTCAACAGTAGATTCTTCTTCCTCTCCGGTTTCTTCGACATTAGATACCATTTATGTATCTTCTTCAATGCAAGCTACCATCTC
TTCGTCCTCATCTTCGAGACAAACAAAAACTTCCTCCTCTTCTCTGTCAACCTCGACCAGTAGCACCGCTACCACAACGGAGAA
CTCTTCAACAACGACCATTGTCAATCTTTTCAATGCTGTTTCGACAGATGAACCACCAACTGTTTTTGACAGATCACCAAATCC
TATGTCATTGGCTGATGGGGTGTCAAACGATGGTCCCATTCAAACAAACAAGTTTTATACAAATTTGATTGTTGGTAGTCAGGA
ATCACCTGCCTTTGTGTATCCTTATTCATTGTGGAAGTACACCTCGAGTTCATATGGTTTTGCTGTTCAACATACAACAGTAGAT
CAATATAGCTACGGTGGTTATGACAGTTCCGGTAATGCAGAATATTTGGTCAATCCTTTGGGAATTGCTCATGTTGTTTTCTCA
GCTTCGAATTTTGATTCAAGTATGACTATGCAAGTTGATGAAATGACCTTGTCTTCAACAAGGGTAGTGCTATCAGAATCCAAT
GATTCTTCGAACTACCTGGAAATACCTTTGGTTCAAGGTATGGGATTTGCAACTGGTATATATCATGGATCTTTAAATGCCAAG
ATTGGTTCCAGCGTCGGATTCAATACCATTGTTTCTGAATCATCAAGCAATTTAGCTCAGGGTATTCTCAAATATCGTATCACC
CTTTTGAATGGAGTCACTTGGTTATGTTACGTGATAGGACCAGACGATTTAACCTCAACGGACTTTTCATTGGAGGTGAGCTCA
GAATATGAAATTAAAGCAAGCGCTAGTGTTGATGGTCTTATCATTCAATTGGCCGTCGCACCTTCTGAAACTGATTACGAGGTC
TTTTATGATCAAGCCGCAGGTATGTACGTTACTAATTTCAAGCTGCAGGGTGTCTCTGACGGTTCTACAGCTACCTATGAATTT
TCTTATACCACCCAAGGTGAGTCGGCATCAGGTAGCACAATGATTTTTGCCTTGCCCCATCATGAATCTTCATTTAGTGATATA
ATGCAAGACTACTATACCGGCATTCAACTCGCTTCTACAACTAAGGGTGTAATGAACGGTTATCTAACAACGAGCCTACAATT
TTCCACATCTCTAAACAGACAAATATCTTGGTTACCATGGTCTTCTCAACTTGGCTCAAACCTTTTAGAGTATTCTAAAGAACA
ATTGCAATTATTAGCTGAAGTTGCCAATTCCGAATTACAGGTTAGTATTTCCGAAAGTATAAGCGGGTTAAACACATATTATTT
GGGTAAAGTTATCGATAAGTATTCTTACATTCTATTAACGGTCTCCGAAATCATCCAAGATGAAGCCAGTACTAAGAGTACTTT
GGAAAACATCAAATCGGCCTTTGATATTCTGCTACAAAATGAACAGACGTATCCACTAATTTACGACACAAAGTTCAACGGTT
TGGTCAGTTCAGGAGATTGGGGTTCCACCAGCACGCAATACGATTTTGGTAACACTTATTACAATGACCATCATTTCCATTATG
GTTACATTATTCATGCAGCTGCTGTCATCGGCTATGTTGACTCTAAACTGAATGGTACTTGGGCTGCCGACAATAAAGATTGGG
TCAATTCATTGGTAAGAGATGTCGCAAATCCATCTGAAAAAGATGAATACTTTGCACAATCGAGAATGTTTGATTGGTTCAAC
GGTCATTCATGGGCAGCTGGACTTTATGAAAACGGTAACGGTAAGAACGAAGAAAGTAGTAGTGAAGATTACAATTTTGCCTA
TGCTATGAAGTTATGGGGGGCCACTATCGGTGACCAGTCAATGGAATTGAGGGGTGACTTGATGATTAGTATAATGAAGGATG
CAATGAACGACTATTTCTATTATCAAAATGACAACACAGTTGAGCCTGAAGAAATCATAGGAAATAAAGTGAGTGGTATTTTA
TTCGATAATATTATCGATTATACTACTTATTTTGGAACAAACACAGAATATATCCACGGTATTCATATGCTACCTATCACACCA
GTTTCTTCTAATATTCGTTCTGAGACTTTCGTCGAAGAAGAATGGCAGACTAAAATCGAGCCAATTATTGAATCGATAGAAAG
CGGCTGGACAGGCATATTGAAGCTGAATCAAGCACTCTTCGACCCAGTAGATTCGTATGCATTTTTCAGTGATTCAACTTTTGA
TTCATCCACATATTTGGATAACGGAATGAGTCGCACATGGGCATTAGCATTTTCAGGGGGACTGGCCAACTCAATTGCTTAG

Data Box 19.2: Tool-results of translation

5′3′ Frame 1

Met Q L Y L T L L F L L S F V E C S Y I S F I S N N A D E I L E T D L I E T L S Y A T L T V G E P Y V A Q S V V V T R V S A A S H
S P L S V S P K N R V S A S P I N S Q D S D S N T R T A V Q L S L S L S N Y A S Q V S Q K I S A Q T N N D P V T V S N I Y A N D
N S K S K S S V H N L S S V S G V A S V Met P S A S T Met R K V T T L L S Q T A S T S T S T L F S S S L S I S G T Q L N G T L L T
S V S K G T I D P L V T Q Met P S Y S S Q E T K I I P S S L T S N K T I Y T I S V R T N A A T A T G E D S F I A S T P A S S T L F Y
P S N S T Q D L V Q T L A S T T A S P A Y P S N R T Q I T L S P S V S L Y S T T S P I Y P S N I T E N G S S P S P S L S S T V S P V
Y P S S S T G N I L L S S L F S T V D S S S S P V S S T L D T I Y V S S S Met Q A T I S S S S S S R Q T K T S S S S L S T S T S S T A
T T T E N S S T T T I V N L F N A V S T D E P P T V F D R S P N P Met S L A D G V S N D G P I Q T N K F Y T N L I V G S Q E S P
A F V Y P Y S L W K Y T S S S Y G F A V Q H T T V D Q Y S Y G G Y D S S G N A E Y L V N P L G I A H V V F S A S N F D S S
Met T Met Q V D E Met T L S S T R V V L S E S N D S S N Y L E I P L V Q G Met G F A T G I Y H G S L N A K I G S S V G F N T I V
S E S S S N L A Q G I L K Y R I T L L N G V T W L C Y V I G P D D L T S T D F S L E V S S E Y E I K A S A S V D G L I I Q L A V A
P S E T D Y E V F Y D Q A A G Met Y V T N F K L Q G V S D G S T A T Y E F S Y T T Q G E S A S G S T Met I F A L P H H E S S F
S D I Met Q D Y Y T G I Q L A S T T K G V Met N G Y L T T S L Q F S T S L N R Q I S W L P W S S Q L G S N L L E Y S K E Q L Q
L L A E V A N S E L Q V S I S E S I S G L N T Y Y L G K V I D K Y S Y I L L T V S E I I Q D E A S T K S T L E N I K S A F D I L L Q
N E Q T Y P L I Y D T K F N G L V S S G D W G S T S T Q Y D F G N T Y Y N D H H F H Y G Y I I H A A A V I G Y V D S K L N G
T W A A D N K D W V N S L V R D V A N P S E K D E Y F A Q S R Met F D W F N G H S W A A G L Y E N G N G K N E E S S S E D
Y N F A Y A Met K L W G A T I G D Q S Met E L R G D L Met I S I Met K D A Met N D Y F Y Y Q N D N T V E P E E I I G N K V S G
I L F D N I I D Y T T Y F G T N T E Y I H G I H Met L P I T P V S S N I R S E T F V E E E W Q T K I E P I I E S I E S G W T G I L K L
N Q A L F D P V D S Y A F F S D S T F D S S T Y L D N G Met S R T W A L A F S G G L A N S I A Stop

5′3′ Frame 2

C N Y I Stop H F F F Y Stop V S S N V H I Stop V S Y R I Met Q T K Y W K L I Stop L K L Y L T L L Stop Q Stop V S L Met L H N R
W L Stop Q E Y L Q L L I A L S Q F R Q K I E Stop A H L Q L I H K T V I L I Q E Q Q F N C H C R Y Q T T Q A K Y L K K Stop A H
K P I Met I L S L F Q I F T Q Met I I R K A K V Q Y T I Stop V R F P A S P Q S C Q V H Q Q C V K S P H Y Y L K L L R L R L R L Y
F L H L Y L F R G H N Stop Met E L Y Stop L L F Q K V Q S I H W Stop L K C L P I L P K K Q K L F L L H Stop P R I K R Y I L F Q
Stop G Q Met Q L Q L P A K I P S L L Q L L L P R L C S I H P T Q P R I Stop S R R S H L Q Q Q A L L T H L I G H K S L F L H L Y H
Y I R Q L V Q F T L Q T L Q K T V L H H H H H Y H R L L V Q F I H R V Q Q E T F F C H H Y F Q Q Stop I L L P L R F L R H Stop I

P F Met Y L L Q C K L P S L R P H L R D K Q K L P P L L C Q P R P V A P L P Q R R T L Q Q R P L S I F S Met L F R Q Met N H Q L
F L T D H Q I L C H W L Met G C Q T Met V P F K Q T S F I Q I Stop L L V V R N H L P L C I L I H C G S T P R V H Met V L L F N I
Q Q Stop I N I A T V V Met T V P V Met Q N I W S I L W E L L Met L F S Q L R I L I Q V Stop L C K L Met K Stop P C L Q Q G Stop C
Y Q N P Met I L R T T W K Y L W F K V W D L Q L V Y I Met D L Stop Met P R L V P A S D S I P L F L N H Q A I Stop L R V F S N I
V S P F Stop Met E S L G Y V T Stop Stop D Q T I Stop P Q R T F H W R Stop A Q N Met K L K Q A L V L Met V L S F N W P S H L L
K L I T R S F Met I K P Q V C T L L I S S C R V S L T V L Q L P Met N F L I P P K V S R H Q V A Q Stop F L P C P I Met N L H L V I
Stop C K T T I P A F N S L L Q L R V Stop Stop T V I Stop Q R A Y N F P H L Stop T D K Y L G Y H G L L N L A Q T F Stop S I L K N
N C N Y Stop L K L P I P N Y R L V F P K V Stop A G Stop T H I I W V K L S I S I L T F Y Stop R S P K S S K Met K P V L R V L W K
T S N R P L I F C Y K Met N R R I H Stop F T T Q S S T V W S V Q E I G V P P A R N T I L V T L I T Met T I I S I Met V T L F Met Q
L L S S A Met L T L N Stop Met V L G L P T I K I G S I H W Stop E Met S Q I H L K K Met N T L H N R E C L I G S T V I H G Q L D F
Met K T V T V R T K K V V V K I T I L P Met L Stop S Y G G P L S V T S Q W N Stop G V T Stop Stop L V Stop Stop R Met Q Stop T T
I S I I K Met T T Q L S L K K S Stop E I K Stop V V F Y S I I L S I I L L I L E Q T Q N I S T V F I C Y L S H Q F L L I F V L R L S S K
K N G R L K S S Q L L N R Stop K A A G Q A Y Stop S Stop I K H S S T Q Stop I R Met H F S V I Q L L I H P H I W I T E Stop V A H

G H Stop H F Q G D W P T Q L L

5'3' Frame 3

A I I S D T S F S I K F R R Met F I Y K F H I E Stop C R R N I G N Stop F N Stop N F I L R Y F D S R Stop A L C C T I G G C N K S I C
S F S Stop P S L S F A K K Stop S E R I S N Stop F T R Q Stop F Stop Y K N S S S T V I V V I K L R K P S I S K N K R T N Q Stop Stop
S C H C F K Y L R K Stop Stop F E K Q K F S T Q F E F G F R R R L S H A K C I N N A Stop S H H I I I S N C F D F D F D F I F F F I F I
Y F G D T I K W N F I D F C F K R Y N R S T G N S N A F L F F P R N K N Y S F F I D L E Stop N D I Y Y F S K D K C S Y S Y R R R
F L H C F N S C F L D F V L S I Q L N P G F S P D A R I Y N S K P C L P I Stop Stop D T N H S F S I C I T I F D N Stop S N L P F K H
Y R K R F F T I T I T I I D C Stop S S L S I E F N R K H S F V I T I F N S R F F F L S G F F D I R Y H L C I F F N A S Y H L F V L I F
E T N K N F L L F S V N L D Q Stop H R Y H N G E L F N N D H C Q S F Q C C F D R Stop T T N C F Stop Q I T K S Y V I G Stop W
G V K R W S H S N K Q V L Y K F D C W Stop S G I T C L C V S L F I V E V H L E F I W F C C S T Y N S R S I Stop L R W L Stop Q
F R Stop C R I F G Q S F G N C S C C F L S F E F Stop F K Y D Y A S Stop Stop N D L V F N K G S A I R I Q Stop F F E L P G N T F G
S R Y G I C N W Y I S W I F K C Q D W F Q R R I Q Y H C F Stop I I K Q F S S G Y S Q I S Y H P F E W S H L V Met L R D R T R R F
N L N G L F I G G E L R I Stop N Stop S K R Stop C Stop W S Y H S I G R R T F Stop N Stop L R G L L Stop S S R R Y V R Y Stop F Q
A A G C L Stop R F Y S Y L Stop I F L Y H P R Stop V G I R Stop H N D F C L A P S Stop I F I Stop Stop Y N A R L L Y R H S T R F Y
N Stop G C N E R L S N N E P T I F H I S K Q T N I L V T Met V F S T W L K P F R V F Stop R T I A I I S Stop S C Q F R I T G Stop Y
F R K Y K R V K H I L F G Stop S Y R Stop V F L H S I N G L R N H P R Stop S Q Y Stop E Y F G K H Q I G L Stop Y S A T K Stop T
D V S T N L R H K V Q R F G Q F R R L G F H Q H A I R F W Stop H L L Q Stop P S F P L W L H Y S C S C C H R L C Stop L Stop T
E W Y L G C R Q Stop R L G Q F I G K R C R K S I Stop K R Stop I L C T I E N V Stop L V Q R S F Met G S W T L Stop K R Stop R
Stop E R R K Stop Stop Stop R L Q F C L C Y E V Met G G H Y R Stop P V N G I E G Stop L D D Stop Y N E G C N E R L F L L S K
Stop Q H S Stop A Stop R N H R K Stop S E W Y F I R Stop Y Y R L Y Y L F W N K H R I Y P R Y S Y A T Y H T S F F Stop Y S F Stop
D F R R R R Met A D Stop N R A N Y Stop I D R K R L D R H I E A E S S T L R P S R F V C I F Q Stop F N F Stop F I H I F G Stop R N

E S H Met G I S I F R G T G Q L N C L

3′5′ Frame 1

L S N Stop V G Q S P Stop K C Stop C P C A T H S V I Q I C G Stop I K S Stop I T E K C I R I Y W V E E C L I Q L Q Y A C P A A F Y
R F N N W L D F S L P F F F D E S L R T N I R R N W C D R Stop H Met N T V D I F C V C S K I S S I I D N I I E Stop N T T H F I S Y
D F F R L N C V V I L I I E I V V H C I L H Y T N H Q V T P Q F H Stop L V T D S G P P Stop L H S I G K I V I F T T T F F V L T V T
V F I K S S C P Stop Met T V E P I K H S R L C K V F I F F R W I C D I S Y Q Stop I D P I F I V G S P S T I Q F R V N I A D D S S C
Met N N V T I Met E Met Met V I V I S V T K I V L R A G G T P I S Stop T D Q T V E L C V V N Stop W I R L F I L Stop Q N I K G R F
D V F Q S T L S T G F I L D D F G D R Stop Stop N V R I L I D N F T Q I I C V Stop P A Y T F G N T N L Stop F G I G N F S Stop Stop
L Q L F F R I L Stop K V Stop A K L R R P W Stop P R Y L S V Stop R C G K L Stop A R C Stop I T V H Y T L S C R S E L N A G I V V
L H Y I T K Stop R F Met Met G Q G K N H C A T Stop C R L T L G G I R K F I G S C R T V R D T L Q L E I S N V H T C G L I I K D
L V I S F R R C D G Q L N D K T I N T S A C F N F I F Stop A H L Q Stop K V R Stop G Stop I V W S Y H V T Stop P S D S I Q K G D
T I F E N T L S Stop I A Stop Stop F R N N G I E S D A G T N L G I Stop R S Met I Y T S C K S H T L N Q R Y F Q V V R R I I G F Stop
Stop H Y P C Stop R Q G H F I N L H S H T Stop I K I R S Stop E N N Met S N S Q R I D Q I F C I T G T V I T T V A I L I Y C C Met L
N S K T I Stop T R G V L P Q Stop I R I H K G R Stop F L T T N N Q I C I K L V C L N G T I V Stop H P I S Q Stop H R I W Stop S V K
N S W W F I C R N S I E K I D N G R C Stop R V L R C G S G A T G R G Stop Q R R G G S F C L S R R Stop G R R D G S L H Stop R
R Y I N G I Stop C R R N R R G R R I Y C Stop K Stop Stop Stop Q K N V S C Stop T R W I N W T N S R Stop Stop Stop W Stop W Stop
R T V F C N V Stop R V N W T S C R I Stop Stop Y R W R K S D L C P I R W V S R A C C C R C E R L D Stop I L G Stop V G W I E Q
S R G S R S Stop S N E G I F A G S C S C I C P Y Stop N S I Y R F I R G Q Stop R R N N F C F L G R I G R H L S Y Q W I D C T F Stop
N R S Q Stop S S I Stop L C P R N R Stop R Stop R K Stop S R S R S R S S L R Stop Stop C G D F T H C Stop C T W H D Stop G D A G
N R T Q I V Y Stop T F A F R I I I C V N I Stop N S D R I I I G L C A Y F L R Y L A C V V Stop Stop R Q Stop Q L N C C S C I R I T V
L Stop I N W R C A H S I F W R N Stop E R A Met R S C R Y S C Y N H R L C N I R L T Y C Q S S V R Stop S F N Stop I S F Q Y F V

C I I R Y E T Y I Stop T F D E T Stop Stop K K K C Q I Stop L H

3′5′ Frame 2

Stop A I E L A S P P E N A N A H V R L I P L S K Y V D E S K V E S L K N A Y E S T G S K S A Stop F S F N Met P V Q P L S I D S I
I G S I L V C H S S S T K V S E R I L E E T G V I G S I Stop I P W I Y S V F V P K Stop V V Stop S I I L S N K I P L T L F P Met I S S
G S T V L S F Stop Stop Stop K Stop S F I A S F I I L I I K S P L N S I D W S P I V A P H N F I A Stop A K L Stop S S L L L S S F L P L
P F S Stop S P A A H E Stop P L N Q S N I L D C A K Y S S F S D G F A T S L T N E L T Q S L L S A A Q V P F S L E S T Stop P Met T
A A A Stop I Met Stop P Stop W K Stop W S L Stop Stop V L P K S Y C V L V E P Q S P E L T K P L N F V S Stop I S G Y V C S F C S
R I S K A D L Met F S K V L L V L A S S W Met I S E T V N R Met Stop E Y L S I T L P K Stop Y V F N P L I L S E I L T C N S E L A
T S A N N C N C S L E Y S K R F E P S Stop E D H G N Q D I C L F R D V E N C R L V V R Stop P F I T P L V V E A S Stop Met P V
Stop Stop S C I I S L N E D S Stop W G K A K I I V L P D A D S P W V V Stop E N S Stop V A V E P S E T P C S L K L V T Y I P A A
Stop S Stop K T S Stop S V S E G A T A N Stop Met I R P S T L A L A L I S Y S E L T S N E K S V E V K S S G P I T Stop H N Q V T P
F K R V I R Y L R I P Stop A K L L D D S E T Met V L N P T L E P I L A F K D P Stop Y I P V A N P I P Stop T K G I S R Stop F E E S
L D S D S T T L V E D K V I S S T C I V I L E S K F E A E K T T Stop A I P K G L T K Y S A L P E L S Stop P P Stop L Y Stop S T V V
C Stop T A K P Y E L E V Y F H N E Stop G Y T K A G D S Stop L P T I K F V Stop N L F V Stop Met G P S F D T P S A N D I G F G D
L S K T V G G S S V E T A L K R L T Met V V V E E F S V V V A V L L V E V D R E E E E V F V C L E D E D E E Met V A C I E E D
T Stop Met V S N V E E T G E E E E S T V E N S D D K R Met F P V E L D G Stop T G L T V D D S D G D G E E P F S V Met F E G
Stop I G L V V E Y S D T D G E R V I C V L L D G Stop A G L A V V D A S V W T K S W V E L D G Stop N K V E E A G V E A Met K
E S S P V A V A A F V L T E I V Y I V L F E V N E E G I I F V S W E E Stop E G I Stop V T S G S I V P F E T E V N K V P F N C V P E
I D K D E E N K V E V E V E A V Stop D N N V V T L R I V D A L G Met T E A T P E T E L K L C T E L L L F E L S F A Stop I F E T
V T G S L L V C A L I F Stop D T W L A Stop F D N D N D S Stop T A V L V L E S L S C E L I G D A L T L F F G E T E R G L Stop E

A A D T L V T T T D C A T Stop G S P T V K V A Stop D K V S I K S V S N I S S A L F D Met K L I Y E H S T K L N R K R S V R Y N C

3'5' Frame 3

K Q L S W P V P L K Met L Met P Met C D S F R Y P N Met W Met N Q K L N H Stop K Met H T N L L G R R V L D S A S I C L S S R
F L S I Q Stop L A R F Stop S A I L L R R K S Q N E Y Stop K K L V Stop Stop V A Y E Y R G Y I L C L F Q N K Stop Y N R Stop Y Y
R I K Y H S L Y F L Stop F L Q A Q L C C H F D N R N S R S L H P S L Y Stop S S S H P S I P L T G H R Stop W P P I T S Stop H R Q
N C N L H Y Y F L R S Y R Y R F H K V Q L P Met N D R Stop T N Q T F S I V Q S I H L F Q Met D L R H L L P Met N Stop P N L Y
C R Q P K Y H S V Stop S Q H S R Stop Q Q L H E Stop C N H N G N D G H C N K C Y Q N R I A C W W N P N L L N Stop P N R Stop
T L C R K L V D T S V H F V A E Y Q R P I Stop C F P K Y S Stop Y W L H L G Stop F R R P L I E C K N T Y R Stop L Y P N N Met C
L T R L Y F R K Y Stop P V I R N W Q L Q L I I A I V L Stop N T L K G L S Q V E K T Met V T K I F V C L E Met W K I V G S L L D
N R S L H P Stop L Stop K R V E C R Y S S L A L Y H Stop Met K I H D G A R Q K S L C Y L Met P T H L G W Y K K I H R Stop L
Stop N R Q R H P A A Stop N Stop Stop R T Y L R L D H K R P R N Q F Q K V R R P I E Stop Stop D H Q H Stop R L L Stop F H I L S S
P P Met K S P L R L N R L V L S R N I T K Stop L H S K G Stop Y D I Stop E Y P E L N C L Met I Q K Q W Y Stop I R R W N Q S W H
L K I H D I Y Q L Q I P Y L E P K V F P G S S K N H W I L I A L P L L K T R S F H Q L A Stop S Y L N Q N S K L R K Q H E Q F P K
D Stop P N I L H Y R N C H N H R S Y I D L L L Y V E Q Q N H Met N S R C T S T Met N K D T Q R Q V I P D Y Q Q S N L Y K T C
L F E W D H R L T P H Q P Met T Stop D L V I C Q K Q L V V H L S K Q H Stop K D Stop Q W S L L K S S P L W Stop R C Y W S R L
T E K R R K F L F V S K Met R T K R W Stop L A L K K I H K W Y L Met S K K P E R K K N L L L K I V Met T K E C F L L N S Met
D K L D Stop Q S Met I V Met V Met V K N R F L Stop C L K G K L D Stop L S N I V I Q Met E K E Stop F V S Y Stop Met G K Q G L
L L Stop Met R A S G L N P G L S W Met D R T K S R K Q E L K Q Stop R N L R R Stop L Stop L H L S L L K Stop Y I S F Y S R S Met
K K E Stop F L F L G K N R K A F E L P V D R L Y L L K Q K S I K F H L I V S P K Stop I K Met K K I K S K S K S K Q F E I I Met W
Stop L Y A L L Met H L A Stop L R R R R K P N S N C V L N F C F S N Y H L R K Y L K Q Stop Q D H Y W F V R L F F E I L G L R S
L I T T Met T V E L L F L Y Stop N H C L V N Stop L E Met R S L Y F L A K L R E G Y E K L Q I L L L Q P P I V Q H K A H L L S K
Stop R K I K F Q L N Q F P I F R L H Y S I Stop N L Y Met N I R R N L I E K E V S D I I A

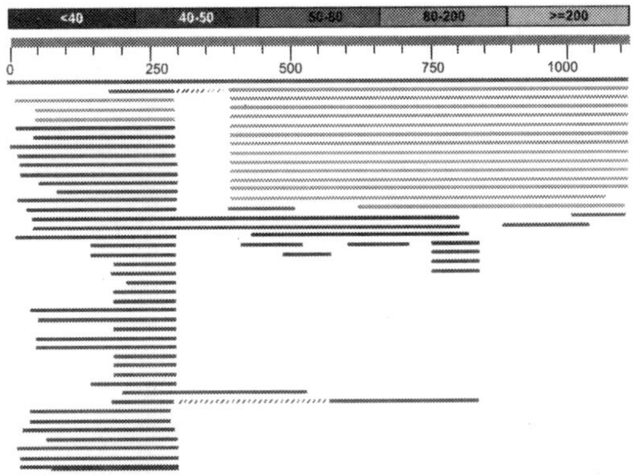

Sequences producing significant alignments:	Score * (bits)		E** Value
gi|1302597|emb|Z71682.1|SCYNR067C	*S. cerevisiae* chromosome X...	1885	0 . 0
gi|44985656|gb|AE016906.1|	*Eremothecium gossypiichromosome...*	755	0 . 0
gi|47075130|ref|NM_211521.1|	*Eremothecium gossypii* AGL208Cp...	755	0 . 0
gi|6562337|emb|AJ251504.1|CAL251504	*Candida albicans* eng1 g...	744	0 . 0
gi|6562333|emb|AJ251464.1|CAL251464	*Candida albicans* eng1 g...	744	0 . 0
gi|1360567|emb|Z73316.1|SCYLR144C	*S.cerevisiae* chromosome X...	520	e - 1 4 7
gi|1262303|gb|U53879.1|YSCL9634	Saccharomyces cerevisiae ch...	520	e - 1 4 7
gi|47075177|ref|NM_211568.1|	*Eremothecium gossypii* AGL161Cp...	510	e - 1 4 4
gi|6562335|emb|AJ251465.1|CAL251465	*Candida albicans* eng2 g...	496	e - 1 4 0
gi|1039338|emb|Z64354.1|SPAC23D3	*S.pombe chromosome* I cosmi...	485	e-136

Fig. 19.6: Output of BLASTN

Data Box 19.3

```
>gi|1302597|emb|Z71682.1|SCYNR067C  S. cerevisiae chromosome XIV reading frame ORF YNR067c
          Length = 5031
Score = 1885 bits (4883), Expect = 0.0
Identities = 967/1117 (86%), Positives = 967/1117 (86%)
Frame = -2
Query:     1   MQXXXXXXXXXSFVECSYISFISNNADEILETDLIETLSYATLTVGEPYVAQSVVVTRVS      60
               MQ         SFVECSYISFISNNADEILETDLIETLSYATLTVGEPYVAQSVVVTRVS
Sbjct:  4064   MQLYLTLLFLLSFVECSYISFISNNADEILETDLIETLSYATLTVGEPYVAQSVVVTRVS 3885
Query:    61   AASHSPLSVSPKNRVSASPINSQDSDSNTRTAVQLSLSLSNYASQVSQKISAQTNNDPVT     120
               AASHSPLSVSPKNRVSASPINSQDSDSNTRTAVQLSLSLSNYASQVSQKISAQTNNDPVT
Sbjct:  3884   AASHSPLSVSPKNRVSASPINSQDSDSNTRTAVQLSLSLSNYASQVSQKISAQTNNDPVT    3705
Query:   121   VSNIYANDXXXXXXXXXXXXXXXGVASVMPSASTMRKVXXXXXXXXXXXXXXXXXXXXXX     180
               VSNIYAND               GVASVMPSASTMRKV
Sbjct:  3704   VSNIYANDNSKSKSSVHNLSSVSGVASVMPSASTMRKVTTLLSQTASTSTSTLFSSSLSI    3525
```

```
Query:   181   XXXXXNGTLLTSVSKGTIDPLVTQMPSYSSQETKIIPSSLTSNKTIYTISVRTNAATATG   240
               NGTLLTSVSKGTIDPLVTQMPSYSSQETKIIPSSLTSNKTIYTISVRTNAATATG
Sbjct:  3524   SGTQLNGTLLTSVSKGTIDPLVTQMPSYSSQETKIIPSSLTSNKTIYTISVRTNAATATG   3345
Query:   241   EDSFIASTPASSTLFYPSNSTQDLVQTLASTTASPAYPSNRTQITLSPSVSLYSTTSPIY   300
               EDSFIASTPASSTLFYPSNSTQDLVQTLASTTASPAYPSNRTQITLSPSVSLYSTTSPIY
Sbjct:  3344   EDSFIASTPASSTLFYPSNSTQDLVQTLASTTASPAYPSNRTQITLSPSVSLYSTTSPIY   3165
Query:   301   PSNITENGXXXXXXXXXXXXXXXXXXXGNIXXXXXXXXXXXXXXXXXXXXXTIYVXXXX   360
               PSNITENG                   GNI                     TIYV
Sbjct:  3164   PSNITENGSSPSPSLSSTVSPVYPSSSTGNILLSSLFSTVDSSSSPVSSTLDTIYVSSSM   2985
Query:   361   XXXXXXXXXXXXXXXXXXXXXXXXXXXXXXXXXXXXIVNLFNAVSTDEPPTVFDRSPN   420
                                  IVNLFNAVSTDEPPTVFDRSPN
Sbjct:  2984   QATISSSSSSRQTKTSSSSLSTSTSSTATTENSSTTTIVNLFNAVSTDEPPTVFDRSPN   2805
Query:   421   PMSLADGVSNDGPIQTNKFYTNLIVGSQESPAFVYPYSLWKYTSSSYGFAVQHTTVDQYS   480
               PMSLADGVSNDGPIQTNKFYTNLIVGSQESPAFVYPYSLWKYTSSSYGFAVQHTTVDQYS
Sbjct:  2804   PMSLADGVSNDGPIQTNKFYTNLIVGSQESPAFVYPYSLWKYTSSSYGFAVQHTTVDQYS   2625
Query:   481   YGGYDSSGNAEYLVNPLGIAHVVFSASNFDSSMTMQVDEMTLSSTRVVLSESNDSSNYLE   540
               YGGYDSSGNAEYLVNPLGIAHVVFSASNFDSSMTMQVDEMTLSSTRVVLSESNDSSNYLE
Sbjct:  2624   YGGYDSSGNAEYLVNPLGIAHVVFSASNFDSSMTMQVDEMTLSSTRVVLSESNDSSNYLE   2445
Query:   541   IPLVQGMGFATGIYHGSLNAKIGSSVGFNTIVSESSSNLAQGILKYRITLLNGVTWLCYV   600
               IPLVQGMGFATGIYHGSLNAKIGSSVGFNTIVSESSSNLAQGILKYRITLLNGVTWLCYV
Sbjct:  2444   IPLVQGMGFATGIYHGSLNAKIGSSVGFNTIVSESSSNLAQGILKYRITLLNGVTWLCYV   2265
Query:   601   IGPDDLTSTDFSLEVSSEYEIKASASVDGLIIQLAVAPSETDYEVFYDQAAGMYVTNFKL   660
               IGPDDLTSTDFSLEVSSEYEIKASASVDGLIIQLAVAPSETDYEVFYDQAAGMYVTNFKL
Sbjct:  2264   IGPDDLTSTDFSLEVSSEYEIKASASVDGLIIQLAVAPSETDYEVFYDQAAGMYVTNFKL   2085
Query:   661   QGVSDGSTATYEFSYTTQGESASGSTMIFALPHHESSFSDIMQDYYTGIQLASTTKGVMN   720
               QGVSDGSTATYEFSYTTQGESASGSTMIFALPHHESSFSDIMQDYYTGIQLASTTKGVMN
Sbjct:  2084   QGVSDGSTATYEFSYTTQGESASGSTMIFALPHHESSFSDIMQDYYTGIQLASTTKGVMN   1905
Query:   721   GYLTTSLQFSTSLNRQISWLPWSSQLGSNLLEYSKEQLQLLAEVANSELQVSISESISGL   780
               GYLTTSLQFSTSLNRQISWLPWSSQLGSNLLEYSKEQLQLLAEVANSELQVSISESISGL
Sbjct:  1904   GYLTTSLQFSTSLNRQISWLPWSSQLGSNLLEYSKEQLQLLAEVANSELQVSISESISGL   1725
Query:   781   NTYYLGKVIDKYSYILLTVSEIIQDEASTKSTLENIKSAFDILLQNEQTYPLIYDTKFNG 840
               NTYYLGKVIDKYSYILLTVSEIIQDEASTKSTLENIKSAFDILLQNEQTYPLIYDTKFNG
Sbjct:  1724   NTYYLGKVIDKYSYILLTVSEIIQDEASTKSTLENIKSAFDILLQNEQTYPLIYDTKFNG 1545
Query:   841   LVSSGDWGSTSTQYDFGNTYYNDHHFHYGYIIHAAAVIGYVDSKLNGTWAADNKDWVNSL   900
               LVSSGDWGSTSTQYDFGNTYYNDHHFHYGYIIHAAAVIGYVDSKLNGTWAADNKDWVNSL
Sbjct:  1544   LVSSGDWGSTSTQYDFGNTYYNDHHFHYGYIIHAAAVIGYVDSKLNGTWAADNKDWVNSL   1365
Query:   901   VRDVANPSEKDEYFAQSRMFDWFNGHSWAAGLYENGNGKNEESSSEDYNFAYAMKLWGAT   960
               VRDVANPSEKDEYFAQSRMFDWFNGHSWAAGLYENGNGKNEESSSEDYNFAYAMKLWGAT
Sbjct:  1364   VRDVANPSEKDEYFAQSRMFDWFNGHSWAAGLYENGNGKNEESSSEDYNFAYAMKLWGAT   1185
Query:   961   IGDQSMELRGDLMISIMKDAMNDYFYYQNDNTVEPEEIIGNKVSGILFDNIIDYTTYFGT   1020
               IGDQSMELRGDLMISIMKDAMNDYFYYQNDNTVEPEEIIGNKVSGILFDNIIDYTTYFGT
Sbjct:  1184   IGDQSMELRGDLMISIMKDAMNDYFYYQNDNTVEPEEIIGNKVSGILFDNIIDYTTYFGT   1005
Query:  1021   NTEYIHGIHMLPITPVSSNIRSETFVEEEWQTKIEPIIESIESGWTGILKLNQALFDPVX 1080
               NTEYIHGIHMLPITPVSSNIRSETFVEEEWQTKIEPIIESIESGWTGILKLNQALFDPV
Sbjct:  1004   NTEYIHGIHMLPITPVSSNIRSETFVEEEWQTKIEPIIESIESGWTGILKLNQALFDPVD 825
Query:  1081   XXXXXXXXXXXXXXXLDNGMSRTWALAFSGGLANSIA 1117
                              LDNGMSRTWALAFSGGLANSIA
Sbjct:   824   SYAFFSDSTFDSSTYLDNGMSRTWALAFSGGLANSIA     714
```

INTERPRETATION OF THE DATA

An important goal in any mapping-sequencing project is the identification of gene, since these represents the functional elements of genomes. In the first step frame 1, 5′ to 3′ could only be selected for BLAST in the given sequence. Rest of the frames are rejected because of occurrence of stop codons in the beginning. BLASTn is done to identify the given sequence. Most of the entries with E-value = 0.0 are fungal genes belonging to *S. cerevisiae* chromosome, *Eremothecium gossypii* or *C. albicaus* genome. A portion of alignment of the best matched sequence i.e., *S. cerevisiae* chromosome xiv reading frame ORF is shown in Data Box 19.3. The gaps inserted by the program are seen as xxx in the query sequence. The problems associated with gene finding in fungi are very different from that in other eukaryotes.

SUMMARY

The fungal kingdom includes some important food material and some devastating pathogens. Fungi, originally classified with plants because of presence of cell walls, now are divided into kingdom fungi, stramenopila and protists. Study of fungal bioinformatics is important for comparative genomics, evolutionary studies, eukaryotic biology and medical studies. *Saccharomyces*, *Schizosaccharomyces*, *Neurospora* and *Ashbya* are model fungal organisms and their genome databases are already complied. First fungal chromosome was sequenced in 1992 and by 2004 complete yeast genome was sequenced. Genome annotation of fungi involves their steps (i) similarity searching (ii) comparative genomics (iii) gene prediction. In Experiment - 19A yeast genome database and fungal genome annotation methods are studied.

EXERCISE

1. Can a program written for finding ORF in *E. coli* be used for *Saccharomyces*?
2. Would it be possible to find introns in EST database of fungi?
3. Explain in details about the databases specially made for fungi. How they are different from prokaryotes and eukaryotes?
4. Write a short note on gene annoation in fungi.
5. Write a short note about special status of fungi as per database formulation is concerned.
6. Give in details the significant discoveries in fungal bioinformatics.
7. Look at the web sites for databases of following fungi and find the given sequence belongs to which fungi:
   (*a*) *Neurospora crassa*
   (*b*) *Sacharomyces cereviseae*
   (*c*) *Ashbya gossypii*
8. Using the sequence in Experiment 19A to find out various ORFs and decide which one is most likely coding for.
9. The yeast proteome database contains an entry for CDC6, the protein that regulates the introtion of DNA replication:
   (*a*) On what chromosome is the gene for yeast CDC6 located?
   (*b*) What post-translational modification does this protein undergo to reach its nature active site?
   (*c*) What are the closest relatives of this protein in other species?
   (*d*) With what other proteins is yeast CDC6 known to interact?
   (*e*) What is the effect of distnycin A on the activity of yeast CDC6?
   (*f*) What is the effect of actinonycin D on the activity 9 yeast CDC6?

20

PLANT GENOME DATABASES AND INFORMATICS

Biologists classify all living things into three domains: bacteria, archaea and eukarya. Domain eukarya is composed of four kingdoms: protista, planta, fungi and animals. Kingdom planta consists of organisms that have cell walls of cellulose and obtain energy by photosynthesis. Kingdom animalia however, contains organisms that lack cell walls and obtain energy by first ingesting and then digesting them internally. Eukaryotic genomes despite being so diverse share several characteristic in terms of genomes. The difference may lie not in the sequence but in the expression. Many gene sequences may be shared between plant and animals. There are only four nucleotides both in animals and plant that makes up genes. There always has been not much difference in plant and animals genetics. But expression of these genes is entirely different in plants and animals, which makes it different from prokaryotes.

20.1 CHARACTERS OF EUKARYOTIC GENOMES

Characters of eukaryotic genome which are shared by both animals and plants are given below:

20.1.1 Organisation of DNA

Most prokaryotic cells contain their genetic material in the form of a single circular piece of double stranded DNA, usually less than 5 mb long. In addition they may also contain plasmids.

In eukaryotes, the DNA is compacted into true chromosomes, which contain several proteins and which are large enough to be visible by light microscopy during the mitotic phase of the cell cycle. The DNA-protein complex of eukaryotic chromosomes is called chromatin. The protein component of chromatin consists primarily of five distinct proteins: his tones H1, H2A, H2B, H3 and H4. The last four histone aggregate to form an octameric protein that contains two molecules of each. DNA is wrapped around the histone octamer, forming a structure called a nucleosome core particle. This wrapping is the first level of compaction of the DNA in chromosomes. Each nucleosome unit contains about 200 nucleotide pairs, of which about 145 are in contact with the core particle. The remaining 55 nucleotide pairs link adjacent nucleosomes. Histone H1 binds to the linker segment and draws the nucleosomes nearer to one another. The DNA in its nucleosome form is further compacted into a helical fiber, the 30-nm fiber. In forming a visible chromosome, this unit undergoes several additional levels of folding, producing a highly compact visible chromosome. The result is that a eukaryotic DNA molecule, whose length and width are about 50,000 mm and 0.002 mm, respectively, is folded to form a chromosome with a length of about 5 mm and a width about 0.5 mm.

In eukaryote loops of chromatin are attached to the nuclear matrix by relatively short sequence (100–1000 bp) sequence called matrix attachment region (MARS) or scaffold associated regions (SARS). These regions are considered to be indicator of presence of an expressed gene.

20.1.2 Genes in Genomes

Typical prokaryotic genome contains only a relatively small amount of noncoding DNA in comparison with eukaryotes, distributed throughout the sequences. This is reflected in number of genes contained in the entire genome. In Table 20.1 number of genes and size of genome for several prokaryotes and eukaryotes is given. Eukaryotes in comparison to prokaryotes have much bigger genome and lesser number of genes. This is attributed to presence of a large amount of junk DNA in eukaryotes which includes repeat sequences, pseudogenes, transposable elements besides other regions.

Table 20.1: Genome size and number of genes found in several eukaryotic and prokaryotic organisms

Species	*Genome* ($\times 10^6$ *bp*)	*Genes*
Prokaryotes		
Mycoplasma genitatium	0.58	470
Hemophilus influenzae	1.83	1,743
Bacillus subtilis	4.2	4,100
E. scherichia coli	4.6	4,288
Eukaryotes		
Saccharomyces cerevisiae	13.5	6,034
Arabidopsis thalliana	119	25,498
Oryza sativa	466	~ 40,000
Drosophila melanogaster	165	13,601
Caenorhabditis elegans	97	18,424
Homo sapiens	3300	20,000 to 25,000
Fugu rubripes (puffer fich)	3890	30,000
Triticum aestivum	16000	30,000

20.1.3 Interrupted Genome

Eukaryotic genomes both in plants and animals have much more DNA content than prokaryotic genomes. Most eukaryotic genes are interrupted by large introns. Even with these introns included, however, genes comprise only about twenty-five per cent of the human genome. Bacteria have small, compact genomes, rich in genes. These genes have fewer noncoding regions and no introns. While eukaryotes generally have more genes than bacteria, the difference in gene content is not as great as the difference in DNA content: there is much more noncoding DNA in eukaryotes. Because of presence of introns in eukaryotes, mechanism of gene expression specifically mRNA splicing makes it more complicated. In fact, gene-coding regions comprise only about two percent of the human genome.

20.1.4 Sequence Repeats

Repetitive sequence refers to a DNA sequence that is present more than once per haploid genome. In most eukaryotic genomes, the entire set of DNA sequences consists of two major components: Single copy sequences and repetatitive DNA sequence. Several types of repeat sequences are shown in Fig. 20.1. Because of the skewed base composition of regions that have sequence repeats they may

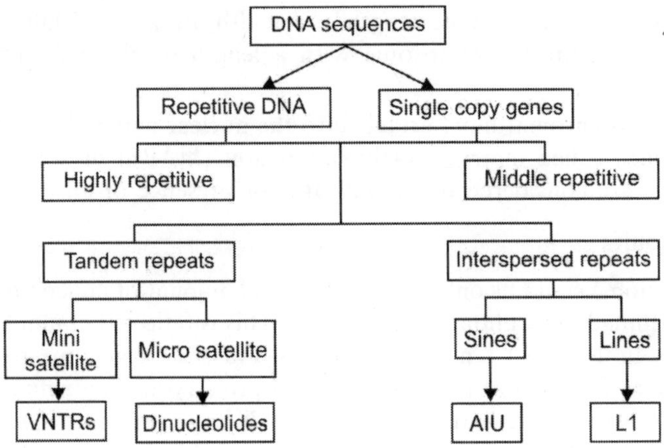

Fig. 20.1: Various types of DNA sequences in eukaryotes

be purified by virtue of having different buoyant densities and are known as satellite DNA. Most of this repetitive DNA is found near the centromere. In 1969 Pardue and Gale discovered satellite DNA using technique of *in situ* hybrisation. AlSo found in eukaryotic genomes are minisatellites made up of repeat units of up to 25 bp and microsatellites composed of repeat units of 4 bp or less. Microsatellite repeats are found at the ends of eukaryotic chromosomes at the telomeres, which in humans comprise hundreds of copies of a 6 bp repeat TTAGGG. The nucleotide sequence composition of many eukaryotic genomes has been examined via analysis of the renaturation kinetics of DNA. The principal finding is that eukaryotic organisms differ widely in the proportion of the genome that consists of repetitive DNA sequences and in the types of repetitive sequences that are present.

Besides, satellite DNA, micro and minisatellite repeat sequences are also present as tandem repeats. Tandem repeat refers to a short nuceotide sequence that is repeated many a times in a row. Just opposite to tandem repeats are interspersed repeat sequences. Lines and sines are two types of interspersed repeat sequences found in mammals. Lines are long interspersed elements that are usually 1 to 5 kbp in length and are found in 20,000 to 100,000 copies per genome. Sines are short interspersed sequences that are less than 500 bp in length. A specific example of sine is the ALU sequence present in 500,000 to 1000,000 copies in human genome. L1 is the example for lines in mammals.

VNTRs are also included in example of middle repetitive DNA. VNTR refers to variable number of tandem repeats, which is 15 to 100 bp, may be found be found within or between the genes. VTNRs are used in DNA fingerprinting. They are also called as minisatellites.

1. **Unique, or single-copy, sequences:** This is usually the major component, typically constituting 30-75 per cent of the chromosomal DNA in most organisms.
2. **Highly repetitive sequences:** This component usually constitutes 5-45 per cent of the genome, depending on the species. The sequences in this class are typically 5-300 nucleotide pairs per repeat and present in as many as 10,000 to millions copies per genome.
3. **Middle-repetitive sequences:** This component typically constitutes 1-30 per cent of a eukaryotic genome; it includes sequences that are usually present in 10-1000 copies per genome.

20.1.4.1 Unique Sequences: Most gene sequences and the adjacent nucleotide sequences required for their expression are contained in the unique-sequence component. With minor exceptions (for example, the repetition of one or a few genes), the genomes of viruses and prokaryotes are composed entirely of single copy sequences; in contrast, such sequences constitute only about 40 per cent of the total genome in some sea urchin species, a little more than 50 per cent of the human genome, and about 70 percent of the *D. melanogaster* genome.

20.1.4.2 Highly Repetitive Sequences: Many highly repetitive sequences are localized in blocks of tandem repeats, whereas others are dispersed throughout the genome. Satellite DNA an example of highly repetitive sequences. Dispersed highly repetitive sequences include some SINE and LINE retrotransposable elements that are found at high copy number in certain genomes, including the human genome. Among the localized highly repetitive sequences, most are fairly short. They can differ greatly in abundance from one genome to the next: 6 per cent in the *Drosophila melanogaster* genome, but 45 percent in the *D. virilis* genome. One of the simplest possible repetitive sequences makes up 25 per cent of the genomes of certain species of land crabs; it is composed of an alternating AT sequence (... ATATATAT ...) with about 3 per cent G + C interspersed. These are present in the centromere and regions where hetrochromatin is present.

20.1.4.3 Middle-Repetitive Sequences: Middle-repetitive sequences constitute about 12 per cent of the *D. melanogaster* genome and 40 percent or more of the human and other eukaryotic genomes. These sequences differ greatly in the number of copies and in their locations in the genome. They comprise many families of related sequences and include several groups of genes. For example, the two major types of ribosomal RNA are transcribed from a tandem pair of genes that is repeated several hundred times in most eukaryotic genomes. The genes for tRNA molecules as well as those for the histone proteins are repeated in the genomes of all eukaryotes. The copies of the tRNA genes are scattered, whereas those for the histones are usually clustered in tandem repeats. Each histone gene is repeated about 10 times per genome in chickens, 20 times in mammals, about 100 times in *drosophila,* and as many as 600 times in some species of sea urchins.

Most of the dispersed middle-repetitive DNA in the *D. melanogaster* genome consists of transposable elements. There are about 50 different families of such elements—with names such as Alu, L1, P, hobo, copia, jockey, Helena, roo, 412, 297, and mdg—each of which is present in 10-50 copies that occur at widely scattered locations throughout the chromosomes.

20.1.5 Transposable Elements

These elements can comprise a large proportion of the eukaryotic genome as repetitive sequence. Transposable elements (TEs) are thought to play an important role in the evolution of these genomes. TEs are DNA sequences that can move from one chromosomal location to another faster than the chromosome can replicate. Hence, TEs have the potential to increase in number until they comprise a large proportion of the genome sequence, a feature already observed in many plants and animals. They remain detectable in the genome until they blend into the background sequence by mutation. The presence of these elements may be demonstrated using programs for detection of low-complexity regions in sequences. The percentage of genomes that are composed of TEs is given in Table 20.2. For example, more than one-third of the human genome consists of interspersed repetitive sequences derived from TEs.

Eukaryotic TEs fall into two main classes according to sequence similarity and the mechanism of transposition. Class I elements encode a reverse transcriptase and use RNA-mediated mechanisms of transcription. There are three main subclasses of these TEs—the long terminal repeat (LTR) retrotransposons, and retrovirus-like elements. The LTR are related by genetic structure to retroviruses. The retroposons include short (80-300 bp long) interspersed nuclear elements (SINES) and long (6-8 kbp long) interspersed nuclear elements (LINES). Ten per cent of the human genome comprises one particular family of the SINE element, designated Alu (1.2 million copies) and 14.6 per cent of one particular LINE designated LINE1 (593,000 copies). Verbebrate chromosomes have long (>300 kb) regions of distinct GC richness, repeat content, and gene density, designated isochores in a model of genome organisation proposing that genomes are made up of distinct segments of unique composition. Human and mouse chromosomal regions that have a low density of genes are AT-rich and have more Alu or B1/B2 (SINES) than LINE1 elements, whereas the reverse is true for regions that have a high gene density, and those regions are more GC-rich. The other class of TEs, class II, is made up of elements that employ a DNA-based mechanism of transposition. The human genome contains about 200,000 copies of this class of elements that probably predate human evolution. Class II elements also include the Activation-Dissociation (Ac-Ds) family in maize and the P element in *Drosophila*. A third category of TEs has features of both class I and class II TEs. These miniature, inverted repeat TEs (MITES) are 400 bp in length and were discovered in diverse flowering plants where they are frequently associated with regulatory regions of genes. Hence, they could be exerting an influence on regulation of gene expression.

Several feature of SINES make them useful for phylogentic studies:

1. A SINE is either present or absent, Presence of SINES indicates similarity.
2. SINES one inserted at random in the non coding portion of genome. Therefore appearance of similar SINES at same locus indicates similarity and common ancesstory.
3. Not only do SINES show relationships, but also implies which species came first.

Table 20.2: Percentage of transposable elements in different genomes

Name of organism	*% transposable*
Homo sapiens	35%
Zea Mays	50%
Drosophila melanogaster	15%
Arabidopsis thalliana	2%
Caenorhabditis elegans	1.8%
Saccharomyces cereviseae	3.1%

20.1.6 Pseudogenes

Pseudogenes are very common in eukaryotes. Pseudogenes are DNA sequences that are derived from a functional copy of a gene but which have become non-functional because of mutation. Pseudogene also comprise a major amount of junk DNA in enkaryotes. The other type of pseudogene found in eukaryotes is called proccessed pseudogene. It is also derived from a functional gene but it does not contain intron and lack a promoter hence are not expressed.

In the present chapter major emphasis is on plant genomes that are sequences and annotated.

20.2 *ARABIDOPSIS THALLIANA* AS A MODEL ORGANISM

For the genetic and genomic studies of the differentiation of cells, tissues and organs as well as development of body forms it becomes necessary to study more complex organisms. These organisms must be easy to culture under controlled conditions have a short life cycle to allow the breeding experiments over several generations and to be small enough to make the production of a large number of individuals.

Arabidopsis thalliana (English name Wall cress or mouse-ear cress) is a model organism in plants because it is a small flowering plant that can be cultured in a large numbers in the greenhouse or laboratory (Fig. 20.2). It has a small genome contained in only five chromosomes. *Arabidopsis* is a member of the mustard (Brassicaceae) family, which includes cultivated species such as cabbage and radish. *Arabidopsis* is not of major agronomic significance, but it offers important advantages for basic research in genetics and molecular biology.

Arabidopsis thalliana was discovered by Johannes Thal (hence, thalliana) in the Harz Mountains in the sixteenth century, though he called it *Pilosella siliquosa* (and it has gone through a number of name changes since). The earliest report of a mutant was in 1873 (by A. Braun). F. Laibach first summarized the potential of *Arabidopsis thalliana* as a model organism for genetics in 1943—he did some work on it much earlier though, publishing its correct chromosome number in 1907. Laibach's student E. Reinholz made the first collection of induced mutants.

Fig. 20.2: Drawing of *arabidopsis* plant

It is ideal for studying the development of higher plants and for comparison of animal and plant development and genome structure. Its entire genome consists of a relatively small set of genes that dictate when the weed will bud, bloom, sleep or seed. Those functional genes have their counterparts in plants with much larger genomes, such as wheat, corn, rice, cotton and soyabean. Unlike the human genome, *Arabidopsis* has few "junk" DNA sequences that contain no genes. The plant is practical for scientists because it matures quickly, is small and reproduces abundantly. All these physical and genetic traits add up to an especially useful organism whose genome is now catalogued for the public's benefit.

The *Arabidopsis* Genome Initiative (AGI) is the name of the multinational effort to sequence the *Arabidopsis* genome. It was established in 1996 and consists of 6 groups. Each group initiated sequencing in a defined region of the genome using BAC, YAC, cosmid, or P1 clones. TIGR, SPP and Genoscope also sequenced the ends of ~18,300 BAC clones, representing 14-fold coverage, for mapping purposes. One critical objective was to obtain very accurate sequence. The minimum standard was that > 97% of the sequence would be double-stranded and the remaining <3% would have double chemistry (pseudo double-stranded). The AGI completed the entire genome by the end of year 2000, much earlier than the projected date of 2004.

Table 20.3: Genome statistics for *Arabidopsis*

Character	*TAIR7 (4/07)*
Protein coding genes	26,819
Transposons and pseudogenes	3,889
Alternatively spliced genes	3,866
Gene density (Kb per gene)	4.44
Avg. exons per gene	5.79
Avg. exon length	268
Avg. intron length	165

20.2.1 Summary of Gene Functional Annotation

TAIR is the *Arabidopsis* genome resource. The latest release of TAIR, TAIR7 reports that *Arabidopsis* genome contains 27,029 protein coding genes, 3889 pseudogenes or transposable elements and 1123 ncRNAs i.e. 32,041 genes in all. The latest data as reported in TAIR7 is summarised in Table 20.3.

The haploid chromosome number of *arabidosis* is five. Besides it has genome present in chloroplast and mitochondria also. In Table 20.4 the details of gene annotation in nuclear and extra nuclear genome of *Arabidoposis* is summarized.

Table 20.3: Arobidoposis genome statistics as released by TAIR7

Chromosome	*Protein coding*	*pre-tRNA*	*rRNA*	*snRNA*	*snoRNA*	*miRNA*	*Other*	*Pseudo*	*Total*
1	6,930	240	0	2	18	30	64	773	8,057
2	4,198	96	2	0	15	21	40	844	5,216
3	5,340	93	2	7	15	20	48	857	6,382
4	4,085	79	0	0	11	15	33	664	4,887
5	6,266	123	0	4	12	28	36	751	7,220
Total	26,819	631	4	13	71	114	221	3,889	31,762
Chloroplast	88	37	8	0	0	0	0	0	133
Mitochondria	122	21	3	0	0	0	0	0	146
Total	27,029	689	15	13	71	114	221	3,889	32,041

(*): Includes transposons and pseudogenes.

20.2.2 Arabidopsis Genome Databases

1. *TAIR: The Arabidopsis Information Resource:* TAIR is a collaborative effort between the Carnegie Institution of Washington Department of Plant Biology at Stanford, California and the National Center for Genome Resources (NCGR) at Santa Fe, New Mexico. TAIR will expand and improve on the AtDB to include comprehensive sequence and map information, community information (people, labs, companies, and databases), expression data, and experimental data from the literature.
2. *Arabidopsis Genome Resource (AGR):* The *Arabidopsis* Genome Resource (AGR) is an information resource being developed at NASC as part of the UK-CropNet project. The aim is to integrate sequence data with physical and genetic maps of *Arabidopsis*, within the UK CropNet project and provide information for identifying gene function and crop plant orthologues of *Arabidopsis* genes.
3. *Arabidopsis Information Management System (AIMS):* AIMS is intended to support the database needs for all *Arabidopsis* Stock Centers around the world.
4. *AraClean:* AraClean is a corrected and redundancy reduced database of *Arabidopsis thalliana* sequences extracted from GenBank.

20.2.3 Analysis of the genome sequence

Analysis of the genome sequence of *Arabidopsis* reveals that a region cover 115.4 megabases of the 125-megabase genome is sequenced and it extends into centromeric regions.

The evolution of *Arabidopsis* involved whole-genome duplication, followed by subsequent gene loss and extensive local gene duplications, giving rise to a dynamic genome enriched by lateral gene transfer from a cyanobacterial-like ancestor of the plastid. The genome contains 25,498 genes encoding proteins from 11,000 families, similar to the functional diversity of *Drosophila* and *Caenorhabditis elegans*—the other sequenced multicellular eukaryotes. *Arabidopsis* has many families of new proteins but also lacks several common protein families, indicating that the sets of common proteins have undergone differential expansion and contraction in the three multicellular eukaryotes. This is the first complete genome sequence of a plant and provides the foundations for more comprehensive comparison of conserved processes in all eukaryotes, identifying a wide range of plant-specific gene functions and establishing rapid systematic ways to identify genes for crop improvement

20.2.4 Implications and Bioinformatics

Genetics reached a major milestone the day the international research team announced that it has completed the first plant genome sequence. The species *Arabidopsis thalliana* has emerged as the plant counterpart of the laboratory mouse, offering clues to how all sorts of living organisms behave genetically, with potentially widespread applications for agriculture, medicine and energy. 2010 Project has the objective to determine the function of 25,000 *Arabidopsis* genes over the next decade. The NSF 2010 Project is part of a worldwide *Arabidopsis* functional genomics effort and will be coordinated in a similar manner to the *Arabidopsis* genome sequencing project.

1. The complete sequence of *Arabidopsis* is directly relevant to human biological functions, because many fundamental life processes at the molecular and cellular levels are common to all higher organisms. Some of those processes are easier to study in *Arabidopsis* than in human or animal models. *Arabidopsis* contains numerous genes equivalent to those that prompt disease in humans —ranging from cancer and premature aging, to ailments such as Wilson's disease, in which the human body's inability to excrete copper can be fatal.
2. Because every molecular function of plants is dictated by DNA, an understanding of the genome could also help scientists develop crops that grow faster and larger, are more disease-resistant, and produce useful chemicals more efficiently.
3. *Arabidopsis* researchers have also identified genes that determine whether the growing shoot of a plant will develop into a flower. By inserting a certain gene into poplar shoots, scientists have shortened that tree's flowering time from six years to only six months.
4. Plants hold great potential as sources of renewable energy, although they currently represent just three per cent of U.S. energy resources. Completion of the *Arabidopsis* genome sequence is revealing how photosynthesis converts solar energy and carbon dioxide into biomass, helping scientists develop better plants for fuel and chemical uses.
5. The *Arabidopsis* initiative also demonstrates how information technology (IT) is revolutionizing biology research. Without advanced computers and networks, the genome's usefulness would not be as extensive. Biologists are the fastest-growing segment of IT users in the scientific community, with the field of "bioinformatics" producing more-powerful databases and pattern recognition tools for DNA analysis.

EXPERIMENT 20A

To Locate Potential Protein Coding Genes in a Recently Sequenced Part of *Arabidopsis* Chromosome 1 using a Program called GENSCAN

About GENSCAN: This server is for the program "GENSCAN", developed by Chris Burge at Stanford University, not to be confused with the Applied Biosystems sequencing software called "GENESCAN". GENSCAN is freely available for academic use. Executables are currently available for the following Unix platforms: Intel/Linux, Sun/Solaris, Intel/Solaris, SGI/Irix, DEC/Tru64, and IBM/AIX.

Once the amino acid sequences of the predicted genes is found program BLAST would be used to identify one of them.

About the sequence: Present sequence is selected from *Arabidopsis* genome. This sequence contains over 102,000 base pairs of DNA is located on chromosome 1. A portion of the sequences is shown in Data Box 20.1.

TESTING THE HYPOTHESIS

Computer set up: A normal PC with internet connection in windows environment:

1. The given sequence is saved as expt_20A.doc is shown in Data Box 20.1. http. www.genes.nit.edu/ Genscan.html

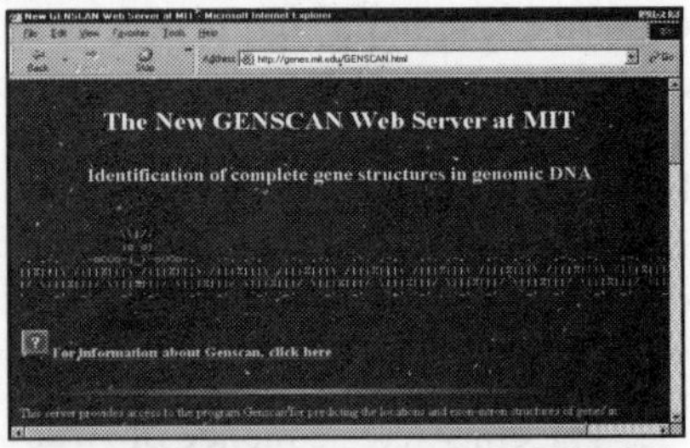

Fig. 20.3: Web view of GENSCAN

2. Log on to the GENSCAN server at MIT and scroll down to the large white box. Click in the box and then press "Alt V" or "Ctrl V" to paste the sequence into the box.
3. Notice the several pull-down menus immediately above the sequence box. Pull down the one that reads Vertebrate (the default) and select *Arabidopsis*.

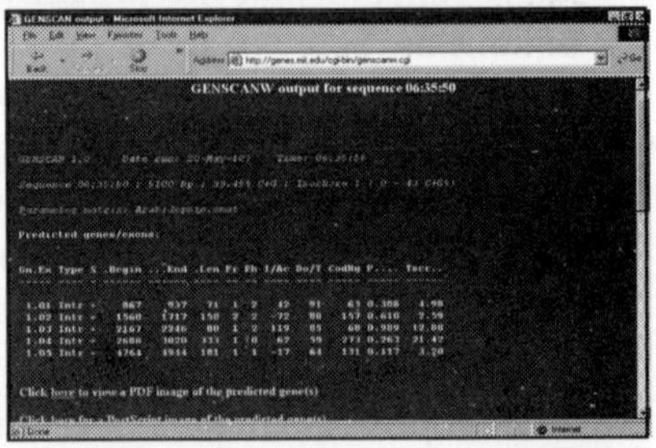

Fig. 20.4: Result obtained through GENSCAN

4. Now click Run GENSCAN. This analysis may take a few moments. Study the output thoroughly. Find the amino acid sequence of gene 14, select it, and copy it into the computer's memory as described above. Now do a BLAST search.

DATA

Data Box 20.1

```
1 gaattcatag tatatattta taaaaatgca aattcttaca tgcataagga tgaggagaga
61 aaataaatca caatcctaat ggaacccata aattacgtta ccacccttta aacaaaacta
121 tttcataaca aaagtaaaca tgcgttgtga aataatgata tgataggcat atattaagga
181 gtggtattag aataaagaag cgattaattg atattgagaa gatagcttat catagcttat
241 cataatcgaa tttagatgtt gcattgtggt accaaagtgg tggtatatat taccttatac
301 atatcagcat ttcgcatgaa ccatatatga gtattaatta ggtgccatat gcagctagtc
361 tttctactac acgccacaga agtaacacgt aattaattat gtgtatcggt ttatgtaatg
421 ccgtaaacta tgtgtttaat gtgtccaatc acgtcttatc atctatagta tgcgtttgtt
481 ctttttgcat tatatgtgta ttaattaggt agacactgag attcgtatga tacgttcaac
541 cgaatcgcat aactctatat gaatctaaag tccacttact cactacaatt aatagtttat
601 ggtagctgtt gatgcacgaa tcacacaagt acgaaaatgg gatctctagg gaaggaagaa
661 gaatctttct attaatgacg agcccgcgac ttaggcggat tggacggatt acaaatgatg
721 atttcttcaa cgagcataaa tggagtaaga acgagccgat tgtagatgaa agacgagctt
781 gaggccggaa gacgatcttc cgaggaagaa tagagtttat tcgttttata gctgaacctt
841 gtaaaaggtt gcctacgtac ccttagaaag ggatcaagcc acacgtagtt catattgtga
901 taaacgagct aggccattat tggaggaagg tcggcaggta attgacgagc cggcagatgt
961 gcgacgagct ggatggcttc tcgagacgtg cgagtaggtt agttattttg ggcgggcagg
1021 tcggagagac acttcggacg gacgagtaaa cgagttactc aagatggatg ggcagacgag
1081 gtactttgga cgggtgagct gacgagctgc ttgatacggg tgagctgacg agctgcttga
1141 gacgggtgag ctgacgagct gcttgagacg ggtgagctga cgagctgctt gagacgggtg
1201 agctgacgag ctgcttgaga cgaaccggct ggcgggatac ttcggacggg tgagctgacg
1261 agctgtttga gacggaccgg ctggcgggat acttcggagg ggagagctga cgagctgatt
1321 gagacgggtg agctgacgag ctgcttgaga cggaccggct ggcgggatac ttcggagggg
1381 agagctgacg agctgcttga gacgggtgag ctgacgatct gcttgagacg ggtgagctga
1441 cgagctgctt gagacggacc ggctggcggg atacttcgga cgggtgagct gacgagctgc
1501 ttgagacggg tgagctgacg agctgtttga gacggaccgg ctggcgggat acttcggagg
1561 ggagagctga cgagctgctt gagacgggtg agctgacgag ctgcttgata cggaccggct
1621 ggcgggatac ttcggagggg agagctgacg agctgcttga gacgggtgag ctgacgagct
```

```
1681 gcttgagacg gaccggctgg cgggatactt cggacgggtg agctgacgag ctgctctgtg
1741 tctctgcgtg tccggcggaa tcccctttg acaagacatg ctattccgta tttatagcct
1801 tctgcgtgta ggacactcct ccaggatttt gacgagccta ccctcttgac ttaacattct
1861 gcttaggcct ttgacttgga tttgggcctt gggccaagca agcccaaatc ttacccctaa
1921 cagtagctga aatggatcac aacgtaccta attccgattt ctatatctag tgtacgtatt
1981 gatattcttt agcaatagtt ttggaaaata taatctgagt ttagtgtact cgttggcttt
2041 attacttttg aattttaata atagtctatt ttaaccataa gtactagcat atactaaata
2101 acttgatcta cacaataaaa tacagaagat tgattctcaa actatttctc tctctctctc
2161 tcacagagaa tctccaatgg cggctcccgt agtggtcgct cctgcgataa atcttcagca
2221 tatggaccta cctgaaggaa atgaaggtaa taaaaaacga ttatattctc agaagcaaga
2281 tataggggta gcgaatttgg ggtaataaag aaattgaaat tgaaagtatc aagaaacatt
2341 ggtgagtttg ggaattggag gaatgtattt aaaattagag tagatctgtg aatttgggaa
2401 atcgaggcta caatctatgg aattggtaac agtgtagaaa tcttagaaaa gttatgatct
2461 atggaatctt agaacagtgt agaaatcgta gtacagtgtt acagatctat ggaatcttag
2521 aacagtgtag aaatcgtagc acagtgttac agatctatgg aatcttagaa tagtctccta
2581 gatctatgga atttgtacaa ttaaggcaag atttaaaaat ttcagaataa aacagttgct
2641 tataatatat atataaatct catatgtttg gttgctattt ttgacagatg tagcagctgt
2701 gcttcaaaca tcaacacagc gctggctcaa cgctgctgag tggagcattc tttataacaa
2761 ccacaaccta ttcccagaat caaactcatt taccatacct attgaagtag agggacttta
2821 cagagttgac aaccgcttta atcaagacag caatacgtgg agaaacacaa gtgctccctc
2881 cacttgcaaa aatgctatca acttctacaa tgttaggatc gccactggct tgtacaggtt
2941 tgatgatgat ggggacattg tgatggacgc tgaagttcaa ggagggtggg tcctttgcag
3001 aaaaaccgtg aggaacagag gtggggggt acttactttt gcacactaca tgtgggtttt
3061 agtgtgagta aaaactatgc ttgagtaaaa actatgcttg agtaaaaact atgtttgagt
3121 tgatactgga aaaatctaaa cgaacttgta gtataaactt ctatgcttat aaacataagc
3181 ccaaattagc taaatcattt ttaattcata tttttaagat gctacaggta aatgctacag
3241 gtaaagtcat agctacatta tttacttcac tatactatcc aattgagaga ataaatgatc
3301 aaaaacaaag aggcgcaaaa aaactaatat taaaatatca agctaggaaa gataaacaaa
3361 gttcaaagac tcagatgtta atgaaacgaa tgagactgat ggcccaaata atccaaatag
3421 atcctagatt atatatgtcg gaagactaat ttcaataagt ataaacacaa gatatttgcg
3481 gatttatatt gttagactaa aatgcaaatg aagtgtgaaa aggtcaagaa tatgcaacac
3541 ttgcgtagtc gtataagaag acattagagt ggtcatataa atgggatatt aattcagaga
3601 agttttaagg gtaggaaaaa caaactgttt aaacaaaaaa tcatgcttaa aacatctgtt
3661 acatgtaaaa aaccgagaaa ttaaaaccaa gacaaaaaaa acagagtatg tgaacttgcc
3721 ttcaattggg taaaaggctg gatctgagat aatctctaaa agtaaagcca gatctgtaaa
3781 atataataac aaagaacaga ttaaattgat gaaaccggac aacacaagaa gaacaatcca
3841 tctttttcag atctgaagta gagatggaaa aaaaacttgc cttttcctat gaacagatga
3901 ctgcggatag cagatctgga gaaaatatat aaaaaatctc aggttaaaat cgtgaaacaa
3961 gaacaaactg atcaaataag aaagaaggac ataacaaaaa acttagcttt tttaatgcgg
4021 atctgaagca tatgggtaga gagccagata gggtttgaga aatcgaagag aggtaatgat
4081 tatagcagtt tagtgtttct cgatggagag agagagagag agattgagtt ttttattttg
4141 gtgagagaca gagaggcgga cgagtggtgt ttatttggag ctttctcgat taacttattt
4201 ttattctgag ttcctgtttt acccttgtaa ccctatcaca tctcataagg tttcgaagac
4261 ttggcttttg tcttaaaaag ttctcggatt ctcttttta ttcgaggaaa caaaacaaat
```

Data Box 20.2: GENSCAN output for sequence

GENSCAN 1.0 Date run: 20-May-07 Time: 06:35:55

Sequence 06:35:50 : 5100 bp : 39.45% C+G : Isochore 1 (0 - 43 C+G%)

Parameter matrix: *Arabidopsis*.smat

Predicted genes/exons:

Gn.Ex	Type	S	.Begin	...End	.Len	Fr	Ph	I/Ac	Do/T	CodRg	P....	Tscr..
1.01	Intr	+	867	937	71	1	2	42	91	63	0.308	4.98
1.02	Intr	+	1560	1717	158	2	2	-72	88	157	0.610	2.59
1.03	Intr	+	2167	2246	80	1	2	119	85	68	0.989	12.88
1.04	Intr	+	2688	3020	333	1	0	62	59	273	0.263	21.42
1.05	Intr	+	4764	4944	181	1	1	-17	64	131	0.117	3.20

Click here to view a PDF image of the predicted gene(s)

Click here for a PostScript image of the predicted gene(s)

Predicted peptide sequence(s):

```
>06:35:50|GENSCAN_predicted_peptide_1|275_aa
XRDQATRSSYCDKRARPLLEEGRQGRADELLETGELTSCLIRTGWRDTSEGRADELLETG
ELTSCLRRTGWRDTSDGESPMAAPVVVAPAINLQHMDLPEGNEDVAAVLQTSTQRWLNAA
EWSILYNNHNLFPESNSFTIPIEVEGLYRVDNRFNQDSNTWRNTSAPSTCKNAINFYNVR
IATGLYRFDDDGDIVMDAEVQGGWVLCRKTVRNRGHCGTRLSHLYKPVFDRQFKMLKIHM
MEDMIKDGEEKSLDTFVISRQLTTFTDLTVCLSTX
```

Data Box 20.3: Explanation of terms in genscan output

Gn.Ex: gene number, exon number (for reference)
Type: Init = Initial exon (ATG to 5' splice site)
Intr = Internal exon (3' splice site to 5' splice site)
Term = Terminal exon (3' splice site to stop codon)
Sngl = Single-exon gene (ATG to stop)
Prom = Promoter (TATA box / initation site)
PlyA = poly-A signal (consensus: AATAAA)
S: DNA strand (+ = input strand; - = opposite strand)
Begin: beginning of exon or signal (numbered on input strand)
End: end point of exon or signal (numbered on input strand)
Len: length of exon or signal (bp)
Fr: reading frame (a forward strand codon ending at x has frame x mod 3)
Ph: net phase of exon (exon length modulo 3)
I/Ac: initiation signal or 3' splice site score (tenth bit units)
Do/T: 5' splice site or termination signal score (tenth bit units)
CodRg: coding region score (tenth bit units)
P: probability of exon (sum over all parses containing exon)
Tscr: exon score (depends on length, I/Ac, Do/T and CodRg scores)

The SCORE of a predicted feature (e.g., exon or splice site) is a log-odds measure of the quality of the feature based on local sequence properties. For example, a predicted 5' splice site with score > 100 is strong; 50-100 is moderate; 0-50 is weak; and below 0 is poor (more than likely not a real donor site).

The PROBABILITY of a predicted exon is the estimated probability under GENSCAN's model of genomic sequence structure that the exon is correct. This probability depends in general on global as well as local sequence properties, e.g., it depends on how well the exon fits with neighbourring exons. It has been shown that predicted exons with higher probabilities are more likely to be correct than those with lower probabilities.

INTERPRETATION OF DATA

GENSCAN looks for features of the DNA sequence that are consistent with protein coding genes. It then removes predicted introns, translates the sequence into protein and then compiles all of the predicted protein sequences into the output file that should now appear in that window is shown in Data Box 20.2. An explanation to several terms in output window in given in Data Box 20.3. As you scroll down the file you will notice that each predicted gene is composed of several exons. In the third column a + or - indicates the strand of DNA that is used as the template. The rest of the columns indicate where each exon is, how long it is, and stuff like that.

The *Arabidopsis* genome contains about 130 megabases of DNA. The present gene contains a very small amount compared to *Arabidopsis* genome. The output shows the amino acid sequences for each of the predicted genes. Between the exon data and the amino acid sequence data are two links, one of which reads "Click here to view a PDF image of the predicted gene(s)". Click on the link a graphical display of the genes along the length of the DNA.

20.3 MAIZE AS A MODEL ORGANISM FOR GENOME DYNAMICS

Maize also known as corn is actually *zea mays* is a member of the grass family. Grasses including rice, wheat and barley are the most important source of calories for humanity. Diploid chromosome number of maize is 20. Genome size of maize is 4.5×10^6 NTP. In 1940 Barbara McClintock while studying the chromosome breakage and reunion in maize discovered the genetic elements that move from one location to another. These genetic elements are called as transposable elements and are also called as controlling elements, jumping genes, mobile genes, genetic elements and transposons. Barbara McClintock was recognized for her seminal discovery by being awarded the 1983 Nobel prize for medicine or physiology (Historical Profile 22).

HISTORICAL PROFILE 22

Barbara McClintock for Gene Finding Software

Barbara McClintock

Edward Barbara McClintock was born on June 16, 1902, in Hartford, Connecticut. McClintock earned her B.S. and M.S. degrees in Botany at Cornell University, and received her Ph.D. in the same subject at Cornell in 1927. She was a highly influential women member of a small group who studied maize (corn) cytogenetics, the genetic study of maize at the cellular level.

Throughout her long and distinguished career, McClintock's work focused on the genetics of maize and, in particular, the relationship between plant reproduction and subsequent mutation. Beginning in the late 1920s, she studied how genes in chromosomes could "move" during the breeding of maize plants. She did groundbreaking research on this phenomenon, where she determined the physical correlate of genetic crossing-over. Later, during the 1940s and 1950s, McClintock showed how certain genes were responsible for turning on or off physical characteristics, such as the color of leaves or individual corn kernels. She developed theories to explain the suppression or expression of genetic information from one generation of maize plants to the next that defied the common wisdom of molecular biology prevalent during the 1950s.

In 1957, McClintock received funding from the National Science Foundation and the Rockefeller Foundation to study different varieties, or races, of maize in South and Central America. In the early 1960s, she traveled extensively, collected maize samples that demonstrated interesting evolutionary characteristics, and mentored junior scientists and young graduate students in maize genetics. McClintock and her colleagues spent two decades assembling data on differences in South American maize, which were finally published in 1981 as *The Chromosomal Constitution of Races of Maize*.

McClintock was recognized throughout her career as one of the most distinguished scientists of the 20th century. In 1944, she became the third woman elected to the National Academy of Sciences. She was the first woman to become president of the Genetics Society of America, to which she was elected in 1945. In 1971, President Richard M. Nixon awarded McClintock the National Medal of Science. In 1981, McClintock became the first recipient of a MacArthur Foundation Grant, now known informally as the "genius" grant, which was awarded for her lifetime. In that same year, she was given the Albert and Mary Lasker Award. In 1983, at the age of 81, she received the Nobel Prize in Physiology or Medicine for her work on "mobile genetic elements", that is, genetic transposition, or the ability of genes to change position on the chromosome. McClintock was the first woman to receive an unshared Nobel Prize in that category. McClintock died on September 2, 1992, at the age of 90.

DNA sequencing of genomes from a variety of microbes, plants, and animals indicate that transposable elements exits in virtually all organisms. Surprisingly, they are by far the largest component of the human genome, accounting for almost 50 per cent of our chromosomes. Despite their abundance, the normal genetic role of these elements is not known with certainty. What is becoming clear is that both plants and animals use epigenetic mechanisms to regulate the proliferation of their transposable elements.

As explained earlier eukaryotic transposable elements fall into one of two classes, called class 1 and class 2 elements, that are distinguished by whether the DNA element transposes through an RNA copy [class 1] or directly [class 2]. Class I elements are also called RNA elements because the DNA element in the genome is transcribed into an RNA copy. In contrast, class II elements are called DNA elements because a class II element itself moves from one site in the genome to another. Unlike class I elements, class II elements can excise from the donor site, which means that, if insertion into a gene has created a mutation, excision of the element can lead to the reverstion of the original mutation. The first transposable element discovered genetically in maize were class II elements that led to unusual spotted kernels due to their excision from genes reponsible for pigment producton in the kernel.

Two example of phenotypic effects of transposable elements were:

1. Mosaicism through the activity of a transposable element in maize. Insertion of transposable element disrupts pigment production, resulting in yellow (colorless) kernels. Excision of te transposable element restores pigment production. The large the spot, the earlier the excisions in kernel development.
2. Mosaicism through the excision of transposable elements in snapdragons (*Antirrhinum*). Insertion of a transposable elements disputes pigment production, resultng in white flowers. An excision of the transposable element restores pigment production, resulting in red floral tissue sectors.

20.3.1 Maize Databases

1. MaizeDB Maize Genome Database: This project is part of the National Agricultural Library's Plant Genome Research Program. This provides access to maize-related information in a variety of formats.
2. National Corn Genome Initiative NCGI: Links to the main maize genome sites.

20.4 RICE GENOME DATA

Rice is a cereal foodstuff which forms an important part of the diet of many people worldwide. Diploid chromosome of rice number is 24. Genome of rice is 466 MB in size and is estimated to contain about 40,000 genes. Rice is an important commercially grown plant whose genome size is little higher than *Arabidopsis thalliana* and is much lesser than that of wheat.

Two different groups of investigators from Beijing, Tokyo, and Washington have sequenced the rice genome. Their work continued for more than a decade. Stephen A. Goff and coworkers from USA, sequenced *Oryza sativa* L. ssp. *Japonica* and Jun Yu and his group from China sequenced subspecies *Indica* of rice. *Science* April 2002 issue published these two sequences for the rice genome. Rice genome is made available to public as Rice Genome Research Program (RGP) by NIAS DNA data.

The draft sequence of the rice genome from both sub-species is produced by whole-genome shotgun sequencing. The genome of indica ssp. was 466 megabases in size, with an estimated 46,022 to 55,615 genes. Functional coverage in the assembled sequences was 92.0%. About 42.2% of the genome was in exact 20-nucleotide oligomer repeats, and most of the transposons were in the intergenic regions between genes. Although 80.6% of predicted *Arabidopsis thalliana* genes had a homolog in rice, only 49.4% of predicted rice genes had a homolog in *A. thalliana*. The large proportion of rice genes with no recognizable homologs is due to a gradient in the GC content of rice coding sequences.

The genome of the japonica sub-species of rice, an important cereal and model monocot, was sequenced and assembled by whole-genome shotgun sequencing. The assembled sequence covers

93% of the 420-megabase genome. Gene predictions on the assembled sequence suggest that the genome contains 32,000 to 50,000 genes. Homologs of 98% of the known maize, wheat, and barley proteins are found in rice. Synteny and gene homology between rice and the other cereal genomes are extensive, whereas synteny with *Arabidopsis* is limited. Assignment of candidate rice orthologs to Arabidopsis genes is possible in many cases. The rice genome sequence provides a foundation for the improvement of cereals, our most important crops.

20.5 GENOME EVOLUTION IN GRASSES

The genomes of grass species vary enormously in size. The smallest, at 400 Mb, is found in rice; the largest, at 16,000 Mb, is found in wheat. Although some of the difference in genome size results from the fact that wheat is an allohexaploid whereas rice is a diploid, a far more important factor is the large variation from one species to the next in types and amount of repetitive DNA sequences present. Each chromosome in wheat contains approximately 25 times as much DNA as each chromosome in rice. For comparison, maize has a genome size of 2500 Mb; it is intermediate in size among the grasses and approximately the same size as the human genome.

Table 20.5: Genome size variation in different grasses

Common name	*Scientific name*	*Monoploid genome size (Mb)*
Rice	*Oryza sativa*	400
Wheat	*Triticum aestivum*	16,000
Maize	*Zea mays*	2500
Foxtail millet	*Setaria itallica*	500
Sugar cane	*Saccharum officinarum*	3500
Sorghum	*Sorghum vulgare*	1000

In spite of the large variation in chromosome number and genome size in the grass family, there are a number of genetic and physical linkages between single-copy genes that are remarkably conserved amid a background of very rapidly evolving repetitive DNA sequences. In particular, each of the conserved regions can be identified in all the grasses and referred to a similar region in the rice genome. Each of such conserved genete and physical linkages is called a synteny group. Synteny groups are found in other species comparisons as well. For example, the human and mouse genomes share many synteny groups. These synteny groups are often useful in identifying the mouse homolog of a human gene.

20.6 OTHER PLANT GENOME DATABASES

There are 49 eukaryotic genome sequences going one of these 6 are complete and 5 are at assembly stage. A small description about other website is given below:

1. *Dendrome:* A Genome Database for Forest Trees. This is a collection of specialized forest tree genome databases which are being developed by the Institute of Forest Genetics. The initial focus of the project is to develop a genome mapping database for loblolly pine (*Pinus taeda L.*).
2. *Demeter's Genomes:* Demeter's Genomes is primarily a site for communication of Plant Genome data among scientists worldwide. The medium is a set of genome databases, similar to the ones for Human genome mapping and sequencing. They include data about plant genes, chromosome maps, DNA fingerprints, beneficial traits, and crop varieties that possess those traits. This is a semi-private research consortium, managed and supported by MAFF, NIAR, a consortium of some twenty Japanese companies and the Japan Racing Association (JRA).
3. *Mendel-Plant Gene Nomenclature Database:* The Mendel database project aims to develop a common nomenclature for sequenced genes, based on gene families, for all photosynthetic

organisms, the organelle genomes of both photosynthetic and non-photosynthetic organisms (fungi, algae, and protozoa) and plant viruses. The Mendel database contains information on: Gene Name Gene synonym Accession number Gene product name and synonym information on expression.

4. *Medicago Genome Initiative: Medicago truncatula* is a model system for studying crop species in the family Fabaceae. Currently, the database contains only cDNA sequences from *Medicago truncatula* that were submitted from the Noble Foundation Center for *Medicago* Genomics Research. NCGR plans to form collaborations that make this resource useful for researchers interested in other legume species by adding publicly available *Medicago* sequences and sequences from other species in the same family. Enhancements to the functionality of the system are planned: Storage and analysis capabilities for genomic sequences, sequence clustering and assembly methods for estimating transcript diversity in cDNA libraries, heuristics for distinguishing species of origin of clones from mixed libraries sequence, similarity searching against sequences in the database and links to other data types (e.g., maps, expression data).

SUMMARY

Plant genomes have all the characters of typical eukaryotic genomes which include: interrupted genome, sequence repeats, transposable elements, pseudogenes and organisation of DNA as chromosomes. *Arabidopsis thalliana* is a model organism in plants because it is a small flowering plant that can be cultured in a large number in laboratory and has a small genome of five chromosoms. *Arabidopsis* genome was completely sequence in 2000 and is reported to contain 27,029 protein coding genes. Maize is another model plant, important for study of transposable elements. Other completely sequenced plant genomes are *Oryza saliva*, *Ostreococcus* and *Pseudendoclonium*.

EXERCISE

1. A cloned gene from *Arabidopsis* is used a radioactive probe against DNA samples from cabbage (which is in the same plant family) digested by three different restriction enzymes. For enzyme 1, there were three radioactive bands on the autoradiogram; for enzyme 2, there was one band; and, for enzyme 3, there were two bands. How can these results be explained?
2. Five YAC clones of maize DNA were tested for sequence-tagged sites (STSs): STS1 through STS7. (An STS is a short unique sequence that is PCR amplified by a specific primer pair.) The results are shown in the following table, in which a plus sign shows that the YAC contains that STS:

	STS						
YAC	1	2	3	4	5	6	7
A	+	–	+	+	–	–	–
B	+	–	–	–	+	–	–
C	–	–	+	+	–	–	+
D	–	+	–	–	+	+	–
E	–	–	+	–	–	–	+

   (*a*) Draw a physical map showing the STS order.

   (*b*) Align the YACs into a contig.

3. Compare and contrast *Arabidopsis* genome with *Drosophila* genome?
4. Write down vital statistics of maize genome?

21

BIOINFORMATICS OF ANIMALS

Studies on animal genomes are much more elaborate as compared to plant genomes. Animal genome sequencing began in 1998, when *Caenorhabditis elegans,* a round worm genome was deciphered by an international team led by Bob Waterson and John Sulston. Next breakthrough achievement was sequencing of *Drosophila* genome in 2000, which is represented in four chromosomes, in a haploid organism. Travelling through mouse, human and chimpanzee, we have reached to the level of fossil human genome or personal genome sequencing. Personal genome project require a time shorter than 70 days and costs similar to a Mercedes. In May 2007 scientists at Baylor College of Medicine Human Genome Sequencing Center in Houston, using data obtained from a Connecticut based sequenced company named 454 Life Sciences sequenced the complete genome of Dr. James D. Watson. Also, they are proposing to sequence fossil human Neanderthal genome sequencing within 2 years.

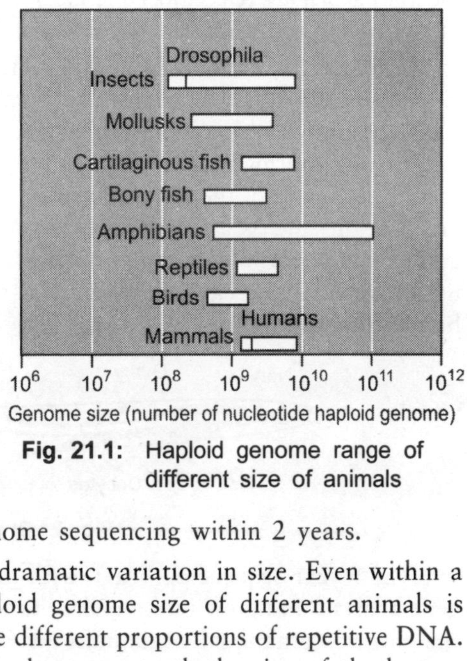

Fig. 21.1: Haploid genome range of different size of animals

When comparing genome size in animals, there is a dramatic variation in size. Even within a group of animals it is varying a lot. In Fig. 21.1 haploid genome size of different animals is depicted. Genomes differ in size, in part because they have different proportions of repetitive DNA. For example, the total genome size of the puffer fish is about one-tenth the size of the human genome. However, the puffer fish genome has about the same number of genes as the human genome, and the genes appear to have the same functions. This is partially because it contains only about fifteen per cent repetitive DNA, while more than half the human genome is repetitive DNA. Because most human genes are present in the puffer fish and repetitive DNA less clutters the puffer fish genome, this model organism may help scientists identify the genes responsible for human diseases.

Animal Genome Size Database, Release 2.0, is a comprehensive catalogue of animal genome size data. Haploid DNA contents (C-values, in picograms) are currently available for 4344 species based on 5762 records from 621 published sources. The largest animal genome reported remains

133pg for *Paranthropus aethiopicus (*an ancient human), followed by the amphibians *Necturus lewisi* and *N. punctatus* at 120pg. At present, the range in published eukaryote genome sizes is between about 0.0023pg in the parasitic microsporidium *Encephalitozoon intestinalis*, and 1,400pg in another protist, the free-living amoeba *Chaos chaos*—a greater than 600,000-fold difference. The largest genome size is often attributed to *Amoeba dubia* at 700pg, conveniently 200 times larger than a human's and sometimes *Encephalitozoon cuniculi* is considered the smallest at about 0.0029pg for a more than 200,000-fold range. Many previous publications have taken brewer's yeast *Saccharomyces cerevisiae* at 0.008pg to be the lowest eukaryotic C-value, for a range of about 80,000-fold. Smallest animal genome size is 0.03pg, for *Meloidogyne graminicola*, a root-knot nematode.

21.1 *CAENORHABDITIS ELEGANS* GENOME

For the first time, scientists have the nearly complete genetic instructions for an animal that, like humans, has a nervous system, digests food, and reproduces sexually. The 97-million-base genome of the tiny roundworm *Caenorhabditis elegans* was deciphered by an international team led by Bob Waterston (Washington University School of Medicine, St. Louis) and John Sulston (Sanger Centre, Cambridge, England)(Historical Profiles 23 and 24). The work was reported in a special issue of the journal *Science* (December 11, 1998) that featured six articles describing the history and significance of the accomplishment and some early sequence-analysis results.

C. elegans is a model organism (Fig. 21.2) with total of only few thousand adult cells. Male *C. elegans* has 11 chromosomes and female has 12 chromosomes. These form a nervous system, a digestive tract with a mouth, pharynx and anus and a reproductive system that can produce eggs and sperms. During its 2- to 3-week life span in the dirt of temperate regions, the benign *C. elegans* carries out many of the same processes as humans. Unlike the much smaller microbes sequenced so far, it begins life as a single fertilized cell that undergoes a series of divisions as it grows into an adult animal, forming complex tissues and organ systems. Researchers have found it particularly useful for studying early development, neurobiology, and aging—processes that have parallels in human biology.

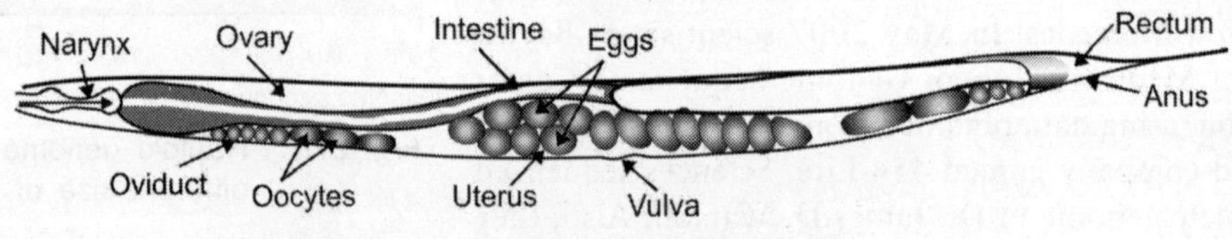

Fig. 21.2: Diagram of *C. elegans* body

The 9-year sequencing began with the development of a clone-based physical map to facilitate gene analysis and grew into a collaboration among *C. elegans* sequencing consortium members and the entire international community of *C. elegans* researchers. In addition to the nuclear genome-sequencing effort, other researchers sequenced it 15 kb mitochondrial genome and carried out extensive cDNA analyses that facilitated gene identification. Free data exchange and immediate data release have been hallmarks of the project, which has been a model for cooperation and sharing among Human Genome Project researchers. The first of a two-part sequencing process used to parse the *C. elegans* genome was the "shotgun" sequencing of randomly chosen subclones (each only a small piece of a much larger cloned DNA molecule). The finishing phase used a more ordered (directed) sequencing strategy to close specific remaining gaps and resolve ambiguities.

The *C. elegans* genome consists of about 97 million base pairs of DNA, approximately one-thirtieth the number in the human genome. It is packaged into 11 chromosomes containing about 19,000 genes, several times the number originally predicted by classical genetics experiments. About

40% of identified genes match those of other organisms, including humans. Like the human genome, *C. elegans* contains large amounts of repeated DNA that does not encode proteins but probably plays a role in chromosome function, gene organization, or regulation of gene activity.

Table 21.1: Vital statistics of genome of *C. elegans*

Genome size	97 Mb
Chromosomes	5 autosomes (2 n = 10), X chromosome
Number of genes	19,000
Percentage with human homologs	25%
Average gene size	5 kb, 5 exons/gene
Transposons	Several types, active in some strains
Genome sequenced in	1998

The *C. elegans* sequence presents scientists with a treasure trove of information and an extremely powerful analytical tool. *Caenorhabditis elegans* is known for its advanced genetic information and complete descriptions of its cellular anatomy, cell lineage, and neuronal wiring diagram. Furthermore, recently developed methods for studying the effects of inhibiting any endogenous gene and for isolating deletions in any gene will greatly facilitate sequence-based gene analysis. The combination of the complete genome sequence with these and other methods will allow investigation of a host of problems at new levels of sophistication.

Comparative analysis of predicted protein sequences encoded by the genomes of *Caenorhabditis elegans* and *Saccharomyces cerevisiae* suggests that most of the core biological functions are carried out by orthologous proteins (proteins of different species that can be traced back to a common ancestor) that occur in comparable numbers. The specialized processes of signal transduction and regulatory control that are unique to the multicellular worm appear to use novel proteins, many of which re-use conserved domains. Major expansion of the number of some of these domains seen in the worm may have contributed to the advent of multicellularity. The proteins conserved in yeast and worm are likely to have orthologs throughout eukaryotes; in contrast, the proteins unique to the worm may well define metazoans.

Despite its humble 302-cell nervous system, the nematode possesses genes for most of the known molecular components of vertebrate brains, which suggests that it provides a valid minimal model for the nervous systems of all animals. The complete sequence reveals the full extent of gene families and shows which developmental control genes are universal and which are missing in different phylogenetic lineages. This will lead to a better understanding of features that make multicellular organisms similar and different.

21.2 THE *DROSOPHILA* GENOME

The *Drosophila melanogaster* genome is 165 Mb, with about 120 Mb of this being euchromatic. The genome is organized in 4 chromosome pairs (Fig. 21.3) and is estimated to contain roughly 13,000 to 14,000 genes. *Drosophila* a fruitfly is a model multicellular animal. In larval stage the chromosome have a well-marked pattern of banding that makes it possible to observe physical changes such as deletion and duplications which can be correlated with genetic changes in morphology and biochemistry. The *Drosophila* research community—both through the efforts of genome projects and of individual investigators—has accumulated a vast amount of data on the genetic and molecular organization of the *Drosophila* genome as well as on the structure, expression and function of individual genes.

The genome of *D. melanogaster*, the largest yet sequenced in full, was described in the 24 March 2000 issue of Science magazine, in a series of articles jointly authored by hundreds of scientists, technicians, and students from 20 public and private institutions in five countries. The collaboration was led by Gerald Rubin. *D. melanogaster* has some 165 million bases in its 80 percent of the genome is located on the large chromosomes labeled 2 and 3. It's genome contain 13000 genes in its genome which is much lower than human beings Vital statistics for *Drosophila* genome is given in Table 21.2.

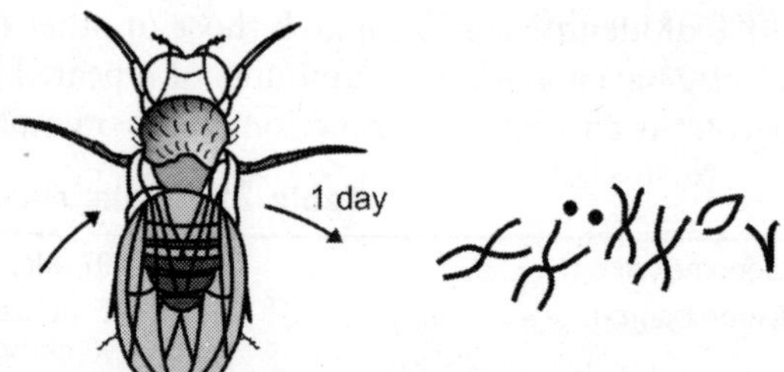

Fig. 21.3: (a) Drosophila body
(b) Metaphase chromosome

Table 21.2: Drossphila genome details

Genome size	165 Mb
Chromosomes	Diploid, 3 autosomes, X and Y (2 n = 8)
Number of genes	13,000 to 14,000
Percentage with human Homologs	~50%
Average gene size	3 kb, 4 exons/gene
Transposons	P elements, among other
Genome sequenced in	2000

D. melanogaster's importance is far greater than as a trial run for the mouse and human genome. In a set of 289 human genes implicated in diseases, 177 are closely similar to fruit fly genes, including genes that play roles in cancers, in kidney, blood, and neurological diseases, and in metabolic and immune-system disorders. The underlying biochemistry of fruit flies and humans is remarkably similar so fruit flies can provide clues to understanding human diseases caused by defective genes. We can find human tumor-suppressing genes in flies easier than we can in the mouse.

21.2.1 Databases

There are two main databases for *Drosophila* genome information: FlyBase and the BDGP database. In addition there are numerous specialized databases dealing with many aspects of *Drosophila* anatomy, gene expression and gene function; a list of these other resources can be found at http://www-leland.stanford.edu/~ger/drosophila.html. FlyBase is the central database for *Drosophila* genetic information. It captures information from the literature, from the major genome projects and through bulk data provided by sequence and bibliographic data banks. The major genomic data sets in FlyBase include genetic information on genes, alleles, chromosomal aberrations and transposons, as well as molecular information on contigs, chromosomal walks, transcripts and proteins. FlyBase data can be accessed through the World Wide Web (http://morgan.harvard.edu/ or http://www.embl-ebi.ac.uk/flybase/) or by gopher server (flybase.bio.indiana.edu). Data available from the BDGP home page (http://www.fruitfly.org/) include data on the P-element gene-disruption project and monthly updates of the P1-based physical map. A sequence display is under development that presents genomic sequence determined by the BDGP, annotated with the results of homology searches, gene prediction programs and cDNA sequence and expression analyses.

21.2.2 Biological Annotation of the Genomic Sequence

A key use of the sequence information from the model organisms, such as *Drosophila*, will be to help interpret the sequence of the human genome. If the model organism genome projects are to be maximally useful in assigning functions to human DNA sequences, they will need to utilize the powerful tools for determining gene function that are available to them so that not only the sequences of the genes, but also their biological functions, are determined. Among the model organisms, *Drosophila* is particularly well-suited for this role. In terms of evolutionary conservation of sequence similarity, *Drosophila* is the closest of the invertebrate model organisms to humans. Moreover, it terms of morphological, physiological, and behavioral complexity *Drosophila* is by far the closest to humans of these model organisms, yet its genome is not substantially bigger than the least complex metazoans. Finally, the large *Drosophila* research community—about one researcher per 2 genes—has provided a wealth of information and understanding unusual in its depth and intellectual breadth. These workers have already extensively characterized about 1000, or 10%, of all *Drosophila* genes in terms of sequence, gene structure, expression pattern and biological function.

As part of its efforts to develop and apply tools for large-scale functional analysis, the BDGP has undertaken a novel gene disruption project based on mutagenesis by transposable element insertion. Transposable elements provide a powerful tool for correlating genetic and molecular information because they generate a simple, reproducible lesion upon insertion that can be detected much more easily than damage produced by other mutagens. In *D. melanogaster* the *P* transposable element has been particularly useful because it moves with high frequency but can be tightly controlled by limiting the availability of an element-encoded transposase. The initial goal of the project is to establish a large collection of *Drosophila* strains that each contain a single genetically engineered *P* transposable element insertion that mutates a different gene, in a genome free of other *P* elements. Moreover, the inserted *P* elements in BDGP lines carry enhancer-traps that can be used to efficiently acquire information about the expression pattern of disrupted genes. The strains in the current collection disrupt 20-25 per cent of essential genes, provide information on their expression patterns, and link the genetic, cytogenetic and physical maps of the *Drosophila* genome at ~100 kb intervals.

21.3 THE MOUSE GENOME

Mus cesculus the house mouse, is a model organism for invertebrate. It has been studied to compare the genetic basis of vertebrate and invertebrate development as well as to explore the genetics of antigen-antibody system of maternal-fetal interaction in uterus and in understanding cancer.

As the most powerful model organism in biomedical research, the mouse was the second mammal to be sequenced as part of the Human Genome Project. A high-quality draft of the mouse genome was produced and analyzed in 2002 by the Mouse Genome Sequencing Consortium, including the Broad Institute, Washington University and the Sanger Institute. The genome sequence allows genetic studies, such as identification of disease mutations, on an unprecedented scale. Initial analysis of the differences and similarities between the mouse and human genomes suggested that as much as 5% of the human genome might be highly conserved between mammals based on function—considerably more than the ~1% of the genome known to contain protein coding genes. This finding has sparked the current search for all the highly conserved, potentially functional elements across mammalian genomes.

Among model organisms, mouse is the one whose genome most closely resembles the human genome. The mouse genome is about 14 per cent smaller than that of humans (the human genome is 3000 Mb), but it has approximately the same number of genes. A surprising 99 per cent of mouse genes seem to have homologs in humans (Table 21.3) there, a large proportion of the genome is syntenic with that of humans; that is, there are large blocks containing the genes in the same relative position. Such similarities are the key to the mouse's success as a model.

These similarities allow mice to be treated as "standing" for their human counterparts in many ways. Potential mutagens and carcinogens that we suspect of causing damage to humans, for example are tested on mice, and mouse models are essential in studying a wide array of human genetic diseases.

Table 21.3: Details Genome structure of *Mus*

Genome size	2600 Mb
Chromosomes	19 autosomes, X Y (2 n = 40)
Number of genes	less than 30,000
Percentage with human homologs	99%
Average gene size	40 kb, 8.3 exons/gene
Transposons	Source of 38% of genome
Genome sequenced in	2002

21.4 THE HUMAN GENOME PROJECT

The Human Genome Project (HGP) was the international, collaborative research program whose goal was the complete mapping and understanding of all the genes of human beings a 13-year effort which began in October 1990 and completed in 2003. The HGP was the natural culmination of the history of genetics research. The HGP has revealed that there are probably somewhere between 20,000 and 25,000 human genes. Human genome diploid is made up of 46 chromósomes size of genome is 3300 MB, nucleotide base pairs. The completed human sequence can now identify their locations. This ultimate product of the HGP has given the world a resource of detailed information about the structure, organisation and function of the complete set of human genes. This information can be thought of as the basic set of inheritable "instructions" for the development and function of a human being.

The International Human Genome Sequencing Consortium published the first draft of the human genome in the journal *Nature* in February 2001 with the sequence of the entire genome's three billion base pairs, some 90 per cent complete. A startling finding of this first draft was that the number of human genes appeared to be significantly fewer than previous estimates, which ranged from 50,000 genes to as many as 40,000. The full sequence was completed and published in April 2003. It indicated are number of genes in human is 20,000 to 25,000.

The tools created through the HGP also continue to inform efforts to characterize the entire genomes of several other organisms used extensively in biological research, such as mice, fruit flies and flatworms. These efforts support each other, because most organisms have many homologous genes with similar functions. Therefore, the identification of the sequence or function of a gene in a model organism, for example, the roundworm *C. elegans*, has the potential to explain a homologous gene in human beings, or in one of the other model organisms.

The vital statistics for human genome is given in Table 21.4. The human genome is distributed over 22 chromosome pairs plus the X and Y chromosomes. The DNA contents of autosomes range from 279 mbp down to 48 mbp. The X-chromosome of human contains 163 mbp and α-chromosomes only 51 mbp. Only a small part of the human genome codes for polypeptides; i.e., only about 3% of it encodes exons of mRNA. Exons are typically small, about 50 bases, whereas introns are large extending between 1000 to 100,000 bp. Transcripts are compared of an average of 10 exons, although many have substantially more. On the basis of current cDNA and EST data, 60% of protein coding genes are likely to have two or more splice variants. On average there are about three splice variants. Hence, the number of distinct proteins encoded by human genome is about three times greater than the number of recognised genes.

HISTORICAL PROFILE 23

Genome Project and Sir John Sulston

John Sulston was born in Buckinghamshire on 24 March 1942, the son of a Church of England minister and a school teacher. A childhood obsession with how things worked—whether animate or inanimate—led to a degree in Natural Sciences at the University of Cambridge, specialising in organic chemistry. He stayed on to do a PhD in the synthesis of oligonucleotides, short stretches of RNA. He arrived there in 1969, and joined the laboratory of Sydney Brenner. Brenner had set out to understand the sequence of events from gene to whole, living, behaving organism by studying the tiny nematode worm *Caenorhabditis elegans*.

For more than 20 years Sulston worked on the worm, charting for the first time the sequence of cell divisions that lead from a fertilised egg to an adult worm, identifying genetic mutations that interfere with normal development, and then going on to map and sequence the 100 million letters of DNA code that make up the worm genome. The success of this project of sequencing of *c. elegans genome* carried out jointly with Bob Waterston of Washington University in St Louis, led the Welcome Trust to put Sulston at the head of the Sanger Centre, established in 1993 to make a major contribution to the international Human Genome Project. There he led a team of several hundred scientists who completed the sequencing of one-third of the 3-billion-letter human genome, together with the genomes of many important pathogens such as the tuberculosis and leprosy bacilli.

As the leader of one of the four principal sequencing centres in the world, Sulston was a major influence on the Human Genome Project as a whole, particularly in establishing the principle that the information in the genome should be freely released so that all could benefit. In 2000 Sulston resigned as director of the Sanger Centre (now the Wellcome Trust Sanger Institute), though he retained an office there for a few more years, continuing to work on the Human Genome Project publications and on outstanding problems with the worm genome.

Anxious to promote his views on free release and global inequality, he published his own account of the 'science, politics and ethics' of the Human Genome Project, while adding his voice to influential bodies such as the Human Genetics Commission and an advisory group on intellectual property set up by the Royal Society. The same year he gave the Royal Institution Christmas Lectures for children on the topic 'The secrets of life'. In 2002, John Sulston was awarded the Nobel Prize for Physiology or Medicine jointly with Sydney Brenner and Bob Horvitz, for the work they had done in understanding the development of the worm and particularly the role of programmed cell death.

Repeat sequences comprise over 50% of the genome. These include LINES, SINES and retroposed pseudogenes. Details of transposable elements in human genome is given in Table 21.5.

Table 21.4: The vital statistics of human genome

Genome size	3.2×10^9bp
Chromosome Number	22 pairs autosome x, y (2n = 46)
No. of genes	v 30,000
Average gene size	50 bp exons and introns 1000 to 100,000 bp
Repeat sequences	50%
Coding sequences	5%
Genome sequenced in	2003

Table 21.5: Type of transposable elements in the human genome

Element	*Size (bp)*	*Copy number*	*Fraction of genome %*
Short interspersed Nuclear Elements (SINEs)	100–300	1500,000	13
Long interspersed Nuclear Elements (Lines)	6000–8000	850 000	21
Long Terminal Repeats	15000–110000	450 000	8
DNA Transposon fossils	80–3000	300000	3

21.4.1 Other Related Genome Projects

(*a*) Neanderthal genome: Neanderthal, a hominid (*Homo neanderthalensis* syn *H. sapiens neanderthalensis*) known from skeletal remains in Europe, northern Africa, and western Asia that lived from about 30,000 to 200,000 years ago—also called *Neanderthal man.* The veil of mystery surrounding our extinct hominid cousins, the Neanderthals, has been at least partially lifted to reveal surprising results. Scientists with the U.S. Department of Energy's Lawrence Berkeley National Laboratory (Berkeley Lab) and the Joint Genome Institute (JGI) have sequenced genomic DNA from fossilized Neanderthal bones. Their results show that the genomes of modern humans and Neanderthals are at least 99.5% identical, but despite this genetic similarity, and despite the two species having cohabitated the same geographic region for thousands of years, there is no evidence of any significant crossbreeding between the two. Based on these early results, *Homo sapiens* and *Homo neanderthalensis* last shared a common ancestor approximately 700,000 years ago.

(*b*) Personal human genome: Personal human genome sequencing refers to sequencing of an individual genome. It has really been possible to because of new sequencing technique called pyrosequencing. The Nobel Prize-winning scientist who helped discover the molecular structure of DNA has become the first person to receive his own personal genome map. The map, a breakdown of his DNA that shows illnesses he is predisposed to contracting, is the first step in making the sequencing of individual human genomes quick, affordable and a routine part of medical care, according to researchers. Watson was chosen for the project because of his contributions to the field, and the map was completed after he submitted a blood sample. A review showed he has some variances that could induce cancer—which appeared to mirror his actual health. Watson said that he has had skin cancer and that his sister had breast cancer. The $1 million, two-month project was a collaboration between 454 Life Sciences Corpration, a Connecticut company that specializes in DNA sequencing, and Baylor College of Medicine's Human Genome Sequencing Center. At the moment, there are no plans to complete more maps in the immediate future, though researchers want to eventually map more people.

21.4.2 Comparative Genomics

Watson's genome was sequenced in May 2007. Then in September 2007 genome of Dr. Craig Venter was sequenced. These studies revealed that these genomes are only 90% similar and not 99.9% as predicted earlier. The area of genomics that compares and contrasts genomes from different organisms is called comparative genomics. It provides a powerful method of identifying essential sequences. Retention of an identical sequence in many different groups in years since their divergence is taken as evidence that the sequence is being maintained by natural selection because it plays an essential role.

A comparison of genomes of yeast, fly, worm and human revealed 1308 group of proteins that appear in four. These form a conserved core of proteins for basic functions, including metabolism,

HISTORICAL PROFILE 24

Robert Waterston Deciphers Genetic Blueprints

Studied several aspects of *C. elegans*, whose muscles are surprisingly similar to those of humans. His group has since defined the role of a major protein called myosin in muscle assembly and contraction. It also has located about 25 muscle genes and their proteins. Watching his Maynard alternately struggle and succeed in developing methods for mapping yeast made him think about the genome of the worm. With the aid of a John Simon Guggenheim Fellowship, Waterston returned to Cambridge in 1985 for a sabbatical. There he began to work with John E. Sulston, and AlanR. Coulson, who had begun to create a physical map of the worm's DNA.

In 1986, Waterston found that David T. Burke had found a way to convert large segments of DNA into artificial chromosomes that yeast cells would copy. These yeast artificial chromosomes, or YACs, provided a way to get sufficient amounts of DNA for analysis and, overlapped, would generate a map. The YACs, combined with a previously developed map made from overlapping bacterial clones, led to extensive coverage of a physical map of the worm by 1989. At a meeting at Cold Spring Harbor, N.Y., that summer, Coulson, Sulston and Waterston pinned six long pieces of paper, each corresponding to a *C. elegans* chromosome, across one room's entire wall. Their impressive map persuaded influential geneticists that it was time to spell out the details of the genome with DNA sequence.

The success of the worm project has been so spectacular that Waterston and Sulston convinced their funding agencies to go ahead with human genome sequencing. Bob Waterston rattled the scientific community by urging the Human Genome Project to get on with our most daunting goal: the letter-by-letter sequencing of the entire human genome.In addition to being wise, thoughtful and generous-spirited when it comes to genome matters, he led the most successful DNA sequencing partnership in the world.

DNA replication and repair, and translation. The proteins of worm and fly are built from a structural repertoir containing about three times as many domains as proteins of yeast. Human proteins are build from about twice as many as those of worm and fly. Most of these domains appear also in bacteria and archaea, but some are specific to vertebrate such as defence and immunity proteins and proteins in newous system.

More than ninety-nine per cent of human genes have a related copy in the mouse. As one examines animals that are more distantly related, the proportion of the genes they share decreases; however, despite about 500 million years of evolutionary separation, half the genes in the lowly sea squirt correspond to those found in humans. This remarkable conservation of gene structure is striking considering how much these animals differ in morphology, physiology, and behavior. If they share so many of the same genes, why are different animals so different? Differences among species result largely from differences in the time and location of the genes' expression. Let us consider our closest relative, the chimpanzee. Not only do chimpanzees and humans share nearly all of the same genes, but the DNA sequences of those genes also are very similar between the two species. Svante Pääbo sequenced three million bases of the chimp genome and found that chimps and humans differ overall by less than two per cent at the sequence level. Based on the low sequence divergence, Pääbo hypothesized that the difference between humans and chimpanzees was due mainly to how the genes were expressed in the different species. To test this hypothesis, Pääbo compared the expression pattern of 20,000 human genes in humans and chimps. He found that while expression levels were similar in liver cells and blood, there were larger differences in brain cells. This suggests that the human brain has increased the use of certain genes compared to those same genes in a chimp. So, it not so much the sequence of the genes that is important, but how they are expressed to make the cell's proteins that determines the unique characteristics of each organism.

21.5 BIOINFORMATICS OF ANIMAL GENOMES

Informatics resources are vividly applied to animal genome information. However, there is nothing specific about their use for animal genomes except that they are used in diseases diagnosis and cure here. The several areas where they are applied is given below:

1. Improve content and utility of databases.
2. Develop better tools for data generation, capture and annotation.
3. Develop and improve tools and databases for comprehensive functional studies.
4. Develop and improve tools for representing and analyzing sequence similarity and variation.
5. Create mechanisms to support effective approaches for producing robust, exportable software that can be widely shared.

EXPERIMENT 21A

To Study the Transcription Factor on Chromosome Number 2 of *Drosophila*

About Gene2promoter: Gene2Promoter provides access to promoter sequences of all genes annotated in the available genomes. For each locus the chromosomal location and a short description of the corresponding gene is provided together with the promoters and transcripts. Several tasks for comparative analysis of the selected promoters can be started directly from the Gene2Promoter output. The number of input genes/sequences is limited to 50, since comparative analysis of more than 50 promoters will not yield meaningful results.

Main features of Gene2Promoter are as follows:

1. Listing of all alternative transcripts and their promoters for a locus
2. Easy check box selection of identified, ranked promoters from a list of matches
3. Immediate interactive graphical analysis for common transcription factor binding sites.
4. Full ElDorado analysis at one click for each promoter and its gene.
5. Direct submission of selected promoters to tasks like FrameWorker (analysis for common appearance of TF binding sites).

Gene2Promoter is available as internet installation evaluation account. We need to get a free of charge evaluation account. In the present study we will use the online account.

TESTING THE HYPOTHESIS

1. Log on to web site of genomatix. (http://www. genomatix. deproducts Gene2Promoters in like Fig 21.4.
2. In the organism select *Drosophila* because of its small genome.
3. We intend to study transcription factors. Add transcription factor to the keyword text box.

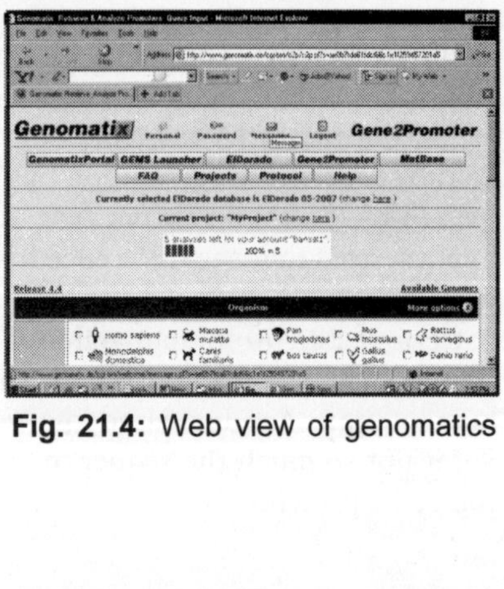

Fig. 21.4: Web view of genomatics

4. Press Submit. The query gave 22 result transcription sequences in *Drosophila*. The view is shown in Fig. 21.5and data is shown in databox 21.1.

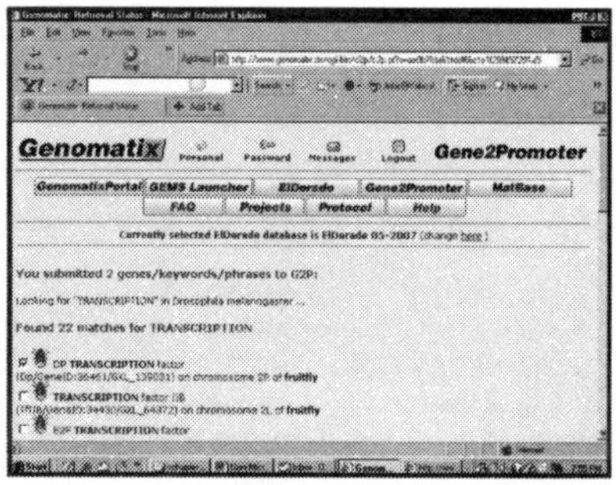

Fig. 21.5: Results of the search

5. Of these select DP "TRANSCRIPTION factor (Dp/GeneID:36461/GXL_139031) on chromosome 2R of fruitfly".
6. Select "Additionally list orthologous genes in output". It gives output and press continue button.

DATA

Data Box 21.1: Found 22 matches for TRANSCRIPTION

DP TRANSCRIPTION factor
(Dp/GeneID:36461/GXL_139031) on chromosome 2R of fruitfly
TRANSCRIPTION factor IIB
(TfIIB/GeneID:34430/GXL_64372) on chromosome 2L of fruitfly
E2F TRANSCRIPTION factor
(E2f/GeneID:42550/GXL_92538) on chromosome 3R of fruitfly
TRANSCRIPTION factor IIA L
(TfIIA-L/GeneID:43284/GXL_231930) on chromosome 3R of fruitfly
TRANSCRIPTION-factor-IIA-S
(TfIIA-S/GeneID:42822/GXL_92655) on chromosome 3R of fruitfly
Another TRANSCRIPTION unit
(Atu/GeneID:40684/GXL_190569) on chromosome 3R of fruitfly
ftz TRANSCRIPTION factor 1
(ftz-f1/GeneID:40045/GXL_196294) on chromosome 3L of fruitfly
E2F TRANSCRIPTION factor 2
(E2f2/GeneID:35381/GXL_66046) on chromosome 2L of fruitfly
Adh TRANSCRIPTION factor 1
(Adf1/GeneID:47082/GXL_3269) on chromosome 2R of fruitfly
TRANSCRIPTION factor IIEbeta
(TfIIEbeta/GeneID:38527/GXL_215101) on chromosome 3L of fruitfly
TRANSCRIPTION factor IIFalpha
(TfIIFalpha/GeneID:40790/GXL_190307) on chromosome 3R of fruitfly
TRANSCRIPTION factor IIEalpha
(TfIIEalpha/GeneID:39313/GXL_108543) on chromosome 3L of fruitfly
TRANSCRIPTION factor TFIIFbeta
(TfIIFbeta/GeneID:41290/GXL_9830) on chromosome 3R of fruitfly
Germ line TRANSCRIPTION factor 1
(Gnf1/GeneID:40607/GXL_224670) on chromosome 3R of fruitfly
Activating TRANSCRIPTION factor-2
(Atf-2/GeneID:37978/GXL_173034) on chromosome 2R of fruitfly
mitochondrial TRANSCRIPTION factor A
(TFAM/GeneID:42433/GXL_250051) on chromosome 3R of fruitfly
Mitochondrial TRANSCRIPTION Factor B1
(mtTFB1/GeneID:39320/GXL_108447) on chromosome 3L of fruitfly
mitochondrial TRANSCRIPTION factor B2
(mtTFB2/GeneID:41228/GXL_190453) on chromosome 3R of fruitfly
Myocardin-related TRANSCRIPTION factor
(Mrtf/GeneID:38338/GXL_187302) on chromosome 3L of fruitfly
Calmodulin-binding TRANSCRIPTION activator
(Camta/GeneID:35974/GXL_8938) on chromosome 2R of fruitfly
mitochondrial TRANSCRIPTION termination factor
(mTTF/GeneID:34837/GXL_110691) on chromosome 2L of fruitfly
Signal-transducer and activator of TRANSCRIPTION protein at 92E
(Stat92E/GeneID:42428/GXL_249955) on chromosome 3R of fruitfly

Data Box 21.2

		Results Locus	Selection	Promoter	Transcript/TSS
TRANSCRIPTION found on:	**Drosophila** **melanogaster** NCBI_build5_1 Chromosome 2R Contig NT_033778	**DP transcription** **factor** Dp (GeneID:36461) GXL_139031 · View this locus in ElDorado · More Gene Info · Comparative Genomics		GXP_167193 (+) 9112329- 9112929 (601 bp)	GXT_1645298 (8 exons) NM_165974 TSS = 501
				GXP_167194 (+) 9109402- 9110002 (601 bp)	GXT_1640909 (9 exons) NM_057691 TSS = 501

INTERPRETATION OF DATA

Our search showed 22 transcription factors present on chromosome number 2 in *Drosophila*. The details of matches is shown in Data Box 21.1. Of these the first entry on chromosome number 2 i.e. DP transcription factor was selected for further study.

Following further analysis is possible with these sequences using gene2promoter :

1. An analysis button to check for the presence of certain transcription factor binding sites in the promoters.
2. Locus of binding site can be viewed in ElDorado (Data Box 21.2). It also provides information about the transcripts.

21.6 APPLICATIONS OF GENOMICS AND BIOINFORMATICS

21.6.1 Molecular Medicine

Clinical application of human genome data range from improved diagnosis of disease, earlier detection of genetic predispositions to disease, rational drug design, gene therapy and control systems for drugs to pharmacogenomics. Technology and resources promoted by the Human Genome Project are starting to have profound impacts on biomedical research and promise to revolutionize the wider spectrum of biological research and clinical medicine. Increasingly detailed genome maps have aided researchers seeking genes associated with dozens of genetic conditions, including myotonic dystrophy, fragile X syndrome, neurofibromatosis types 1 and 2, inherited colon cancer, Alzheimer's disease, and familial breast cancer.

It is generally agreed that the sequencing of the human and other genomes will lead to improvement in the health of mankind. Clinical application of genome projects is given below:

1. Diagnosis of disease and disease risks: DNA sequencing can detect the absence of a particular gene, or a mutation. Identification of specific gene sequences associated with diseases will permit fast and reliable diagnosis of conditions (a) when a patient presents with symptoms, (b) in advance of appearance of symptoms, as in tests for inherited late-onset conditions such as Huntington disease, (c) for in utero diagnosis of potential abnormalities such as cystic fibrosis, and (d) for genetic counselling of couples contemplating having children.

   An example of a risk factor detectable at the genetic level involves α-antitrypsin, a protein that normally functions to inhibit elastase in the alveoli of the lung.

2. Pharmacogenomics: Because people differ in their ability to metabolize drugs, different patients with the same condition may require different dosages. Sequence analysis permits selecting drugs and dosages optimal for individual patients, a fast-growing field called *pharmacogenomics.*
3. Identification of drug targets: A target is a protein the function of which can be selectively modified by interaction by a drug, to affect the symptoms or underlying causes of a disease. Identification of a target provides the focus for subsequent steps in the drug design process. Among drugs now in use, the targets of about half are receptors, about a quarter enzymes, and about a quarter hormones. Approximately 7% act on unknown targets.
4. Gene therapy: If a gene is missing or defective, we would like to replace it or at least supply its product. If a gene is overactive, we would like to turn it off. Direct supply of proteins is possible for many diseases, of which insulin replacement for diabetes and Factor VIII for a common form of haemophila are perhaps the best known.

21.6.2 Energy and Environmental Applications

Despite our reliance on the inhabitants of the microbial world, we know little of their number or their nature: estimates are that less than 0.01% of all microbes have been cultivated and characterized. Microbial genome sequencing will help lay a foundation for knowledge that will ultimately benefit human health and the environment. The economy will benefit from further industrial applications of microbial capabilities. In 1994, taking advantage of new capabilities developed by the genome project, Microbial Genome Program is initiated to sequence the genomes of bacteria useful in energy production, environmental remediation, toxic waste reduction and industrial processing.

Information gleaned from the characterization of complete microbial genomes will lead to insights into the development of such new energy-related biotechnologies as photosynthetic systems, microbial systems that function in extreme environments and organisms that can metabolize readily available renewable resources and waste material with equal facility. Expected benefits also include development of diverse new products, processes and test methods that will open the door to a cleaner environment. Biomanufacturing will use nontoxic chemicals and enzymes to reduce the cost and improve the efficiency of industrial processes. Microbial enzymes have been used to bleach paper pulp, stone wash denim, remove lipstick from glassware, breakdown starch in brewing, and coagulate milk protein for cheese production. In the health arena, microbial sequences may help researchers find new human genes and shed light on the disease-producing properties of pathogens.

Microbial genomics will also help pharmaceutical researchers gain a better understanding of how pathogenic microbes cause disease. Sequencing these microbes will help reveal vulnerabilities and identify new drug targets.

Gaining a deeper understanding of the microbial world also will provide insights into the strategies and limits of life on this planet. Data generated in this young program have helped scientists identify the minimum number of genes necessary for life and confirm the existence of a third major kingdom of life. Additionally, the new genetic techniques now allow us to establish more precisely the diversity of microorganisms and identify those critical to maintaining or restoring the function and integrity of large and small ecosystems; this knowledge also can be useful in monitoring and predicting environmental change. Finally, studies on microbial communities provide models for understanding biological interactions and evolutionary history.

21.6.3 Bioarchaeology, Anthropology, Evolution and Human Migration

Understanding genomics will help us to understand human evolution and the common biology we share with all of life. Comparative genomics between humans and other organisms such as mice already has led to similar genes associated with diseases and traits. Further comparative studies will help determine the yet-unknown function of thousands of other genes. Comparing the DNA sequences of entire genomes of different microbes will provide new insights about relationships among the three kingdoms of life: archaebacteria, eukaryotes and prokaryotes.

21.6.4 DNA Forensics (Identification)

Any type of organism can be identified by examination of DNA sequences unique to that species. Identifying individuals is less precise, although when DNA sequencing technologies progress further, direct characterization of very large DNA segments, and possibly even whole genomes, will become feasible and practical and will allow precise individual identification. To identify individuals, forensic scientists scan about 10 DNA regions that vary from person to person and use the data to create a DNA profile of that individual (sometimes called a DNA fingerprint). There is an extremely small chance that another person has the same DNA profile for a particular set of regions.

21.6.5 Agriculture, Livestock Breeding and Bioprocessing

Understanding plant and animal genomes will allow us to create stronger, more disease-resistant plants and animals—reducing the costs of agriculture and providing consumers with more nutritious, pesticide-free foods. Already growers are using bioengineered seeds to grow insect- and drought-resistant crops that require little or no pesticide. Farmers have been able to increase outputs and reduce waste because their crops and herds are healthier.

Alternate uses for crops such as tobacco have been found. One researcher has genetically engineered tobacco plants in his laboratory to produce a bacterial enzyme that breaks down explosives such as TNT and dinitroglycerin. Waste that would take centuries to break down in the soil can be cleaned up by simply growing these special plants in the polluted area.

SUMMARY

Study of animal genomes are much more elaborate as compared to plant genomes. Ist animal genome to be deciphered was of round worm in 1998, followed by *Drosophila* genome in 2000 mouse genome in 2000, human genome in 2003, Neanderthal genome in 2006 and genome of Dr. James D. Watson in 2007. In Experiment - 21.1A transcription factor is located on chromosome number 2 of *Drosophila* using software Gene-2-promoter. In historical profile section profiles of Bob Waterson and John Sulston are discussed for their contribution to animal genome projects. Applications of genome projects range from clinical implication, bioarcheology, DNA forensic to agriculture and bioprocessing.

EXERCISE

1. Two living genera of elephant are represented by African elephant and Indian elephant. It is possible to sequence the mitochondrial cytochrome *b* from a specimen of a fossil elephant preserved in erratic region. Design a program to find to which modern elephant is this fossil elephant more closely related? Mention databases, and software needed for this purpose. Analyses your result accordingly.
2. How many floppy discs, CDs, or DVDs are required to store entire human genome?
3. Write a brief note on legal aspects of genomics and bioinformatics.
4. A team of researchers has obtained a dinosaur bone (*Tyannosauris rex*) and has attempted to extract ancient DNA from it. Using primers to the 125 rRNA gene, they have used PCR and obtained DNA segments that yields a sequenced homologies to crocodile DNA. Other scientists are skeptical that this sequence is really from dinosaur; they believe instead it may be due to

continuation from more recent DNA, such as the remains of a reptile that lived more recently. What criteria might you use to establish the credibility of the dinosaur sequence?

5. One way to identify human oncogene is to case human DNA from a malignant cell to transform a mouse cell. A mouse cell that has been transformed by human DNA will have a human gene incorporated into to genome; it may also have ALU sequences that are closely linked to human oncogeny. As might be already known that human genome contains ALU sequences interspersed every 5000 to 6000 bp. ALU sequences are not found in the mouse genome. Discuss how ALU sequences can provide away to clone human genome?
6. Compare and contrast the experimental advantages of *Drosophila* and C. *elegans* in the study of genomics.
7. Align the sequences of human neutrophil elastase and C. *elegans* elastase.
   (a) In the optimal alignment, how many identical residues are there?
   (b) Would it be reasonable to build a model of C. *elegans* elastase starting from structure of human neutrophil elastase?

WEB ADDRESS

BCM	http://searchlauncher.bcm.tmc.edu/
BLAST	http://0-www.ncbi.nlm.nih.gov.library.vu.edu.au/BLAST/
BLOCKS	http://blocks.fhere.org/blocks/
CATH	http://www.biochem.ucl.ac.uk/bsm
CINEMA 2.1	http://www.bioinf.manchester.ac.uk/dbbrowser/CINEMA2.1/
ClustalW	http://www.ebi.ac.uk/Tools/clustalw/index.html
CLUSTALX	http://bips.u-strasbg.fr/fr/Documentation/ClustalX/
CMR	http://cmr.jcvi.org.tigr-scripts/cmr/cmr homepage
DCA	http://bibiserv.techfak.uni-bielefeld.de/dca/
DIALIGN	http://bibiserv.techfak.uni-bielefeld.de/dialign/
DDJB	http://www.ddbj.nig.ac.jb/
Dotlet	http://myhits.isb-sib.ch/util/dotlet/doc/dotlet_about.html
Dotplot	http://www.hku.hk/bruhk/gcgdoc/dotplot.html
Dotter	http://www.cgb.ki.se/cgb/groups/sonnhammer/Dotter.html
EBI	http://www.ebi.ac.uk
EcoCyc	http://ecocyc.org/
EMBL	http://www.ibi.ac.uk/embi
Emboss Align	http://emboss.sourceforge.net/apps/
Emotif	http://dna.stanford.edu/emotif/
ExPASY	http://www.expasy.ch/tools/dnot. html
Expasy	(http://ca.expasy.org/tools/dna.html)
FASTA	http://fasta.bioch.virginia.edu/fasta
FGENEH	http://genomic.sanger.ac.uk/gf/gf.shtml
GENBANK	http://www.ncbi.nlm.nih.gov/Entrez
GENEBUILDER	http://www.itba.ni.cnr.it/webgene/
GenelD	http://wwwl.imim.es/geneid.html
GENSCAN	http://genes.mit.edu/genscan.html
GENSCAN	http://genes.mit.edu/GENSCAN.html
Glimmer	http://nbc11.biologie.uni-kl.de/framed/left/menu/auto/right/glimmer2.02/
gorf.html	http/www.ncbi.nlm.nih.gov/gorf/
GRAIL-EXP	http://compbio.ornl.gov/grailexp
GRAIL	http://cornpbio.ornl.gov/tools/index.shtml
GRAOL	http://combio.ornl.gov
HMMgene	http://www.cbs.dtu.dk/services/HMMgene
ITIS	http://www.itis.ugda.gov
JALVIEW	http://www.jalview.org/
KEGG	http://www.genome.ad.jp/kegg/)
LALIGN	http://www.ch.embnet.org/software/LALIGN_form.html
LFASTA	http://envgen.nox.ac.uk/bioinformatics/docs/lfasta.html
MACAW	http://bcr.musc.edu/manuals/MACAW.html
MAFFT	http://align.bmr.kyushu-u.ac.jp/mafft/software/
MZEF	http://www.cshl.org/genefinder
NCBI	http://www.ncbi.nlm.nih.gov
NDB	http://ndbserver.rutgers.edu

ORF Finder	http://www.snake.ne.jp/~yama/nph-docomo.cgi/010000A
PDB	http://www.resb.org/pdb
Pfam	http://www.sanger.ac.uk/Pfam
PHIBLAST	http://www.ncbi.nlm.nih.gov
PILUP	http://www.umdnj.edu/gcg/pileup.html
PIR	http://www.nbrf.georgetown.edu/pirwww/
PRINTS	http://biochem.ucl.ac.uk/bsm/
PROCRUSTES	http://www-hto.usc.edu/software/procrustes
PROSITE	http://www.expasy.ch/tools/senpsit2.html
PROSITE	http://www.expasy.ch/prosite
RASMOL	http://www.umass.edu/microbio/rasmol/
READSEQ	http://iubio.bio.Indiana.edu
RSCB	http://www.rcsb.org
SCOP	http://scop.mrc.imb.cam.ac.uk/scop
SIB	http://www.expasy.ch/
SIM	http://www.expasy.ch/tools/sim-prot.html
SIM4	http://globin.cse.psu.edu/html/docs/sim4.html
Splice Predictors	http://deepc2.psi.iastate.edu/cgi-bin/sp.cgi
SSEARCH	http://helix.nih.gov/docs/gcg/ssearch.html
SWISSPROT	http://www.expasy.ch/cgi-bin/sprotE-search-de
T-cofee	http://www.tcoffee.org/Projects_home_page/t_coffee_home_page.html
TIGR	http://www.tigr.org
Translation	http://www.expasy.ch/tools/dna.html
Webgene	(http://www.itba.mi.cnr.it/webgene)
WITZ	http://igweb.integratedgenomics.com/WIT2/)
YEAST GENOME DATABASE	http://www.yeastgenome.org

APPENDIX

UNIX/LINUX: A QUICK REFERENCE

The present book on bioinformatics was incomplete without an introduction to UNIX or Linux operating system. UNIX is operating systems which was first developed in the 1960s, and has been under constant development ever since. UNIX systems also have a graphical user interface (GUI) similar to Microsoft Windows which provides an easy to use environment. However, knowledge of UNIX is required for operations which aren't covered by a graphical program, or for when there is no windows interface available, for example, in a telnet session. This leaves the students of molecular biology and bioinformatics with a blank face in practical classes. There are many different versions of UNIX, although they share common similarities. The most popular varieties of UNIX are Sun Solaris, GNU/Linux, and MacOS X.

Linux is an operating system that was initially created as a hobby by a young student, Linus Torvalds, at the University of Helsinki in Finland. He began his work in 1991 when he released version 0.02 and worked steadily until 1994 when version 1.0 of the Linux Kernel was released. The kernel is developed and released under the GNU General Public License and its source code is freely available to everyone. Apart from the fact that it's freely distributed, Linux's functionality, adaptability and robustness, has made it the main alternative for proprietary Unix and Microsoft operating systems. IBM, Hewlett-Packard and other giants of the computing world have embraced Linux and support its ongoing development. Well into its second decade of existence, Linux has been adopted worldwide primarily as a server platform. Its use as a home and office desktop operating system is also on the rise. In chapter 2 of this book a small database is created using Linux operating system.

The UNIX/linux operating system is made up of three parts; the kernel, the shell and the programs.

1. **The kernels:** The kernel of UNIX is the hub of the operating system. It allocates time and memory to programs and handles the file store and communications in response to system calls.
2. **The shell:** The shell acts as an interface between the user and the kernel. When a user logs in, the login program checks the username and password, and then starts another program called the shell. The shell is a command line interpreter (CLI). It interprets the commands the user types in and arranges for them to be carried out.
3. **The programs:** The commands themselves are programs. When they terminate, the shell gives the user another prompt (% on our systems). Staff and students in the colleges have the tcsh shell by default.

   As an illustration of the way that the shell and the kernel work together, suppose a user types rm myfile (which has the effect of removing the file myfile). The shell searches the filestore for the file containing the program rm, and then requests the kernel, through system calls, to execute the program rm on myfile. When the process rm myfile has finished running, the shell then returns the UNIX prompt % to the user, indicating that it is waiting for further commands Everything in UNIX is either a file or a process. A process is an executing program identified by a unique PID (process identifier). A file is a collection of data. They are created by users using text editors, running compilers etc. All the files are grouped together in the directory structure. The file-system is arranged in a hierarchical structure, like an inverted tree. The top of the hierarchy is traditionally called root (written as a slash /)

BASIC UNIX COMMANDS

Please note UNIX is case-sensitive. Care is to be taken before writing commands. Some commonly used commands are given below:

1. **ls:** lists your files

2. **ls -l:** lists your files in 'long format', which contains lots of useful information, e.g. the exact size of the file, who owns the file and who has the right to look at it, and when it was last modified.
3. **more filename:** shows the first part of a file, just as much as will fit on one screen. Just hit the space bar to see more or q to quit. You can use /pattern to search for a pattern.
4. **emacs filename:** is an editor that lets you create and edit a file.
5. **mv filename1 filename2:** moves a file (is used to give a different name, or moves it into a different directory).
6. **cp filename1 filename2:** copies a file.
7. **rm filename:** removes a file. It is wise to use the option rm -i, which will ask you for confirmation before actually deleting anything
8. **diff filename1 filename2:** compares files, and shows where they differ
9. **wc filename:** tells you how many lines, words, and characters there are in a file
10. **chmod options filename:** lets you change the read, write, and execute permissions on your files. chmod o+r filename will make the file readable for everyone, and chmod o-r filename will make it unreadable for others again.
11. **gzip filename:** compresses files, so that they take up much less space.
12. **gunzip filename:** uncompresses files compressed by gzip.
13. **lpr filename:** print
14. **mkdir dirname:** make a new directory
15. **cd dirname:** change directory
16. **pwd:** tells you where you currently are.
17. **ff:** find files anywhere on the system
18. **grep string filename(s):** looks for the string in the files.
19. **du filename:** shows the disk usage of the files and directories in filename
20. **rlogin hostname:** lets you connect to a remote host
21. **telnet hostname:** also lets you connect to a remote host. Use rlogin whenever possible.
22. **ftp hostname:** lets you download files from a remote host which is set up as an ftp-server. This is a common method for exchanging academic papers and drafts. If you need to make a paper of yours available in this way, you can (temporarily) put a copy in /user/ftp/pub/TMP.
23. **date:** shows the current date and time.
24. **cal:** shows a calendar of the current month.

INDEX